Phosphatidylcholine Metabolism

Editor
Dennis E. Vance, Ph.D.
Professor and Director
Lipid and Lipoprotein Research Group
and
Department of Biochemistry
University of Alberta
Edmonton, Alberta, Canada

CRC Press, Inc.
Boca Raton, Florida

Library of Congress Cataloging-in-Publication Data

Phosphatidylcholine metabolism / editor, Dennis E. Vance.
p. cm.
Includes bibliography and index.
ISBN 0-8493-6338-1
1. Lecithin—Metabolism. I. Vance, Dennis E.
[DNLM: 1. Phosphatidylcholines—metabolism. QU 93 P575]
QP752.L4P47 1989
612′.397—dc19
DNLM/DLC
for Library of Congress

89-750
CIP

Direct all inquiries to CRC Press, Inc., 2000 Corporate Blvd., N.W., Boca Raton, Florida, 33431.

International Standard Book Number 0-8493-6338-1

Library of Congress Number 89-750
Printed in the United States

DEDICATION

This book is dedicated to my mentors, Charles C. Sweeley, David S. Feingold, and Konrad E. Bloch, my father Russell E. Vance, and my wife Jean. I can only begin to express my gratitude for your kindness and enthusiastic support throughout the years.

PREFACE

The decision to edit and publish a book is not easy. The contributors are active scientists with many constraints on their time. However, not a single monograph on phosphatidylcholine metabolism exists and there are several reasons why a book on this subject would be a valuable contribution. First, this book brings together a complete, cogent, and current summary of the field which should be a most valuable resource for scientists presently engaged in research and would be an important starting reference for future scientists who enter the field. Second, the field is large and our state of knowledge sufficiently developed so that more than enough material is available to be summarized in this volume. Third, with the recent successes in the purification of several of the enzymes involved in phosphatidylcholine metabolism, the field has reached a turning point in the types of experiments that are possible and the types of questions that can be asked. The application of molecular biological, immunological, and genetic approaches to phosphatidylcholine metabolism has already begun. Finally, a single volume on phosphatidylcholine metabolism will hopefully stimulate research in this area.

The book begins with a historical introduction by Eugene Kennedy, who discovered the CDP-choline biosynthetic pathway in the 1950s. The next six chapters discuss the current status of the enzymes involved in the biosynthesis of phosphatidylcholine from animal sources. These are followed by a chapter on phospholipases and phosphatidylcholine catabolism. Chapters 9 to 12 focus on special aspects of phosphatidylcholine metabolism, including a discussion on platelet activating factor. The last chapter summarizes regulatory and functional aspects of phosphatidylcholine metabolism.

I wish to acknowledge and thank the contributors to this volume for their excellent contributions and timely submission of their manuscripts, Sharline Willows for her able assistance in typing several chapters and in the editing process of this book, and the staff of CRC for their cooperation and efforts in publishing this book.

Dennis E. Vance
August 1988
Edmonton, Alberta

THE EDITOR

Dennis E. Vance, Ph.D., is the Director of the Lipid and Lipoprotein Research Group at the University of Alberta. He is also a Heritage Medical Scientist of the Alberta Heritage Foundation for Medical Research and Professor of Biochemsitry at the University of Alberta.

Dr. Vance obtained his Ph.D. in 1968 with Professor Charles Sweeley at the University of Pittsburgh. He worked on glycosphingolipid metabolism and the genetic defect that results in accumulation of a glycosphingolipid in Fabry's disease. Subsequently he studied enzymology for 2 years on an NIH Postdoctoral Fellowship under Professor David S. Feingold in the Department of Microbiology at the University of Pittsburgh. Dr. Vance returned to the lipid field during a postdoctoral fellowship with Professor Konrad Bloch at Harvard University. Dr. Vance's interest in regulation of lipid metabolism was stimulated by this research period in which he studied the mechanism by which polysaccharides stimulate fatty acid synthesis in *Mycobacteria*. In 1972, he received an American Heart Association-British Heart Foundation Fellowship to study phospholipid biosynthesis in virus-infected cells with Professor Derek Burke at the University of Warwick in Great Britain.

In 1973, Dr. Vance accepted an appointment as an Assistant Professor at the University of British Columbia and also held an Established Investigatorship from the American Heart Association from 1973 to 1978. It was at this university that he developed his long-standing interest in the regulation of phosphatidylcholine biosynthesis. From 1978 to 1981, Dr. Vance was an Associate Dean for Research in the Faculty of Medicine. In 1981, he had a sabbatical leave with Professor Daniel Steinberg at the University of California, San Diego, where he was introduced into the field of lipoprotein metabolism. In 1982, Dr. Vance became Professor of Biochemistry and Head of the Department at the University of British Columbia. In 1986, he moved to his present position at the University of Alberta to establish the Lipid and Lipoprotein Research Group.

Dr. Vance has contributed the chapters on lipid structure and metabolism in both editions of the textbook *Biochemistry* edited by Geoffrey Zubay. He has also edited with Dr. Jean Vance an advanced textbook entitled *Biochemistry of Lipid and Membranes*. He is currently a member of the Editorial Board for *Journal of Biological Chemistry, Biochemical Journal*, and *Biochimica Biophysica Acta* and serves on a grant panel for the Medical Research Council of Canada. In 1989, Dr. Vance was awarded the Boehringer Mannheim Canada Prize by the Canadian Biochemical Society.

CONTRIBUTORS

Gilbert Arthur, Ph.D.
Assistant Professor
Department of Biochemistry
University of Manitoba
Winnipeg, Manitoba, Canada

Merle L. Blank
Department of Biological Chemistry
Oak Ridge Associated Universities
Oak Ridge, Tennessee

George M. Carman, Ph.D.
Professor
Department of Food Science
Rutgers University
New Brunswick, New Jersey

Patrick C. Choy, Ph.D.
Professor
Department of Biochemistry
University of Manitoba
Winnipeg, Manitoba, Canada

Rosemary Cornell, Ph.D.
Assistant Professor
Department of Chemistry
Simon Fraser University
Burnaby, British Columbia, Canada

Edward A. Dennis, Ph.D.
Professor of Chemistry and Biochemistry
Department of Chemistry
University of California, San Diego
La Jolla, California

Kozo Ishidate, Ph.D.
Associate Professor
Department of Chemical Toxicology
Medical Research Institute
Tokyo Medical and Dental University
Tokyo, Japan

Julian N. Kanfer, Ph.D.
Professor
Department of Biochemistry
Faculty of Medicine
University of Manitoba
Winnipeg, Manitoba, Canada

Eugene P. Kennedy, Ph.D.
Professor
Department of Biological Chemistry
Harvard Medical School
Boston, Massachusetts

Ten-ching Lee, Ph.D.
Scientist
Department of Biological Chemistry
Oak Ridge Associated Universities
Oak Ridge, Tennessee

Fred Possmayer, Ph.D.
Professor
Departments of Obstetrics and Gynaecology
The University of Western Ontario
London, Ontario, Canada

Neale D. Ridgeway, Ph.D.
Department of Molecular Genetics
University of Texas at Dallas
Dallas, Texas

Mary F. Roberts, Ph.D.
Associate Professor
Department of Chemistry
Boston College
Chestnut Hill, Massachusetts

Fred Snyder, Ph.D.
Vice Chairman
Medical Sciences Division
Oak Ridge Associated Universities
Oak Ridge, Tennessee

Matthew W. Spence, M.D., Ph.D.
Director
Atlantic Research Centre for Mental Retardation
and
Professor
Department of Pediatrics and Biochemistry
Dalhousie University
Halifax, Nova Scotia, Canada

Dennis E. Vance, Ph.D.
Professor and Director
Lipid and Lipoprotein Research Group
and
Department of Biochemistry
University of Alberta
Edmonton, Alberta, Canada

TABLE OF CONTENTS

Chapter 1

DISCOVERY OF THE PATHWAYS FOR THE BIOSYNTHESIS OF PHOSPHATIDYLCHOLINE

Eugene P. Kennedy

TABLE OF CONTENTS

I. INTRODUCTION

In 1952, when I began the study of the biosynthesis of membrane phospholipids, nothing was known about the detailed pathways by which these complex molecules are synthesized in living cells. In fact, at that period little was known about the biosynthesis of any of the major classes of cell constituents — proteins, nucleic acids, carbohydrates, or lipids. New technology, especially the isotope tracer method and the development of new forms of chromatography, had just become available to the biochemist. New insights into the relation between biosynthetic and catabolic systems and the central role of ATP, as well as the need for "common intermediates" that must link endergonic and exergonic systems, had been provided by the seminal papers of Lipmann[1] and of Kalckar.[2] The stage was set, and problems of biosynthesis were about to become central to the work of many laboratories throughout the world. Indeed, in future histories of biochemistry, the decades of the 1950s and 1960s might well be termed the "era of biosynthesis".

My initial interest in the study of the biosynthesis of phospholipids was centered on the phosphodiester bond because it is a striking feature of the structure of nucleic acids also. At that time, just before the dawn of the Watson-Crick double helical model of DNA, we had no well-defined idea of the specific biological functions of nucleic acids, let alone the detailed molecular basis for their replication. It was clear, however, that nucleic acids, both RNA and DNA, must be of central importance in biological processes of living cells of every kind. My original intent was therefore to study the formation of phosphodiester bonds in phospholipids with the hope of using the information so obtained to guide later studies on the more complex problem of the biosynthesis of RNA. It is worth noting that the same notion occurred to Arthur Kornberg at the same time and guided his brief but highly productive foray into the field of the biosynthesis of phospholipids. Unfortunately (or perhaps fortunately), the study of phospholipid biosynthesis and related problems of membrane structure and function has proved so engrossing that I could not abandon it for the study of nucleic acids.

In 1952, there was already a large body of information derived from isotope tracer studies, in which a suspected precursor, usually $^{32}P_i$, was injected into the intact animal and the rate of formation or breakdown of phospholipids in various tissues was determined. This work was summarized in reviews at that period by Artom[3] and by Chaikoff and Zilversmit.[4]

Such studies provided a valuable framework for our understanding of the overall metabolism of phospholipids but could hardly be expected to provide information on the detailed mechanism of biosynthesis. Therefore, I determined to attack the problem at the level of cell-free enzyme systems. I was strongly encouraged by the success of the previous studies of Friedkin and Lehninger,[5] who found that cell-free particulate enzyme preparations from rat liver vigorously incorporated $^{32}P_i$ into phospholipid, phosphoprotein, and nucleic acid fractions. Their work also provided clear evidence that the intermediate formation of ATP was essential for these transformations.

The decision to concentrate first on the biosynthesis of phosphatidylcholine was obviously dictated by the fact that this is the predominant phospholipid of animal tissues, and by the fact that choline is so important in the diet of mammals, and (as acetylcholine) in neurotransmission.

II. ROLE OF *sn*-3-GLYCEROPHOSPHATE

In my first experiments on the *in vitro* conversion of $^{32}P_i$ to phospholipid, particulate enzyme preparations from rat liver were employed that are now known to have contained both microsomes and mitochondria. Because the incorporation of $^{32}P_i$ into [^{32}P]ATP was an essential step in its further transformation to phospholipid, it was at first believed that the mitochondria themselves were the site of the biosynthetic reactions. It is now recognized that most of the lipid biosynthetic reactions are in fact occurring in the microsomal fraction.

It was found[6] that $^{32}P_i$ in such crude systems is principally converted to phosphatidic acid. The incorporation into phospholipid was greatly stimulated by the addition of glycerol, providing the first evidence for the role of glycerophosphate in this process and for the presence in liver of the enzyme glycerokinase, the discovery of which was reported by Bublitz and Kennedy.[7] These experiments prompted the synthesis of ^{32}P-labeled glycerophosphate by chemical methods and the demonstration that it is α-glycerophosphate (now known to be *sn*-3-glycerophosphate) that is the precursor of membrane lipids, and not β-glycerophosphate. The important studies of Kornberg and Pricer[8] had provided strong evidence for the direct acylation of glycerophosphate with the formation of phosphatidic acid in reactions involving acyl CoA as the activated form of fatty acid.

III. THE FUNCTION OF CYTIDINE COENZYMES

Kornberg and Pricer[9] reported experiments in which phosphocholine, doubly labeled with ^{32}P and ^{14}C, was converted to a lipid product by rat liver enzyme preparations with the ratio of ^{32}P to ^{14}C in the lipid closely similar to that of the labeled phosphocholine. On the basis of these results, it was suggested that phosphocholine may be incorporated as a unit into the phosphatidylcholine molecule by the following reaction:

$$\text{Phosphatidic acid} + [^{32}\text{P}]\text{phosphocholine} \longrightarrow [^{32}\text{P}]\text{phosphatidylcholine} + \text{P}_i \quad (1)$$

The conversion of phosphophocholine to phosphatidylcholine was reported to require added ATP.

Unaware of these studies, in 1952 I had also started to explore the origins of the phosphodiester bridge of phosphatidylcholine. In these experiments, I had tested both labeled choline and labeled phosphocholine as possible precursors. In an important, and as it proved crucial, difference from the approach of Kornberg, I had employed oxidative phosphorylation catalyzed by mitochondria present in the preparation to generate ATP, or other intermediates that might be needed for biosynthetic processes, as enunciated by Lipmann.[1] I observed that, under these conditions, free choline, but not phosphocholine, was converted into a lipid form. I was for a time led down the garden path by this observation, until it was learned that free choline is converted in this system to a fatty acyl ester,[10] a previously undiscovered lipid form of choline, the function of which remains unclear to the present day.

In July of 1954, there arrived in my laboratory at the University of Chicago my first postdoctoral fellow, Samuel Weiss, who was to become my gifted collaborator and life-long friend. We decided to undertake a detailed investigation of the difference in conditions between my experiments and those of Kornberg and Pricer. I had already conceived a very considerable respect for the abilities of Kornberg, which was only to increase during subsequent years, and I was not surprised when Weiss was able to reproduce his experimental findings exactly, as well (fortunately!) as my own. We then decided to confront the anomaly that the conversion of phosphocholine to phosphatidylcholine required ATP from the bottle, while ATP generated via oxidative phosphorylation (which we carefully showed was actually occurring in our system) was ineffective.

ATP was needed in the system of Kornberg and Pricer at relatively high levels, giving rise to the suspicion that it might contain some active impurity. Crystalline ATP had just become available from commercial sources, and, through the kindness of Drs. S. A. Morrel, S. Lipton, and A. Frieden of the Pabst Laboratories, the crystalline material and various batches of highly purified but amorphous ATP were tested with the results shown in Table 1. Because the active amorphous material had in fact been highly purified by ion-exchange chromatography, we

TABLE 1
Cofactor Needed for Conversion of Phosphocholine to Phosphatidylcholine

Added cofactor	Phosphatidylcholine synthesized (cpm)
1. 5 m*M* ATP lot 116 (amorphous)	590
2. 5 m*M* ATP lot 122 (crystalline)	20
3. 5 m*M* ATP lot 122 + 0.5 m*M* ITP	0
4. 5 m*M* ATP lot 122 + 0.5 m*M* UTP	50
5. 5 m*M* ATP lot 122 + 0.5 m*M* GDP	57
6. 5 m*M* ATP lot 122 + 0.5 m*M* CTP	1677
7. 5 m*M* ATP lot 122 + 0.5 m*M* CTP + 2.5 m*M* pyrophosphate	750

Note: Each tube contained $MgCl_2$ (10 m*M*); phosphocholine-^{32}P (3 m*M* with specific activity 113,000 cpm/μmol); and 25 mg of particulate rat liver enzyme in a total volume of 1.0 ml. The tubes were incubated at 37° for 60 min. Data from Kennedy and Weiss.[12]

reasoned that the active cofactor might also be a triphosphate. We therefore tested the other ribonucleoside triphosphates and discovered that CTP played a highly specific role in this system.

Khorana[11] had just published his elegant new procedure for the synthesis of substituted pyrophosphates by the dicyclohexylcarbodiimide method. I had noted Khorana's paper because it offered a new approach to the synthesis of ^{32}P-labeled ATP. It occurred to me, however, that the method could also be used for the synthesis of asymmetrical disubstituted pyrophosphates such as cytidine diphosphate (CDP) choline. Both CMP and [^{32}P]phosphocholine, the required starting materials for the synthesis, were already available in the laboratory. Although we had no direct evidence whatever that CDP choline was an intermediate in the reaction, we decided to adopt the stochastic method of arriving at the truth by conjecture, and proceeded with the synthesis, which fortunately yielded pure CDP-choline in surprisingly good yield. In fact, the most formidable part of the synthesis proved to be the preparation of the reagent dicyclohexylcarbodiimide. CDP-choline was isolated from the rather complex mixture of the products of the reaction by chromatography on Dowex®-1 formate and proved to be the direct precursor of phosphatidylcholine in our enzyme system.[12,13] We immediately synthesized CDP-ethanolamine also. Although the yield was poor, presumably because of the unprotected amino group of phosphoethanolamine, pure synthetic CDP-ethanolamine was also obtained and shown to be the precursor of phosphatidylethanolamine.[13]

IV. CENTRAL ROLE OF PHOSPHATIDIC ACID

In these early experiments, the conversion of labeled CDP-choline to phosphatidylcholine took place by transfer of the phosphocholine residue from CDP-choline to an endogenous acceptor present in the membrane preparations used as source of enzyme. We were by no means sure that the endogenous acceptor was in fact *sn*-1,2-diacylglycerol. Weiss attacked the problem with his usual ingenuity and vigor and, in an early example of the use of synthetic detergents to study the enzymic synthesis of lipids, discovered that an emulsion of diacylglycerol derived from egg phosphatidylcholine mixed with the commercially available detergent Tween®-20 (polyoxyethylene sorbitan monolaurate) was a highly effective acceptor for phosphocholine.[13]

Exploitation of a similar approach based on the use of the detergent Tween®-20 rapidly led to the discovery of an enzyme catalyzing the net synthesis of triacylglycerol from diacylglycerol and a long-chain thioacyl ester of CoA.[14]

The pioneering studies of Kornberg and Pricer, as well as my own work outlined above, had clearly demonstrated a rapid synthesis of phosphatidic acid from *sn*-3-glycerophosphate. However, phosphatidic acid had not as yet been demonstrated to be a naturally occurring component of fresh mammalian tissues. No phosphatidic acids were found by Marinetti and Stotz[15] in rat heart or liver. Their failure to find phosphatidic acids led Marinetti and Stotz to question the physiological significance of the enzymic reactions leading to the synthesis of phosphatidic acids. This did not appear to us, however, to be a serious objection. If phosphatidic acids and diacylglycerol are very rapidly metabolized, it is clear that their steady-state concentration may be extremely low while the rate of conversion to final products remains high.

An enzyme catalyzing the dephosphorylation of phosphatidic acid with the specific formation of *sn*-1,2-diacylglycerol was an essential postulate of our formulation. Such an enzyme was soon found.[16] With its discovery, a strong link between the pathways for the synthesis of neutral triglyceride and of phospholipids, and between glycolysis and lipid synthesis via *sn*-3-glycerophosphate, was immediately apparent.

V. THE METHYLATION OF PHOSPHATIDYLETHANOLAMINE

Stekol et al.[17] suggested that the transfer of the methyl group of methionine to ethanolamine in the formation of choline involves a phospholipid form of ethanolamine. Bremer and Greenberg[18] discovered that the microsomal fraction of liver catalyzes the methylation of phosphatidylethanolamine to phosphatidylcholine with *S*-adenosylmethionine as the methyl donor.

There has been considerable confusion in textbooks and in other literature on the relative physiological role of the methylation pathway vs. the cytidine coenzyme pathway for the formation of phospholipids in animal tissues. It is clear that methylation of phosphatidylethanolamine represents the sole known source of choline in nature and is of prime importance for this reason. In mammals, however, most of the choline needed for the synthesis of phosphatidylcholine is obtained from the diet. It was suggested by Bremer et al.[19] that the formation of phosphatidylethanolamine and phosphatidylcholine from phosphatidylserine via the decarboxylation of phosphatidylserine, discovered by Borkenhagen et al.[20] might represent the *de novo* pathway for the synthesis of phosphatidylethanolamine and phosphatidylcholine, while the cytidine nucleotide pathways might represent "salvage" reactions. Indeed, this may be the case in yeast where the synthesis of phosphatidylserine may take place from a reaction of L-serine with CDP-diacylglycerol, a reaction analogous to that found in *Escherichia coli* by Kanfer and Kennedy.[26] In animal tissues, however, no such synthesis of phosphatidylserine takes place. Phosphatidylserine is formed by exchange of the head groups of phosphatidylethanolamine and of phosphatidylcholine with free serine. Thus, in animal tissues, all of the major nitrogen-containing phospholipids are derived from phosphatidylethanolamine and phosphatidylcholine which must be synthesized via the cytidine nucleotide pathway. Indeed, the enzymic methylation of phosphatidylethanolamine occurs in the tissues of higher animals at an appreciable rate only in liver. In all other tissues the methylation of phosphatidylethanolamine makes a negligible contribution to the net formation of phosphatidylcholine.

Although the rate of methylation of phosphatidylethanolamine in nonhepatic tissues is low, considerable attention has recently been directed to the possible role of this methylation system in coupling cellular responses to receptor-mediated signaling. This concept has been put forward and developed by Hirata and Axelrod.[22] Unfortunately, some of the reports on which this hypothesis was based have not been confirmed by later investigations (reviewed by Kennedy).[23] It appears that convincing and definitive evidence is lacking to support the notion that the methylation of phosphatidylethanolamine is linked to the transduction of membrane receptor signals.

VI. THE PROSPECT BEFORE US

Further work with cell-free enzyme systems in the decade of of the 1950s had led by 1961 to a fairly comprehensive view of the integration of the biosynthesis of phospholipids and triacylglycerol in animal tissues as revealed by Figure 1, from a review[24] that appeared in that year. Although it is gratifying that the 1961 formulation has stood the test of time, it must be emphasized that a very large body of work remains to be done on the characterization of the enzymes in animal tissues that catalyze the individual reactions outlined in this figure. Our understanding of the detailed enzymology of the biosynthesis of phospholipids is in fact very limited if compared with the advanced state of comparable studies of *E. coli*. The bacterial system has the intrinsic advantage of the higher rates of biosynthesis associated with very rapid growth, as well as benefiting from the remarkable development of bacterial genetics.

The dramatic development of eukaryotic genetics in recent years, however, will surely prove a great stimulus to studies of phospholipid biosynthesis in animal cells, as well as in yeast.

It will be of particular importance to understand the regulation of the activity of these biosynthetic enzymes and to learn how the complex process of the biogenesis of subcellular organelles is integrated with their activity.

After a period of eclipse, the study of the metabolism of membrane lipids in animal tissues has been resumed with new vigor. Two very dramatic developments have spurred researchers to these new efforts. The first of these developments is the recognition that the agonist-stimulated turnover of phosphatidylinositol, first discovered by Hokin and Hokin,[25] leads to the generation of second messengers that play a vital role in the regulation of metabolism and growth in normal and transformed cells. This area of investigation is now one of the most active in the entire domain of cell biology.

The identification of 1-O-alkyl-2-O-acetyl-*sn*-glycero-3-phosphocholine as a potent hormone (platelet-activating factor, or PAF) that acts not only on blood platelets, but on a number of other target organs is also a most surprising and potentially important revelation of the specific role of phospholipids in regulatory phenomena. As pointed out in an earlier review,[23] the great diversity of individual species of glycerophosphatides found in animal tissues has long been noted and considered by some a good reason for staying out of the field of lipid metabolism. The emergence of so-called minor lipids as cellular constituents of the first importance has certainly changed this picture. The possibility must now be considered that each of these minor lipids may play a highly specific regulatory, as well as structural, role. Such cellular regulatory functions may include not only transmembrane signaling, but also the modulation of activity of specific enzymes.

ACKNOWLEDGMENTS

It is a pleasure to acknowledge the invaluable contributions of Samuel Weiss, Sylvia Wagner Smith, and Louise Fencl Borkenhagen to the early studies on the biosynthesis of phosphatidylcholine reviewed here.

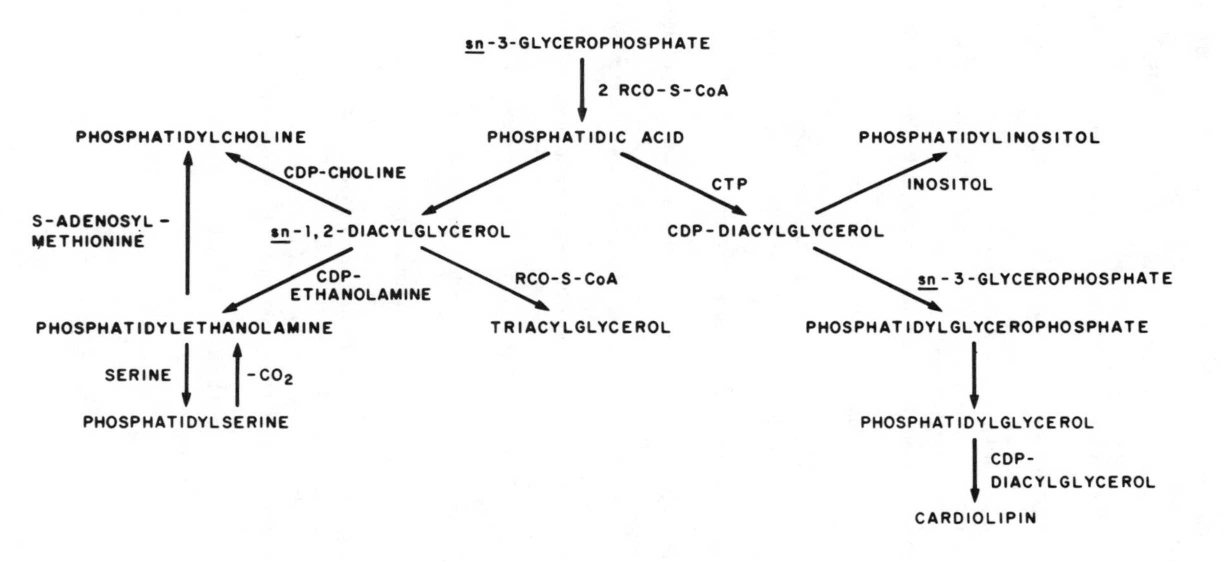

FIGURE 1. Biosynthesis of phospholipids and triacylglcyerol in animal tissues. Redrawn from Reference 24.

REFERENCES

1. **Lipmann, F.,** Metabolic generation and utilization of phosphate bond energy, *Adv. Enzymol.,* 1, 99, 1941.
2. **Kalckar, H. M.,** The nature of energetic coupling in biological syntheses, *Chem. Rev.,* 28, 71, 1941.
3. **Artom, C.,** Lipid metabolism, *Annu. Rev. Biochem.,* 22, 211, 1953.
4. **Chaikoff, I. L. and Zilversmit, D. B.,** Radioactive phosphorus: its application to the study of phospholipid metabolism, *Adv. Biol. Med. Phys.,* 1, 321, 1948.
5. **Friedkin, M. and Lehninger, A. L.,** Oxidation-coupled incorporation of inorganic radiophosphate into phospholipide and nucleic acid in a cell-free system, *J. Biol. Chem.,* 177, 775, 1949.
6. **Kennedy, E. P.,** Synthesis of phosphatides in isolated mitochondria, *J. Biol. Chem.,* 201, 399, 1953.
7. **Bublitz, C. and Kennedy, E. P.,** The enzymatic phosphorylation of glycerol, *J. Biol. Chem.,* 211, 851, 1954.
8. **Kornberg, A. and Pricer, W. E., Jr.,** Enzymatic synthesis of phosphorus-containing lipides, *J. Am. Chem. Soc.,* 74, 1617, 1952.
9. **Kornberg, A. and Pricer, W. E., Jr.,** Studies on the enzymatic synthesis of phospholipides, *Fed. Proc. Fed. Am. Soc. Exp. Biol.,* 11, 242, 1952.
10. **Kennedy, E. P.,** The biological synthesis of phospholipids, *Can. J. Biochem. Physiol.,* 34, 334, 1956.
11. **Khorana, H. G.,** Carbodiimides. VI. A novel synthesis of adenosine di- and triphosphate and P^1, P^2-diadenosine-5′-pyrophosphate, *J. Am. Chem. Soc.,* 76, 3517, 1954.
12. **Kennedy, E. P. and Weiss, S. B.,** Cytidine diphosphate choline: a new intermediate in lecithin biosynthesis, *J. Am. Chem. Soc.,* 77, 250, 1955.
13. **Kennedy, E. P. and Weiss, S. B.,** The function of cytidine coenzymes in the biosynthesis of phospholipides, *J. Biol. Chem.,* 222, 193, 1956.
14. **Weiss, S. B. and Kennedy, E. P.,** The enzymatic synthesis of triglycerides, *J. Am. Chem. Soc.,* 78, 3550, 1956.
15. **Marinetti, G. V. and Stotz, E.,** Chromatography of phosphatides on acid impregnated paper, *Biochim. Biophys. Acta,* 21, 168, 1956.
16. **Smith, S. W., Weiss, S. B., and Kennedy, E. P.,** The enzymatic dephosphorylation of phosphatidic acids, *J. Biol. Chem.,* 228, 915, 1957.
17. **Stekol, J. A., Weiss, S., Hsu, P.-T., and Smith, P.,** Cystathione-C^{14} and cystine-3-C^{14} as sources of methyl groups of choline and creatine in the intact rat, *Fed. Proc. Fed. Am. Soc. Exp. Biol.,* 12, 274, 1953.
18. **Bremer, J. and Greenberg, D. M.,** Methyl transferring enzyme system of microsomes in the biosynthesis of lecithin (phosphatidylcholine), *Biochim. Biophys. Acta,* 46, 205, 1961.
19. **Bremer, J., Figard, J. H., and Greenberg, D. M.,** The biosynthesis of choline and its relation to phospholipid metabolism, *Biochim. Biophys. Acta,* 43, 477, 1960.
20. **Borkenhagen, L. F., Kennedy, E. P., and Fielding, L.,** Enzymatic formation and decarboxylation of phosphatidylserine, *J. Biol. Chem.,* 236, PC 28, 1961.
21. **Kanfer, J. and Kennedy, E. P.,** Metabolism and function of bacterial lipids. II. Biosynthesis of phospholipids in *Escherichia coli, J. Biol. Chem.,* 239, 1720, 1964.
22. **Hirata, F. and Axelrod, J.,** Phospholipid methylation and biological signal transmission, *Science,* 209, 1082, 1980.
23. **Kennedy, E. P.,** The biosynthesis of phospholipids, in *Lipids and Membranes, Past, Present and Future,* Op den Kamp, J. A. F., Roelofsen, B., and Wirtz, K. W. A., Eds., Elsevier, Amsterdam, 1986, chap. 8.
24. **Kennedy, E. P.,** Biosynthesis of complex lipids, *Fed. Proc. Fed. Am. Soc. Exp. Biol.,* 20, 934, 1961.
25. **Hokin, M. R. and Hokin, L. E.,** Enzyme secretion and the incorporation of P^{32} into phosopholipides of pancreas slices, *J. Biol. Chem.,* 203, 967, 1953.
26. **Kanfer, J. and Kennedy, E. P.,** Metabolism and function of bacterial lipids. II. Biosynthesis of phospholipids in *Escherichia coli, J. Biol. Chem.,* 239, 1720, 1964.

Chapter 2

CHOLINE TRANSPORT AND CHOLINE KINASE

Kozo Ishidate

TABLE OF CONTENTS

I. INTRODUCTION

Choline kinase (CK) (ATP:choline phosphotransferase, EC 2.7.1.32) catalyzes the phosphorylation of choline by ATP in the presence of Mg^{2+}, yielding phosphocholine and ADP, as illustrated in Figure 1. This enzymatic step commits choline to the cytidylyl diphosphate (CDP)-choline pathway[1] for the biosynthesis of phosphatidylcholine (PC) in all animal cells.

In 1953, Wittenberg and Kornberg[2] were the first to establish the existence of a cytosolic enzyme in yeast which phosphorylated choline with ATP. This enzyme had an absolute requirement for Mg^{2+} and was recovered in acetone powder extracts of liver, brain, intestinal mucosa, and kidney from various animals. Dimethyl- and diethylethanolamine, monomethyl- and monoethylethanolamine, and ethanolamine could also be phosphorylated, although the affinities of these substrates differed widely. The K_m was 20 μ*M* for choline and 10 m*M* for ethanolamine, the poorest substrate among them. The reaction showed a pH optimum between 8.0 and 9.5.[2] Some requirement for a sulfhydryl agent (cysteine) was also documented.

Since the discovery of this key enzyme in PC biosynthesis, characterization and purification of CK have been the topics of numerous investigations. Only recently, the complete purification of CK has been reported by our laboratory.[3]

An area of considerable interest has been the high inducibility of CK under certain experimental conditions and its possible relation to regulation of net PC biosynthesis. CK is the only enzyme in the PC biosynthetic pathway from an animal source that appears to be regulated at the level of gene expression other than during development. Other areas of interest include cAMP-dependent phosphorylation and dephosphorylation as a mechanism of CK regulation, the physiological significance of multiple forms of CK, and whether or not the same enzyme catalyzes the phosphorylation of choline and ethanolamine. This chapter will focus on the above current areas of interest. However, the first part of the chapter is devoted to choline transport, a process distinct from CK activity, but relevant to understanding the role of CK in PC biosynthesis. Brief reviews on CK are also available.[4-6]

II. TRANSPORT OF CHOLINE INTO CELLS

Choline is a dietary requirement that crosses the cell membrane before being phosphorylated by CK or metabolized in liver or kidney by two mitochondrial enzymes, i.e., choline dehydrogenase (EC 1.1.99.1) and betaine aldehyde dehydrogenase (EC 1.2.1.8) to betaine which can donate a methyl group to homocysteine for the formation of methionine.[4] An adequate supply of choline for acetylcholine (ACh) synthesis is especially critical for cholinergic nerve function, and the nervous system does not synthesize choline *de novo*.[7-9]

A specific, carrier-mediated transport system for the uptake of choline was first demonstrated in kidney cortex[10] and giant axons of *Loligo*[11] in 1965. The oxidation of [^{14}C]choline to [^{14}C]betaine in rat kidney cortex slices was inhibited by 2,4-dinitrophenol (DNP), by hemicholinium-3 (HC-3, an analogue of choline), and by the absence of Na^+ from the incubation medium. However, in kidney homogenates appreciable conversion of choline to betaine occurred and was not inhibited by such agents, suggesting that choline transport in kidney cortex could be an ATP- and Na^+-dependent process.[10] A saturable (K_m: 100 μ*M*), temperature-sensitive uptake of choline occurred in *Loligo,* and HC-3, ACh, and other choline analogues were competitive inhibitors.[11]

It is now well established that there exist two distinct transport mechanisms for choline. One is a high-affinity (K_m or K_t <5 μ*M*), Na^+-dependent, and HC-3-sensitive mechanism tightly coupled to ACh synthesis in cholinergic synaptosomes. The other is a low-affinity (K_t >30 μ*M*), relatively Na^+-independent, and HC-3-insensitive mechanism found in most animal cells as well as in cholinergic nerve tissues.

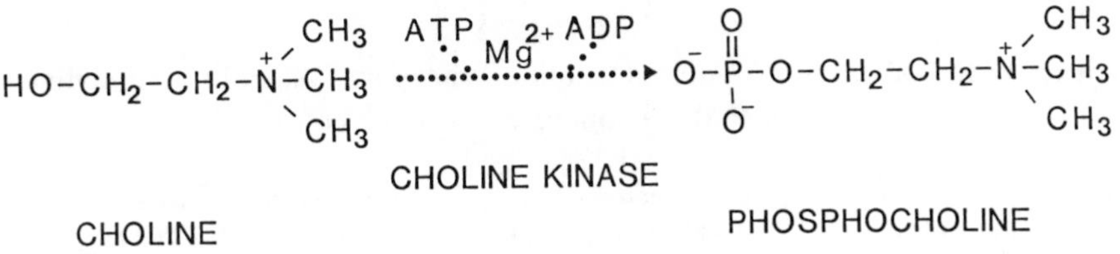

$$HO-CH_2-CH_2-\overset{+}{N}(CH_3)_3 \xrightarrow[\text{CHOLINE KINASE}]{ATP \quad Mg^{2+} \quad ADP} {}^{-}O-\overset{O}{\overset{\|}{P}}(O^{-})-O-CH_2-CH_2-\overset{+}{N}(CH_3)_3$$

CHOLINE PHOSPHOCHOLINE

FIGURE 1. The reaction catalyzed by choline kinase.

A. CHOLINE TRANSPORT IN NONCHOLINERGIC CELLS

The currently available data for carrier-mediated transport systems for choline in noncholinergic mammalian tissues (or cells) are briefly summarized in Table 1. Although there is a wide range in the apparent K_t values (the concentration of choline at the half-maximal transport activity) for choline, these data indicate that choline is taken up by mammalian cells via a specific transport mechanism that is temperature sensitive and metabolic energy dependent. A nonsaturable component for choline transport has also been detected in perfused rat liver,[16] cultured rat hepatocytes,[17] Ehrlich ascites cells,[26] Novikoff hepatoma cells,[28] and human erythrocytes.[31] This simple diffusion mechanism appears to operate only at relatively high extracellular choline concentrations. Since the plasma choline level is normally in the range of 10 to 15 μM,[40-45] choline uptake probably occurs largely through a specific carrier-mediated transport mechanism.

The transport of ethanolamine also occurs through a saturable, high-affinity carrier-mediated mechanism in perfused hamster heart (K_t: 170 μM),[12,46] Krebs II ascites cells (K_t: 34 μM),[23] Y-79 human retinoblastoma (K_t: 41 μM),[25,47] and rabbit photoreceptor cells (K_t: 2 μM). This process could be separate and distinct from choline transport because it was not inhibited by choline. Choline transport in these same systems, on the other hand, was significantly inhibited by ethanolamine in either a competitive or mixed-type manner as shown in Table 1. It has not been clarified, however, whether ethanolamine can also be taken up through a choline transporter in those mammalian cells. Incidentally, the plasma ethanolamine levels have been reported to be in a range of 5 to 20 μM in rats and humans.[49-52]

Other than mammalian systems, a similar carrier-mediated transport of choline has been reported to exist in an anerobic protozoon *Entodinium caudatum*[53] and the yeast *Saccharomyces cerevisiae*.[54] Interestingly, the uptake process of choline was highly coupled to CK in the protozoon system, and free choline was never detected. Thus, in *E. caudatum*, CK activity might be rate-limiting for choline uptake, and the enzyme could play an active role in the choline transport process. Tight coupling between choline transport and CK activity was observed in Krebs II ascites cells.[24]

The choline transport system in yeast required metabolic energy, was temperature dependent, did not require Na^+ ion, and was not inhibited by the presence of ethanolamine.[54] Furthermore, polyamines induced the choline transporter in yeast and caused a parallel stimulation of PC biosynthesis without any significant changes in enzymatic activities involved in the pathway.[55] A yeast mutant lacking choline transport activity facilitated cloning of the yeast choline transporter gene which was responsible for both choline and ethanolamine transport.[56]

The rate of PC biosynthesis does not appear to be influenced significantly by the rate of choline transport in most nonproliferating cells.[4] However, in both Ehrlich ascites[26] and Novikoff hepatoma[28] cells, choline was immediately converted to PC without detectable accumulation of other choline metabolites. Consequently, the apparent K_m values for choline in PC biosynthesis were the same as K_t values for choline in choline transport.[26,28] Thus, choline transport may limit PC biosynthesis in actively proliferating cells.

TABLE 1
Characteristics of Carrier-Mediated Choline Transport System in Mammalian Cells other than Cholinergic Neuronal Tissues

Sources	Apparent *Kt* (μ*M*)	Specific features and notes	Intracellular fate of choline [when incubated at the indicated concentration]	Ref.
Hamster heart	100	Competitive inhibition by ethanolamine and HC-3; stimulation by glycine	Phosphocholine → CDP-choline→PC [10—50 μ*M*]	12—14
Rat kidney cortex		ATP and Na^+ dependent; inhibition by HC-3 and other structural analogues of choline	Betaine (80%); phosphocholine (10%) [200 μ*M*]	10, 15
Rat liver	170	Simple diffusion with choline concentration exceeding 0.3 m*M* in the perfusate	Betaine (60%); phosphocholine (30%); choline (10%) [5—125 μ*M*]	16
Rat hepatocytes	12	Nonsaturable uptake with choline concentration exceeding 40 μ*M* in the culture medium; depression by a cAMP analogue or aminophylline	Betaine (57%); phosphocholine (33%); choline (10%) [40 μ*M*]	17, 18
Rat lens	200—300	Competitive inhibition by ethanolamine	Phosphocholine (75—95%) [3—100 μ*M*]	19
Rabbit photoreceptors	1.4 (high); 128 (low)	Na^+ independent, HC-3 insensitive	Phosphocholine [0.4 μ*M*]	20
Hamster glial cells	0.65—1.52 (high); 24—54 (low)		Not determined	21
Neuroblastoma cells	2.1—4.5 (starved)	Na^+ independent, HC-3 insensitive; temperature sensitive	Phosphocholine (70—95%) [50 μ*M*]	22
	10.2—20.7 (unstarved)	DNP and oubain sensitive	Choline (1—4%)	
Krebs II ascites cells	36—46	Na^+ and temperature sensensitive; competitive inhibition by ethanolamine and HC-3; inhibition by structural analogues of choline; ouabain and iodoacetamide sensitive; DNP insensitive; a cyclic transport mechanism	Phosphocholine (90%) [0.4—1 m*M*]	23, 24
Y-79 retinoblastoma cells	0.93	Mixed-type inhibition by ethanolamine; inhibition by monomethyl- and dimethylethanolamine	Phosphocholine→CDP-choline → PC [21.5 μ*M*]	25
Ehrlich ascites cells	15—59	Inhibition by deoxyglucose, DNP, and choline analogues; uptake process is rate limiting for PC biosynthesis; transport energy: 16 kcal/mol; competitive inhibition by amiloride and demethylamiloride	Phosphocholine →CDP-choline → PC [3—15 μ*M*]	26, 27

TABLE 1 (continued)
Characteristics of Carrier-Mediated Choline Transport System in Mammalian Cells other than Cholinergic Neuronal Tissues

Sources	Apparent *Kt* (μ*M*)	Specific features and notes	Intracellular fate of choline [when incubated at the indicated concentration]	Ref.
Novikoff hepatoma cells	4—7	Simple diffusion with choline concentration exceeding 20 μ*M* in the culture medium; competitive inhibition by phenethylalcohol; transport energy: 16 Kcal/mol; uptake process is rate limiting for PC biosynthesis with extracellular choline concentration below 20 μ*M*	Phosphocholine→CDP-choline→PC [20 μ*M*] Phosphocholine (95%) [100 μ*M*]	28, 29
L1210 leukemia cells	7.9	Noncompetitive inhibition by tetraalkyl ammonium ions	Not determined	30
Human erythrocytes	20—30	Na^+ dependent; ouabain insensitive; competitive inhibition by tetraalkyl ammonium ions; a cyclic transport mechanism; structure affinity-transport activity relationships characterized	Choline (100%)	31—37
Rat diaphragm (muscle fiber)	500	Temperature dependent; against a concentration gradient; inhibition by SH reagents, ouabain, and HC-3	Choline (80—85%) [5 μ*M*]	38, 88
Chick fibroblasts	2.9—3.3	Na^+ dependent, HC-3 sensitive; ouabain, KCN sensitive	Not determined	39
L5178Y lymphoblasts and Yoshida sarcoma cells	12—25	"Uphill" transport against a concentration gradient; competitive inhibition by ethanolamine, HC-3, and nitrogen mustard (HN2); transport of HN2 on the choline transport carrier	Not determined	89, 90

The choline transport system in erythrocytes has been studied despite its unclear physiological function. Thus, Deves and Krupka[36] studied affinity and transport activity in the choline transport system of human erythrocytes with a series of structural analogues of choline. The cation binding site had subsites for the three methyl substituents on the quaternary nitrogen atom, being very specific at two of the three positions. Lengthening of one methyl substituent beyond propyl lead to increased affinity, possibly due to nonpolar structures in the region immediately adjacent to the choline binding site.

The requirements for binding at the carrier site on the surface of the membrane and for the subsequent step of translocation through the membrane were distinctly different. The hydroxymethylene group made an important contribution to choline binding, but did not facilitate translocation.[36] In a subsequent study,[37] it was concluded that a cyclic carrier model in which a substrate site is exposed on only one side at a time could be responsible for choline transport in erythrocytes. A similar conclusion was reached for Ehrlich ascites cells.[23]

B. SODIUM-DEPENDENT HIGH-AFFINITY CHOLINE TRANSPORT SYSTEM IN CHOLINERGIC NERVE ENDINGS

The notion that choline transport is an important aspect of a functioning cholinergic nerve terminal has long been suspected. As early as 1961, Birks and MacIntosh[57] predicted that a choline carrier must be located in a membrane lying between the extracellular fluid and the sites of ACh formation. Early results suggested that choline moved into neuronal tissue by a mechanism that was not highly energy- or Na^+ dependent and only a small portion of the choline was converted to ACh, with most remaining as free choline.[58-62] It was in 1971 when Haga[63] found that the synaptosomal synthesis of radiolabeled ACh from radiolabeled choline was dependent on the presence of Na^+ ion and the conversion of choline to ACh was much more efficient at low concentrations (μM range). Since then the presence of two choline transport systems have been demonstrated in nervous tissues.[64-67] One was a low-affinity system ($K_t > 30$ μM) which was not very energy and Na^+ dependent, relatively insensitive to HC-3, did not parallel the regional distribution of choline acetyltransferase in brain, and was assumed mainly to subserve phospholipid synthesis. The other had a high affinity for choline ($K_t < 5$ μM), strict Na^+ dependency, was highly sensitive to HC-3, transported choline against a concentration gradient, and was associated with an efficient conversion of choline to ACh. It had a regional distribution in brain which paralleled that of choline acetyltransferase. It was suggested that sodium-dependent high-affinity choline transport (SDHACT) had a direct or kinetic coupling to the acetyltransferase[68] and might be rate limiting as well as the regulatory step for ACh synthesis in cholinergic synaptosomes.[69-72] In addition, a precise structure-activity relationship on SDHACT with a variety of synthetic choline analogues has been reported by several groups.[73-76] More information on the SDHACT system is available in previous reviews.[45, 77-79]

"High-affinity transport" in noncholinergic tissues has not been shown to have the properties of SDHACT in cholinergic nerve terminals. It may be that actively dividing cells such as neuroblastoma,[22] Y-79 retinoblastoma,[25] Novikoff hepatoma,[28,29] chick fibroblasts,[39] as well as rat pheochromocytoma[80] and brain embryo cells[81] exhibit a high-affinity choline uptake that is related to their rates of membrane formation.

Specific information is not currently available on the molecular component(s) of the SDHACT carrier nor on the low-affinity choline transport system. However, it has been reported that antibodies (anti-Ub) directed against ubiquitin, a protein found universally in eukaryotes, specifically inhibited the SDHACT activity in rat cerebral cortical synaptosomes.[82] Binding of anti-Ub to synaptosomes was saturable and occurred over the same concentration range at which SDHACT inhibition was observed, indicating a close relationship between ubiquitin and the SDHACT system at the synaptosomal surface. In other studies [^{3}H]HC-3 was shown to bind to crude synaptic membranes from rat brain.[83] This labeled ligand may be a useful probe for investigating the SDHACT system because its affinity for the carrier was approximately 150-fold higher than choline.[69]

Recently, the relationship between CK and choline acetyltransferase activities was investigated in isolated superior cervical ganglia as a function of extracellular choline concentration.[84] Ganglionic CK activity was reduced at low extracellular choline (1 to 5 μM) but rose as the choline concentration was raised to 10 to 50 μM in the incubation medium. Acetyltransferase activity, on the other hand, remained at a significantly high level during incubations at low choline levels (1 to 10 μM), but became inhibited as the extracellular choline concentration was raised to 50 to 100 μM in the medium. From the observed opposite changes in ganglionic CK and acetyltransferase activities as functions of extracellular choline concentration, it was proposed that CK could direct choline taken up by the cell either for storage in PC or for synthesis of ACh in cholinergic nerve terminals.[84] This idea is supported by earlier investigations indicating that the concentrations of ACh and choline in brain following the administration of choline could be modulated by CK activity.[85] How the activation and the inactivation of ganglionic CK are regulated by choline remains to be clarified. A significant amount of CK has been shown to exist in membrane-associated form in rat striata synaptosomes,[86,87] which might be of relevance to the above findings.[84]

III. SUBCELLULAR LOCALIZATION OF CHOLINE KINASE

It is generally believed that CK exists in all eukaryotic cells. In most instances, the activity has been recovered in the high-speed supernatant, indicating a cytosolic origin.

There are a few exceptions in which CK activity was detected on subcellular membranes, for instance, the anaerobic protozoon *E. caudatum.*[53] This location might be connected with a role of CK in trapping and concentrating choline within the cell. An efficient mechanism for trapping choline might be essential for an organism that has an obligatory nutritional requirement for choline,[91] and where the presence of this base in the normal environment of the organism, i.e., the rumen, is likely to be at very low concentration and transitory.[92]

Membrane-associated CK activity has been characterized in rat brain.[86,93-95] In corpus striata, approximately 40% of the total activity was associated with the particulate fraction, of which 18% was with the purified synaptosomal fraction.[86] The same investigators compared the kinetic mechanism of CK associated with cytosolic and membrane fractions of synaptosomes and found no significant differences between the two activities.[87] The authors noted a possible relationship between membrane-bound CK activity and the high-affinity transport of choline (SDHACT)[86] because corpus striatum has been shown to be a rich source of cholinergic synaptosomes in brain tissue.[67,78]

Both CK and ethanolamine kinase (EK) activities were found in highly purified myelin fraction from rat brain, amounting to 17 to 20% when compared to that in brain homogenate.[94,95] Mixing experiments and repeated purification of isolated myelin indicated that CK and EK were intrinsic components of the myelin membrane. Researchers also found that myelin had the other enzymes needed to convert diacylglycerol to both PC and phosphatidylethanolamine (PE), i.e., phosphocholine and phosphoethanolamine cytidylyltranferases and choline- and ethanolaminephosphotransferases.[94,95] The authors suggested that such reactions could be involved in a maintenance and/or remodeling role rather than synthesis of bulk myelin constituents during initial myelogenesis.

A report that CK activity was found almost entirely in the microsomal fraction in both liver and brain from rat and mouse[96] appears to be incorrect.[3,4] Finally, the majority of CK activity could be detected in the mitochondrial faction in the filaments of *Cuscuta reflexa,* whereas no significant activity was recovered in the high-speed supernatant of the homogenate.[97] Thus, with a few exceptions, CK appears to be a cytosolic enzyme.

IV. ASSAY METHODS

The most widely used methods of analysis of the reaction products of CK can be classified as follows:

1. Unreacted free [^{14}C]choline is precipitated as an iodine complex[112] or as a reineckate salt.[98] The [^{14}C]phosphocholine, which is not precipitated by this reaction, is measured.
2. Extraction is performed of unreacted [^{14}C]choline by an organic cation exchanger, sodium tetraphenylboron in allyl cyanide[99,100] (or 3-heptanone),[101] and determination of the remaining [^{14}C]phosphocholine in the aqueous phase.
3. One of the CK reaction products, ADP is coupled to pyruvate kinase (1) and lactate dehydrogenase (2) reactions in the presence of phosphoenoylpyruvate and NADH. The disappearance of NADH can be monitored either spectrophotometrically[2] or spectrofluorometrically.[102]

Choline + ATP → phosphocholine + ADP
ADP + phosphoenoylpyruvate → pyruvate + ATP (1)
Pyruvate + NADH → lactate + NAD^{+2}

4. Separation of [^{14}C]phosphocholine from unreacted [^{14}C]choline is carried out by means of either cationic[103,128] or anionic[104,105] ion-exchange resin, by paper chromatography,[5,106] by thin-layer chromatography,[81,107] or by high performance liquid chromatography.[108]

In our laboratory, an anion-exchange column (Dowex® 1 × 8, OH^- form) method has been used extensively for the isolation of [^{14}C]phosphocholine from the incubation medium because of the high sensitivity of the method, its reproducibility, and its relative ease. We have simplified[109,110] the method[104] and modification[105] by utilizing a Gilson-type (P-1000) pipette tip with a cotton plug at the bottom as a column in which 0.5 ml of the Dowex® resin is packed. The time required for the assay is about 2 h for 100 samples. The reaction rate is constant for at least 30 min and proportional to the amount of enzyme within the range of 0.5 mg protein when crude tissue high-speed supernatant is the enzyme source.[110,111]

V. PURIFICATION OF CHOLINE KINASE

The results of major purification studies of CK from various enzyme sources are summarized in Table 2 in chronological order.

In 1976, Brophy and Vance[114] introduced choline-Sepharose® affinity chromatography for the purification of rat liver CK and reported a 500-fold purification in one step. This approach made a significant contribution to the first complete purification of CK by our laboratory[3] in 1984. The purification was started with freshly prepared kidney cytosol from 130 Wistar rats. The steps involved pH 5.1 acid precipitation, ammonium sulfate fractionation, DEAE-cellulose chromatography, and Sephadex® G-150 gel filtration. The partially purified preparation was applied on a small choline-Sepharose® affinity column.[3] The activity of CK was eluted in two major fractions, I and II, with approximately 80% recovery. Both fractions were analyzed by sodium dodecylsulfate polyacrylamide gel electrophoresis. Only one major protein band was detectable in fraction II which migrated at the approximate molecular size of 42 kDa. To evaluate whether the 42-kDa band in fraction II was actually a form of rat kidney CK, we performed native polyacrylamide gel electrophoresis and compared the location of the protein band with that of CK activity. The only detectable protein band was located at exactly the same position as the highest CK activity. Thus, we concluded that the 42- kDa polypeptide in fraction II was rat kidney CK. On the other hand, the molecular size of the intact form of kidney CK was estimated as 75 to 80 kDa by Sephadex® G-150 gel filtration,[3] which indicated that the enzyme in rat kidney existed in a dimeric form.

We also applied choline-Sepharose® affinity chromatography and similar initial steps in an attempt to purify CK from both rat liver and lung.[3] However, only one main peak of CK activity was recovered after the affinity column.[3] The overall elution pattern of the activity from both lung and liver were similar to each other and also to that of fraction I for the kidney preparation. These findings strongly suggested that a 42- kDa form of CK does not appear to exist significantly in either liver or lung cytosols of the rat.

In our initial studies,[3] we suggested that both fractions I and II of the kidney enzyme contained an identical CK protein. We now have evidence that there exist at least two isoforms of CK in fraction I, both of which appear to be distinct from the 42-kDa CK in fraction II. The purification of each of the isoforms in fraction I from rat kidney as well as the form(s) in liver and lung is now under way in this laboratory.

Other reports of purification of CK to near homogeneity exist for monkey lung,[5,106] germinating soybean seeds,[117] rooster liver,[118] and chicken liver.[120] In all of these studies, however, conclusive evidence for the purity of their final preparations was not provided. In

TABLE 2
Purification of Choline Kinase from Various Sources

Source	Purification index (recovery)	Estimated molecular size	Kinetic constants and optimal pH	Ref.
Cuscuta reflexa (mitochondria)	283-fold (33%)		K_m (choline): 2.5 m*M* (allosteric nature) K_m (ATP): 5 m*M* Optimal pH: 8.5	97
Brewer's yeast (cell extract)	308-fold (5%)	67,000 (gel filtration and sedimentation constant)	K_m (choline): 15 μ*M* K_m (ATP): 0.14 m*M* Optimal pH: 8.0—9.5	2, 112
Rabbit brain (acetone powder)	203-fold (10%)		K_m (choline): 32 μ*M* (low); 0.31 m*M* (high) K_m (ATP): 1.1 m*M* Optimal pH: 9.0—10.5	113
Rat liver (cytosol)	68-fold (20%)	166,000 (gel filtration)	K_m (choline): 30 μ*M* K_m (ATP): 3.7 m*M* Optimal pH: 8.0	105
	550-fold (19%)		K_m (choline): 33 μ*M*	114, 115
Monkey lung (cytosol)	1000-fold (10%)	80,000 (gel filtration and sedimentation constant)	K_m (choline): 30 μ*M* K_m (ATP): 5 m*M* Optimal pH: 8.5	5, 106, 116
Soybean seeds (cytosol)		36,000 (SDS-PAGE)[a]		117
Rooster liver (cytosol)	780-fold (36%)		K_m (choline): 40 μ*M* (low); 0.2—0.5 m*M* (high) K_m (ATP): 2.1 m*M*	118
Rat kidney (cytosol)	1500-fold (8.6%)	42,000 (SDS-PAGE)[a] 75,000—80,000 (gel filtration)	K_m (choline): 0.1 m*M* K_m (ATP): 1.5 m*M* Optimal pH: 8.5—9.0	3, 119
Chicken liver (cytosol)	601-fold (17%)	36,000 (gel filtration)	K_m (choline): 60 μ*M* K_m (ATP): 13.3 m*M* Optimal pH: 9.0	120

[a] SDS-PAGE = sodium dodecylsulfate-polyacrylamide gel electrophoresis.

addition, these preparations were extremely unstable whereas the 42-kDa CK purified from rat kidney was stable when sterilized and stored at –20°C, without any stabilizing additives such as glycerol, albumin, etc. Most of the purified CK preparations from the above sources appeared to have a significant activity for ethanolamine phosphorylation. On the contrary, although the highly purified CK preparations from soybean seeds still had some EK activity,[117] these activities could be separated, and highly purified EK preparation from the same source did not have any CK activity.[121] The reported minimum molecular size of EK was 17 to 19 kDa whereas that of CK was 38 kDa.[117,121]

VI. GENERAL PROPERTIES

A large number of reports now exist on the kinetic characterization of the CK reaction from various enzyme sources, and typical parameters from these reports are summarized in Table 3.

A. OPTIMAL pH

There is general agreement for an alkaline pH optimum between pH 8.0 and 9.5. In most instances, the CK activity was reduced at neutral pH, and essentially no activity was detected in acidic pH ranges.[2,97,104-106,113,122-124,126,128,134]

TABLE 3
Kinetic Parameters and Inhibitors of Choline Kinase from Various Tissue Sources

Sources	Optimal pH	Apparent K_m		Inhibitors	Ref.
		Choline (μM)	ATP-Mg^{2+} (mM)		
Yeast	8.0—9.5	20	0.29		2
	8.5—9.5	330	0.83		98
		15	0.14	NEM, DTNB, phosphocholine, ADP, 5′-AMP, ACh	112
		12 (low) 370 (high)			106
Spinach	7.5—10.0	50—100		Excess ATP	122
Rapeseed	8.6—10.0	60	9.0	Phosphocholine, ADP, Mn^{2+}, Pb^{2+}, Hg^{2+}, excess ATP	123
Castor bean	10.0	260	0.86	HC-3, ADP	124
Cuscuta reflexa	8.5	2500	5.0	Excess ATP, ADP, phosphocholine, serine, methionine	97
Rhizobium-infected soybean		150 (CK-I) 81 (CK-II)			125
Marine mollusk	9.0	400	26.0		101
Plasmodium-infected erythrocytes	7.9—9.2	79	1.3	Excess ATP, phosphocholine, ADP, ethanolamine, HC-3	126 127
Ehrlich ascites cells	8.5—10.5	120		Mn^{2+}	128
Krebs II ascites cells		350		HC-3	138
CHO cells	8.0—9.5	40	1.3		129
Chick embryo muscle cells		69	3.6		130
L6 myoblast cells		110	3.2		131
Brain					
(Rabbit)	10.2—10.9	5000	1.5	Serine, ethanolamine, KF	104
		32 (low) 310 (high)	1.1	Excess ATP, ADP, 5′-AMP, CDP-choline, tetramethyl- and tetraethylammonium	113
(Rat)		1700 (low) 5000 (high)		HC-3	103
		2600		Excess ATP, HC-3	132
		120			139
(Rat, soluble)		520	0.23	Excess ATP, HC-3, thiocholine	86, 87
(Rat, particulate)		580	0.32	Phosphocholine, MgAMP-PNP	
(Mouse)		550			140
Spinal cord (rat)		160 (low)		HC-3	133
	560 (high)				
Mammary tissues (bovine)	9.2	250		Ethanolamine, chlorocholine	134, 13
Calvaria (mouse)		380			136
Liver (rat)	8.0	30	3.7	Excess ATP, ethanolamine	105
		33		Ethanolamine, phosphocholine, betaine, HC-3, stearoyl CoA	115
		110	6.7		110
		25—43	3.6		111

TABLE 3 (continued)
Kinetic Parameters and Inhibitors of Choline Kinase from Various Tissue Sources

Sources	Optimal pH	Apparent K_m Choline (μM)	ATP-Mg^{2+} (mM)	Inhibitors	Ref.
Kidney (rat)		31	8.3—10.8		111
		20—25			141
	8.5—9.0	110	1.5	Excess ATP, ethanolamine	119
Lung					
(Monkey)	8.5	30	5.0	HC-3, butyrylcholine,	5
(Rat)		100	2.0	propionylcholine	137
		28	4.2		111
Intestine (rat)		36—89	5.7—10.0		111

B. KINETIC PARAMETERS

There have been wide differences in the values presented for the apparent K_m for choline ranging from micromolar to millimolar as can be seen in Table 3. It was reported that double-reciprocal plots of the initial velocity vs. choline concentration were nonlinear for CK from brewer's yeast at relatively high choline concentrations.[112] This nonlinearity was apparently not recognized in early investigations and possibly the cause of conflicting reports for the apparent K_m for choline. More than one K_m has been reported in highly purified CK preparations from rabbit brain,[113] rooster liver,[118] and primate lung cytosols.[5] These observations could be accounted for either by the presence of two catalytic species (or sites) with different affinities for choline or by a mechanism of negative cooperativity between choline binding sites;[112] choline bound to one site would increase the K_m of the enzyme for choline at another site on the same protein. In the case of the first possibility, there is evidence for the existence of multiple forms of the enzyme,[111,115,125,133] most likely physiologically relevant isoenzymes, or less likely,[111,115] artifacts of proteolysis during preparation. For the second possibility a homogeneous CK preparation from rat kidney has been shown to have two binding sites for choline, probably one catalytic and the other regulatory.[119] Thus, more precise kinetic characterization using homogeneous preparations of each isoform will be required to distinguish the two possibilities.

The apparent K_m values for ATP-Mg^{2+} also have varied widely. Several investigations indicated that the highest activity was obtained with an ATP/Mg^{2+} ratio of 1.0 and that excess ATP was inhibitory.[105,119,122,123,126,132,142] In other studies excess Mg^{2+} besides ATP-Mg^{2+} complex was required for maximal activity.[124,132,143] We observed that the apparent K_m value of a homogeneous rat kidney CK for ATP-Mg^{2+} was extremely high (10 mM) when estimated in the presence of equivalent amounts of ATP and Mg^{2+}, whereas when the value was estimated in 1.5-fold higher concentration of Mg^{2+} than ATP, an apparent K_m for ATP of 1.5 mM was obtained.[119] Thus, the affinity of rat kidney CK for ATP was highly dependent on free Mg^{2+} concentration. Several other investigators report similar results for the CK[132,143] and the EK[105] reactions. For crude enzyme preparations endogenous hydrolytic activities toward ATP must be considered.[99,100]

C. PHOSPHATE DONOR AND Mg^{2+} REQUIREMENT

ATP is specifically required as a phosphate donor for the reaction. No detectable phosphocholine was formed in the presence of CTP, GTP, ITP, or UTP,[104,122,123] with the exception of one report in which GTP gave 50% of the activity with ATP and UTP 20%.[97]

Like other kinase reactions, CK activity is totally dependent on Mg^{2+} ion. Other divalent cations so far examined — Mn^{2+}, Hg^{2+}, Co^{2+}, Ca^{2+}, Zn^{2+}, Ni^{2+}, Cd^{2+}, Cs^{2+}, Cu^{2+}, Ba^{2+}, and Fe^{2+} — could not effectively substitute for Mg^{2+}.[122,126,128,144]

D. REACTION MECHANISM

The early initial velocity, product, and inhibitor studies suggested a random equilibrium mechanism rather than an ordered one.[105,112] Subsequent studies suggested that the forward reaction followed a sequentially ordered mechanism with ATP-Mg^{2+} (or choline) binding to the enzyme first, followed by choline (or ATP-Mg^{2+}), and then activation of the ternary complex by free Mg^{2+}.[87,124,143,145] The release of phosphocholine occurred prior to that of ADP-Mg^{2+}. Thus, the overall rate of the reaction was probably limited by the release of ADP-Mg^{2+} from the complex. The presence of a reverse, ATP-synthesizing reaction was demonstrated in castor-bean endosperm.[124]

E. SUBSTRATE SPECIFICITY AND STRUCTURE-ACTIVITY RELATIONSHIP

The initial studies by Wittenberg and Kornberg[2] demonstrated that the yeast CK preparation could phosphorylate dimethyl- and monomethylethanolamine, diethyl- and monoethylethanolamine, and, though very weakly, ethanolamine. Similar results were reported with partially purified CK preparations from rabbit brain,[113] *Phormia regina* larvae,[149] and rapeseed seeds.[123] CK preparations lacking EK activity have also been obtained from soybean seeds,[121] *E. caudatum*,[146] spinach leaves,[147] *Culex pipiens fatigans*,[148] and rat liver.[115] On the other hand, the highly purified CK preparations from primate lung,[116] rat liver,[114] rat kidney,[119] and chicken liver[120] were all shown to have a considerable catalytic activity toward ethanolamine phosphorylation.

Recently, the substrate specificity of CK partially purified from brewer's yeast was examined with more than 50 structural analogues of choline.[150] Yeast CK had relatively stringent requirements for its substrate, when compared to choline transport in erythrocytes described in an earlier section of this review:

1. None of the substitutents on the quaternary nitrogen could be longer than three atoms (carbon or oxygen) with the exception of the hydroxyalkyl chain which could be no longer than four atoms.
2. There could be no substituent on the β-carbon atom of the hydroxyalkyl chain and no more than one on the α-carbon.

Dimethylethanolamine was as effective a substrate as choline for rabbit brain CK,[113] but a poor substrate for yeast CK.[150] Another reported difference between yeast CK and CK from other systems is that a functional SH could be involved in the former[2,112] but not in the latter systems.[104,113,122,123,128,149]

F. INHIBITORS

There is general agreement that CK is inhibited weakly by ethanolamine and strongly by HC-3. The inhibition by ethanolamine was competitive with choline in most instances,[104,105,115,119,121,127,134,149] whereas the inhibition by HC-3 was reported as competitive,[124,138] noncompetitive,[87,132,133] uncompetitive,[103,142] or mixed-type[138] inhibition vs. choline. The case that ethanolamine was not an effective inhibitor for CK reaction has also been reported.[128,146,148] In a partially purified enzyme preparation from *P. regina* larvae, the closer the structure of the compound was to choline, the more effectively it inhibited CK, e.g., dimethylethanolamine > monomethylethanolamine > ethanolamine.[149]

On the other hand, there is little information on the inhibition of CK by structural analogues of ATP. ADP, a product of CK reaction, has been reported to be a competitive,[97,112] noncompetitive,[123,145] or uncompetitive[124] inhibitor with respect to ATP. Both 5′-AMP and AMP-PNP, an ATP analogue, were found to be competitive vs. ATP in some systems.[87,112]

With a partially purified rat kidney CK preparation, the most efficient inhibition was obtained by adenosine, followed by 3′,5′-cAMP > 5′-ADP > 3′-AMP > 2′,3′-cAMP, and no significant inhibition was observed by 5′-AMP, 2′, 5′-ADP, 3′, 5′-ADP, and 2′-AMP when they were added to the CK assay mixture up to 5 m*M*. The inhibition by either adenosine or 3′,5′-cAMP appeared to be competitive vs. ATP, and the estimated K_i was 0.26 or 4 m*M*, respectively (Ishidate, K. et al., unpublished observation).

G. ACTIVATORS

Polyamines stimulated CK activity severalfold in a partially purified enzyme preparation from rat liver cytosol by increasing the affinity of the enzyme for ATP-Mg^{2+}.[151] This effect was strongest with spermine followed by spermidine, cadaverine, and putrescine. Similar *in vitro* results have been reported with other enzyme systems.[110,111,139,152] The possible physiological role of polyamines in modulating CK activity has been discussed.[139,144,153,154]

There is one report in which several structural analogues of choline caused a significant stimulation of CK activity in an enzyme preparation from Ehrlich ascites cells.[128]

VII. ENZYME INDUCTION

A significant and intensely studied aspect of CK has been its inducibility in various systems. CK activity has been induced by a number of experimental manipulations (Table 4). Administration of estrogen analogues to young male chickens caused a three- to fivefold increase in hepatic CK activity.[161,162] The increase in CK activity could be accounted for entirely by an increase in the amount of enzyme[118] and that the hormone treatment caused a corresponding increase in translatable CK mRNAs in chicken liver.[162] We found that the administration of polycyclic aromatic hydrocarbon carcinogens such as 3-methylcholanthrene (3-MC) and 3,4-benzo(a)pyrene caused a significant and specific induction of hepatic CK activity in the rat.[109,157] Actinomycin D (AD) or cycloheximide (CH) completely blocked the induction of CK activity by these carcinogens.[157] In a subsequent study, we found that a more hepatotoxic agent, carbon tetrachloride (CCl_4) caused a more rapid and drastic induction of CK activity in rat liver.[159] When AD was administered with CCl_4, the induction of CK activity was almost completely blocked. When AD was injected 3 h after CCl_4 treatment, the rapid increase of CK activity during 3 to 6 h was not significantly affected, indicating that the messenger for CK, which had been stimulated by CCl_4 treatment, continued as a template for additional enzyme synthesis during 3 to 6 h. An inhibitor of protein synthesis, CH almost completely blocked the induction of CK activity by CCl_4 even when injected 3 h after the CCl_4.[159]

The induction of CK in rat liver caused by 3-MC or CCl_4 was further characterized by immunological cross-reactivity with rabbit polyclonal antibodies raised against the pure rat kidney CK.[3] While the CK activity in rat kidney, lung, intestine, and uninduced normal liver cytosols was found to be blocked almost completely by the addition of small amounts of antiserum, it blocked only partially CK in 3-MC-treated rat liver. Most of the activity in CCl_4-induced liver could not be immunoprecipitated by antiserum.[3] Approximately 40% of CK activity in 3-MC-induced liver and more than 80% of the activity in CCl_4-induced liver was not immunoprecipitable. These results indicated that induced form(s) of CK were different from forms existing in normal untreated rat liver cytosol. Further biochemical characterization of the inducible form(s) of CK by 3-MC and CCl_4 has been reported.[110,111,153]

Similar findings were reported in soybean plants infected with *Rhizobium*,[125,170] in which a new form of CK (termed CK-II) was induced in the host-cell cytoplasm. Since *Rhizobium* lacked CK, the newly formed enzyme was thought to be plant derived. CK-II was separated from CK-I, the normal enzyme in plant tissue. Thus, *Rhizobium* has a gene or set of genes which send a

TABLE 4
Modulation of Intracellular Choline Kinase Activity by Various Experimental Manipulations

Tissue sources	Factors	Proposed mechanism	Degree of change (%)[a]	Ref.
		Stimulation		
Liver (rat)	Essential fatty acid deficiency		+250	155
	Ethionine		+30—260	156
	3-Methyl-cholanthrene	Enzyme induction	+80—105	157, 158
	Carbon tetra-chloride	Enzyme induction	+200—300	153, 159
Liver (mouse)	Excess choline or ethanolamine	Enzyme induction	+50—100	160
Liver (rooster)	Diethylstilbestrol	Enzyme induction	+200	118, 161
Liver (chicken)	17β-Estradiol	Enzyme induction	+324	162
Hepatocytes (rat)	cAMP analogue		+21	163
3A2 liver cells	Insulin	Enzyme induction	+120	152
Kidney (rat)	Uninephrectomy		+23—61	141, 164
Lung (rat)	Insulin		+27	165
Mammary gland (mouse)	Insulin + cortisol + prolactin	Activation by polyamines	+270—460	144
Superior cervical ganglion (rat)	Axotomy		+100	166
3T3 fibroblasts	10% serum		+100—200	167, 168
Nb2 lymphoma cells	Prolactin	Enzyme induction	+110—340	169
Soybean	*Rhizobium* infection	Enzyme induction	+250	125, 170
		Suppression		
Liver (rat)	Fasting-refeeding		−70	171
		Malathion	−20	172
Liver (chicken)	cAMP	Phosphorylation	−45	120, 173
Kidney and brain (mouse)	Ethanol		−40 ~ −60	174
L6 myoblast cells	25-Hydroxycholesterol	−58	131	

[a] The values are expressed by percent changes in specific activity with that in untreated animals or cells as 100%.

message to the eukaryotic partner inducing the formation of CK-II. The possibility still exists that CK-II may be posttranslationally modified from CK-I.

VIII. COVALENT MODIFICATION

Apart from the mechanism for enzyme induction, there has been some evidence for the short-term regulation of CK via covalent modification. cAMP caused a significant inhibition of *de novo* phospholipid synthesis in chicken liver slices.[173] An analysis of [^{32}P]incorporation into the precursors of phospholipids suggested the formation of phosphocholine and phosphoethanolamine to be the main sites of inhibition by cAMP. In subsequent studies,[120] researchers incubated [γ-^{32}P]ATP with chicken liver cytosol in the presence or absence of cAMP, then purified the enzyme through a choline-Sepharose® column. The results suggested that CK could be phosphorylated when the cytosol was incubated in the presence of ATP together with cAMP and that such phosphorylated enzyme was less active.

Contrary to the avian system, cAMP analogues have been reported to cause either a stimulation[163] or no significant change[17] in CK activity when added to the medium of primary cultured rat hepatocytes. Furthermore, the effect of cAMP *in vitro* on CK activity in both rat liver and kidney has been thoroughly examined in our laboratory. At this point, however, there is no evidence that AMP-dependent protein kinase or any other kind of phosphorylation-dephosphorylation mechanism, could be involved in regulation of CK activity in a mammalian system (Ishidate, K. et al., unpublished observation).

On the other hand, Hosaka et al.[175] recently cloned a yeast (*S. cerevisiae*) CK gene (termed *CKI*) and found that there were two putative phosphorylation sites (Arg-Arg-His-Ser and Arg-Arg-Ala-Ser)[176] near the N-terminal region of a deduced amino acid sequence of CK. It is, however, not known whether any covalent modification mechanism is involved in the regulation of yeast CK activity.

IX. ISOZYMES AND THEIR MOLECULAR CHARACTERIZATION

The first successful separation of CK activity into two components was reported with the enzyme preparation from spinach leaves.[147] Four peaks of CK and three peaks of EK were resolved for the activities from rat liver.[115] The existence of two forms of EK and their physical separation through either a DEAE-cellulose column or gel filtration has been reported for the enzyme preparations from rat[105] and primate[5] livers and from soybean seeds.[121] Recently, multiple forms of CK in rat tissues were extensively characterized in our laboratory, and we concluded that CK does not exist in one particular active form but instead in several isoforms (probably isozymes) which differ from each other in electrophoretic mobility, molecular size, and kinetic properties.[111]

Relatively larger molecular form(s) of CK existed in liver, particularly in CCl_4-induced liver, and the form in both kidney and lung appeared to be intermediate in size, with the smallest form in the intestinal preparation. There appear to be three distinct forms of CK in rat liver, termed CK-I, -II, and -III.[153] CK-II appeared to be the main form in normal rat liver cytosol, having an apparent molecular size of 130 to 140 kDa and an apparent *pI* of 5.3 to 5.5. CK-I was a relatively minor form with a molecular mass of 95 kDa and an apparent *pI* of 4.8. CK-III was not present significantly in normal rat liver, but it was the most abundant form in CCl_4-treated rat liver. CK-III had an apparent molecular size of 175 to 195 kDa and an estimated *pI* of 6.0. In addition, while CK-I and -II were shown to have common antigenicity, CK-III was immunologically distinct from the other two forms.[3,153]

More recently, we have utilized immunoprecipitation techniques and identified three major antibody-specific bands for rat kidney cytosol, with estimated molecular sizes of 52, 42, and 35 kDa. Since the form we previously purified from rat kidney showed a molecular size of 42 kDa, the other form(s) of kidney CK remaining unpurified must consist of components having 52 and/or 35 kDa as their minimum molecular size. Likewise, the possible molecular forms or subunits of CK in other rat tissues were estimated as follows: 65 and 52 kDa in both lung and intestine with some 42 and 35 kDa as minor forms and 65, 52, and 44 kDa in normal untreated liver (Ishidate, K. et al., unpublished data).

X. IDENTITY WITH ETHANOLAMINE KINASE

Evidence for the existence of separate kinases for the phosphorylation of choline and ethanolamine has been reported in some plant tissues,[117,121,123,147] rumen protozoon,[146] *C. pipiens fatigans*,[148] and in animal liver.[5,105,115] Individual control mechanisms of the two kinase activities have been investigated under certain physiological states[177,178] as well as under dietary or experimental manipulations such as fasting-refeeding,[171] choline deficiency,[179] essential fatty acid deficiency,[155,180] excess choline or ethanolamine,[160] and ethanol[174] treatments of animals. All of these findings indicated that CK and EK were either separate enzymes or these two kinase activities could be mediated by two distinct active sites on the same enzyme.[105,132,134,142,155,180,181]

On the other hand, a single enzyme catalyzing the phosphorylation of choline and ethanolamine has been suggested from several purification studies. The initial purification study from rat liver suggested that both CK and EK activities may be associated with the same protein.[114] In subsequent studies,[115] however, both activities could be resolved by gel electrophoresis, suggesting that both kinase activities must have resided on separate proteins in rat liver. A highly purified CK from primate lung still had considerable EK activity.[116] When the enzyme was inactivated by heat, HC-3, trypsin digestion, or *p*-hydroxymercuribenzoate, both CK and EK activities were destroyed at the same rate.[116] Thus, in primate lung tissue, there appeared to be only one enzyme for the phosphorylation of choline and ethanolamine. In the liver from the same animal, at least two distinct kinases were detected: one that phosphorylated choline and ethanolamine and appeared similar in properties to the CK of the lung, and a second enzyme that phosphorylated only ethanolamine. A single enzyme appeared to phosphorylate choline and ethanolamine in chicken liver.[120]

Results from our laboratory suggest that, although the CK protein appears to exist in multiple forms in rat tissues, both CK and EK activities reside on the same enzyme in all rat tissues studied.

1. Both kinases were copurified from rat kidney cytosol to apparent homogeneity.[119] The purification index was essentially the same in which 1500-fold purification was achieved for CK with 8.6% recovery of the total activity whereas 1400-fold purification with 8.0% recovery for EK was achieved.[119]
2. Rat polyclonal antibodies were raised against the highly purified rat kidney CK.[3] CK and EK activities not only in kidney, but also in liver, lung, and intestinal cytosols were almost completely blocked by the addition of small amounts of antiserum. In all cases, the inhibition by antiserum was dose dependent and the titration curves for both kinase activities appeared to overlap completely.[119]
3. Crude CK and EK preparations from various rat tissue sources were assayed after gel electrophoresis.[111] In the four tissue preparations examined, both kinase activities were found to comigrate in the gel, contrary to an earlier report.[115]
4. CK and EK activities in rat liver were both highly inducible by the treatment of rats with several hepatotoxins such as 3-MC and CCl_4.[109]

There is some evidence that both kinase activities do not share a common catalytic site on a single protein.[119] The homogeneously purified kidney enzyme showed apparent K_m values of 100 and 585 μM for choline and ethanolamine, respectively. Ethanolamine appeared to be a weak competitive inhibitor (K_i: 4.8 mM) for choline in the CK reaction, and choline appeared to be a very strong competitive inhibitor (K_i: 7.5 μM) for ethanolamine in the EK reaction. Although the exact meaning of the difference in K_m and K_i values for each base has not been fully understood, this could suggest that CK and EK activities do not have a common active site on the same enzyme protein.[119,134]

XI. GENE CLONING

Recently, yeast mutants defective in CK were isolated.[184] The structural gene for CK (*CKI*) was isolated from a *S. cerevisiae* genomic library by means of genetic complementation using one of the mutants.[175] The cloned DNA was subcloned into a 2.7- Kb DNA fragment. The *CKI* gene on the multicopy plasmid caused the overproduction of not only CK, but also EK activities in yeast strains carrying the plasmid. The nucleotide sequence of the subcloned DNA which contained *CKI* gene was determined. The sequence contained an open reading frame capable of encoding a protein of 582 amino acid residues with a calculated molecular size of 66,316, which was in good agreement with the reported molecular weight of 67,000 for yeast CK in its native form.[112] The primary translation product contained two putative phosphorylation sites[176] in the N-terminal region of the protein. When placed under the control of the *lac* promoter on plasmid pUC19 vector, the *CKI* gene was expressed in *Escherichia coli* which does not possess CK. The cell-free extract of *E. coli* cells carrying the recombinant plasmid exhibited phosphorylation activity for both choline and ethanolamine. When the *CKI* locus in the wild-type yeast genome was inactivated by replacement with the *in vitro* disrupted *CKI* gene, the yeast cells lost virtually all of the CK activity and most of the EK activity. Thus, researchers concluded that CK appears to be monocistronic and the EK activity could be a second activity of CK in the yeast.[175] The cloning of CK from any other source has not yet been reported.

XII. FUTURE STUDIES

While there is evidence for the induction of CK in various experimental systems, the physiological meaning has not been fully clarified. A regulatory role for CK in PC biosynthesis, as well as in ACh synthesis in cholinergic synaptosomes,[84,85] has been demonstrated in some model systems such as CHO cells,[129] 3T3 fibroblasts,[167,168] rooster liver,[161,162,181] mouse calvaria,[136] and also in rat liver.[155,182] On the other hand, a number of recent studies have suggested phosphocholine cytidylyltransferase (EC 2.7.7.15) to be rate limiting and/or regulatory in overall PC biosynthesis in various animal cells, as will be discussed in Chapter 3. However, a long-term regulatory mechanism, i.e., an increased or decreased synthesis of the enzyme, has not been established for the cytidylyltransferase. It may be that PC biosynthesis is regulated under both long- and short-term mechanisms in which CK can be involved in the former case, whereas cytidylyltransferase is involved in the latter. More precise studies are required into the actual relationship between CK induction and the rate of PC biosynthesis.

The induction of CK by several means suggests that more than one mechanism could be involved in this process. Actually, distinct mechanisms of CK induction were demonstrated between rooster liver by estrogens and rat liver by some hepatotoxins. A new form of CK appeared to be preferentially induced in the latter case. Thus, the mechanism of the induction should be characterized further at the level of gene expression.

It now appears that there exists more than one form of CK in most animal cells. Although this apparent multiplicity might be caused partly by the action of endogenous proteolytic activities, several biochemical and immunological studies indicated that at least two forms of CK could be present in rat tissues. Each form of CK needs to be purified and the function of the various forms elucidated.

Finally, it now seems very likely that CK and EK are the same enzyme in most mammalian systems. A genetic study with cultured mammalian cells would further prove their identity. There is one interesting report[183] that early deaths of two siblings (10 and 17 months) were due to a genetic mutation in EK. Although CK activity was not checked, it may be that CK activity was also deficient. Insufficient production of PC could be a primary cause of early death rather than the proposed insufficiency in PE formation, because PE can be formed through phosphatidylserine decarboxylation. It is clear that the cDNA for mammalian CK needs to be cloned and the sequence of CK determined. This should facilitate studies on the active sites for the CK and EK activities.

ACKNOWLEDGMENTS

The work of our laboratory discussed in this review was supported in part by Grants-in-aid for Scientific Research from the Ministry of Education, Science, and Culture, Japan, the Tokyo Biochemical Research Foundation, the Naito Foundation, Japan, and the Meiji Milk Products Co. Ltd., Japan.

I wish to thank Drs. Yasuo Nakazawa and Dennis E. Vance for their useful discussions on this review and Drs. Satoshi Yamashita and Kohei Hosaka for providing their unpublished observations on yeast CK gene cloning. I also extend my sincere thanks to many collaborators who have participated in the research emanating from this laboratory. Also, I wish to thank Mrs. Ritsuko Matsuo and Mr. Kerry Ko for their excellent assistance in the preparation of the manuscript.

REFERENCES

1. **Kennedy, E. P.,** Metabolism of lipides, *Annu. Rev. Biochem.,* 26, 119, 1957.
2. **Wittenberg, J. and Kornberg, A.,** Choline phosphokinase, *J. Biol. Chem.,* 202, 431, 1953.
3. **Ishidate, K., Nakagomi, K., and Nakazawa, Y.,** Complete purification of choline kinase from rat kidney and preparation of rabbit antibody against rat kidney choline kinase, *J. Biol. Chem.,* 259, 14706, 1984.
4. **Pelech, S. L. and Vance, D. E.,** Regulation of phosphatidylcholine biosynthesis, *Biochim. Biophys. Acta,* 779, 217, 1984.
5. **Ulane, R. E.,** The CDP-choline pathway: choline kinase, in *Lung Development: Biological and Clinical Perspectives,* Vol. 1, Farrell, P. M., Ed., Academic Press, New York, 1982, 295.
6. **Cornell, R. B., Ishidate, K., Ridgway, N. D., Sanghera, J. S., and Vance, D. E.,** The enzymes of phosphatidylcholine biosynthesis, in *Enzymes of Lipid Metabolism II,* Freysz, L., Dreyfus, H., Massarelli, R., and Gatt, S., Eds., Plenum Press, New York, 1986, 47.
7. **Bremer, J. and Greenberg, D. M.,** Methyl transferring enzyme system of microsomes in the biosynthesis of lecithin (phosphatidylcholine), *Biochim. Biophys. Acta,* 46, 205, 1961.
8. **Ansell, G. B., and Spanner, S.,** The metabolism of labeled ethanolamine in the brain of the rat *in vivo, J. Neurochem.,* 14, 873, 1967.
9. **Browning, E. T. and Schulman, M. P.,** [^{14}C]Acetylcholine synthesis by cortex slices of rat brain, *J. Neurochem.,* 15, 1391, 1968.
10. **Sung, C.-P. and Johnston, R. M.,** Evidence for active transport of choline in rat kidney cortex slices, *Can. J. Biochem.,* 43, 1111, 1965.
11. **Hodgkin, A. L. and Martin, K.,** Choline uptake by giant axons of *Loligo, J. Physiol.,* 179, 26, 1965.
12. **Zelinski, T. A., and Choy, P. C.,** Ethanolamine inhibits choline uptake in the isolated hamster heart, *Biochim. Biophys. Acta,* 794, 326, 1984.
13. **Zelinski, T. A., Savard, J. D., Man, R. Y. K., and Choy, P. C.,** Phosphatidylcholine biosynthesis in isolated hamster heart, *J. Biol. Chem.,* 255, 11423, 1980.
14. **Hatch, G. M. and Choy, P. C.,** Enhancement of choline uptake by glycine in hamster heart, *Biochim. Biophys. Acta,* 884, 259, 1986.
15. **Sung, C. P. and Johnstone, R. M.,** Evidence for the existence of separate transport mechanisms for choline and betaine in rat kidney, *Biochim. Biophys. Acta,* 173, 548, 1969.
16. **Zeisel, S. H., Strory, D. L., Wurtman, R. J., and Brunengraber, H.,** Uptake of free choline by isolated perfused rat liver, *Proc. Natl. Acad. Sci. U.S.A.,* 77, 4417, 1980.
17. **Pelech, S. L., Pritchard, P. H., and Vance, D. E.,** Prolonged effects of cyclic AMP analogues on phosphatidylcholine biosynthesis in cultured rat hepatocytes, *Biochim. Biophys. Acta,* 713, 260, 1982.
18. **Pritchard, P. H. and Vance, D. E.,** Choline metabolism and phosphatidylcholine biosynthesis in cultured rat hepatocytes, *Biochem. J.,* 196, 261, 1981.
19. **Jernigan, H. M., Kador, P. F., and Kinoshita, J. H.,** Carrier-mediated transport of choline in rat lens, *Exp. Eye Res.,* 32, 709, 1981.
20. **Masland, R. H. and Mills, J. W.,** Choline accumulation by photoreceptor cells of the rabbit retina, *Proc. Natl. Acad. Sci. U.S.A.,* 77, 1671, 1980.

21. **Massarelli, R., Ciesielski-Treska, J., Ebel, A., and Mandel, P.**, Choline uptake in glial cell cultures, *Brain Res.*, 81, 361, 1974.
22. **Lanks, K., Somers, L., Papirmeister, B., and Yamamura, H.**, Choline transport by neuroblastoma cells in tissue culture, *Nature*, 252, 476, 1974.
23. **Ribbes, G., Hamza, M., Chap, H., and Douste-Blazy, L.**, Carrier-mediated choline uptake by Krebs II ascites cells, *Biochim. Biophys. Acta*, 818, 183, 1985.
24. **Lloveras, J., Hamza, M., Chap, H., and Douste-Blazy, L.**, Action of hemicholinium-3 on phospholipid metabolism in Krebs II ascites cells, *Biochem. Pharmacol.*, 34, 3987, 1985.
25. **Yorek, M. A., Dunlap, J. A., Spector, A. A., and Ginsberg, B. H.**, Effect of ethanolamine on choline uptake and incorporation into phosphatidylcholine in human Y79 retinoblastoma cells, *J. Lipid Res.*, 27, 1205, 1986.
26. **Haeffner, E. W.**, Studies on choline permeation through the plasma membrane and its incorporation into phosphatidylcholine of Ehrlich-lettre-ascites tumor cells *in vitro*, *Eur. J. Biochem.*, 51, 219, 1975.
27. **Doppler, W., Hoffmann, J., Maly, K., and Grunicke, H.**, Amiloride and 5-*N*, *N*-dimethylamiloride inhibit the carrier-mediated uptake of choline in Ehrlich ascites tumor cells, *Biochem. Pharmacol.*, 36, 1645, 1987.
28. **Plagemann, P. G. W.**, Choline metabolism and membrane formation in rat hepatoma cells grown in suspension culture. III. Choline transport and uptake by simple diffusion and lack of direct exchange with phosphatidylcholine, *J. Lipid Res.*, 12, 715, 1971.
29. **Plagemann, P. G. W. and Roth, M. F.**, Permeation as the rate-limiting step in the phosphorylation of uridine and choline and their incorporation into macromolecules by Novikoff hepatoma cells: competitive inhibition by phenethyl alcohol, persantin, and adenosine, *Biochemistry*, 8, 4782, 1969.
30. **Naujokaitis, S. A., Fisher, J. M., and Rabinovitz, M.**, Tetraalkylammonium ions: protection of murine L1210 leukemia and bone marrow progenitor cells *in vitro* against mechlorethamine cytotoxicity and inhibition of the choline transport system, *Chem. Biol. Interactions*, 40, 133, 1982.
31. **Askari, A.**, Uptake of some quaternary ammonium ions by human erythrocytes, *J. Gen. Physiol.*, 49, 1147, 1966.
32. **Martin, K.**, Active transport of choline into human erythrocytes, *J. Physiol.*, 191, 105, 1967.
33. **Martin, K.**, Concentrative accumulation of choline by human erythrocytes, *J. Gen. Physiol.*, 51, 497, 1968.
34. **Martin, K.**, Effects of quaternary ammonium compounds on choline transport in red cells, *Br. J. Pharmacol.*, 36, 458, 1969.
35. **Martin, K.**, Extracellular cations and the movement of choline across the erythrocyte membrane, *J. Physiol.*, 224, 207, 1972.
36. **Deves, R. and Krupka, R. M.**, The binding and translocation steps in transport as related to substrate structure: a study of the choline carrier of erythrocytes, *Biochim. Biophys. Acta*, 557, 469, 1979.
37. **Krupka, R. M. and Deves, R.**, An experimental test for cyclic *versus* linear transport models: the mechanism of glucose and choline transport in erythrocytes, *J. Biol. Chem.*, 256, 5410, 1981.
38. **Adamic, S.**, Accumulation of choline by the rat diaphragm, *Biochim. Biophys. Acta*, 196, 113, 1970.
39. **Barald, K. F. and Berg, D. K.**, High-affinity choline uptake by spinal cord neurons in dissociated cell culture, *Dev Biol.*, 65, 90, 1978.
40. **Wang, F. L. and Haubrich, D. R.**, A simple, sensitive, and specific assay for free choline in plasma, *Anal. Biochem.*, 63, 195, 1975.
41. **Zeisel, S. H., Epstein, M. F., and Wurtman, R. J.**, Elevated choline concentration in neonatal plasma, *Life Sci.*, 26, 1827, 1980.
42. **Bligh, J.**, The level of free choline in plasma, *J. Physiol.*, 117, 234, 1952.
43. **Zeisel, S. H. and Wurtman, R. J.**, Developmental changes in rat blood choline concentration, *Biochem. J.*, 198, 565, 1981.
44. **Cohen, E. L. and Wurtman, R. J.**, Brain acetylcholine: control by dietary choline, *Science*, 191, 561, 1976.
45. **Tucek, S.**, Regulation of acetylcholine synthesis in the brain, *J. Neurochem.*, 44, 11, 1985.
46. **Zelinski, T. A. and Choy, P. C.**, Phosphatidylethanolamine biosynthesis in isolated hamster heart, *Can. J. Biochem.*, 60, 817, 1982.
47. **Yorek, M. A., Rosario, R. T., Dudley, D. T., and Spector, A. A.**, The utilization of ethanolamine and serine for ethanolamine phosphoglyceride synthesis by human Y79 retinoblastoma cells, *J. Biol. Chem.*, 260, 2930, 1985.
48. **Pu, G. A.-W. and Anderson, R. E.**, Ethanolamine accumulation by photoreceptor cells of the rabbit retina, *J. Neurochem.*, 42, 185, 1984.
49. **Sundler, R. and Akesson, B.**, Regulation of phospholipid biosynthesis in isolated rat hepatocytes: effect of different substrates, *J. Biol. Chem.*, 250, 3359, 1975.
50. **Milakofsky, L., Hare, T. A., Miller, J. M., and Vogel, W. H.**, Rat plasma levels of amino acids and related compounds during stress, *Life Sci.*, 36, 753, 1985.
51. **Kruse, T., Reiber, H., and Neuhoff, V.**, Amino acid transport across the human blood-CSF barrier: an evaluation graph for amino acid concentrations in cerebrospinal fluid, *J. Neurol. Sci.*, 70, 129, 1985.

52. **Baba, S., Watanabe, Y., Gejyo, F., and Arakawa, M.**, High-performance liquid chromatographic determination of serum aliphatic amines in chronic renal failure, *Clin. Chim. Acta*, 136, 49, 1984.
53. **Bygrave, F. L. and Dawson, R. M. C.**, Phosphatidylcholine biosynthesis and choline transport in the anaerobic protozoon, *Entodinium caudatum*, *Biochem. J.*, 160, 481, 1976.
54. **Hosaka, K. and Yamashita, S.**, Choline transport in *Saccharomyces cerevisiae*, *J. Bacteriol.*, 143, 176, 1980.
55. **Hosaka, K. and Yamashita, S.**, Induction of choline transport and its role in the stimulation of the incorporation of choline into phosphatidylcholine by polyamines in a polyamine auxotroph of *Saccharomyces cerevisiae*, *Eur. J. Biochem.*, 116, 1, 1981.
56. **Nikawa, J., Tsukagoshi, Y., and Yamashita, S.**, Cloning of a gene encoding choline transport in *Saccharomyces cerevisiae*, *J. Bacteriol.*, 166, 328, 1986.
57. **Birks, R. and MacIntosh, F. C.**, Acetylcholine metabolism of a sympathetic ganglion, *Can. J. Biochem. Physiol.*, 39, 787, 1961.
58. **Potter, L. T.**, Uptake of choline by nerve endings isolated from the rat cerebral cortex, in *The Interaction of Drugs and Subcellular Components of Animal Cells*, Campbell, P. N., Ed., J. & A. Churchill, London, 1968, 293.
59. **Diamond, I. and Kennedy, E. P.**, Carrier-mediated transport of choline into synaptic nerve endings, *J. Biol. Chem.*, 244, 3258, 1969.
60. **Marchbanks, R. M.**, The uptake of [^{14}C]choline into synaptosomes *in vitro*, *Biochem. J.*, 110, 533, 1968.
61. **Schuberth, J., Sundwall, A., Sorbo, B., and Lindell, J.-O.**, Uptake of choline by mouse brain slices, *J. Neurochem.*, 13, 347, 1966.
62. **Schuberth, J., Sundwall, A., and Sorbo, B.**, Relation between Na^+-K^+ transport and the uptake of choline by brain slices, *Life Sci.*, 6, 293, 1967.
63. **Haga, T.**, Synthesis and release of [^{14}C]acetylcholine in synaptosomes, *J. Neurochem.*, 18, 781, 1971.
64. **Yamamura, H. I. and Snyder, S. H.**, Choline: high-affinity uptake by rat brain synaptosomes, *Science*, 178, 626, 1972.
65. **Kuhar, M. J.**, Neurotransmitter uptake: a tool in identifying neurotransmitter-specific pathways, *Life Sci.*, 13, 1623, 1973.
66. **Haga, T. and Noda, H.**, Choline uptake systems of rat brain synaptosomes, *Biochim. Biophys. Acta*, 291, 564, 1973.
67. **Yamamura, H. I. and Snyder, S. H.**, High-affinity transport of choline into synaptosomes of rat brain, *J. Neurochem.*, 21, 1355, 1973.
68. **Meyer, E. M., Engel, D. A., and Cooper, J. R.**, Acetylation and phosphorylation of choline following high- or low-affinity uptake by rat cortical synaptosomes, *Neurochem. Res.*, 7, 749, 1982.
69. **Guyenet, P., Lefresne, P., Rossier, J., Beaujouan, J. C., and Glowinski, J.**, Inhibition by hemicholinium-3 of [^{14}C]acetylcholine synthesis and [^{3}H]choline high-affinity uptake in rat striatal synaptosomes, *Mol. Pharmacol.*, 9, 630, 1973.
70. **Atweh, S., Simon, J. R., and Kuhar, M. J.**, Utilization of sodium-dependent high-affinity choline uptake *in vitro* as a measure of the activity of cholinergic neurons *in vivo*, *Life Sci.*, 17, 1535, 1975.
71. **Barker, L. A. and Mittag, T. W.**, Comparative studies of substrates and inhibitors of choline transport and choline acetyltransferase, *J. Pharmacol. Exp. Ther.*, 192, 86, 1975.
72. **Simon, J. R., Atweh, S., and Kuhar, M. J.**, Sodium-dependent high affinity choline uptake: a regulatory step in the synthesis of acetylcholine, *J. Neurochem.*, 26, 909, 1976.
73. **Hemsworth, B. A., Darmer, K. I., and Bosmann, H. B.**, The incorporation of choline into isolated synaptosomal and synaptic vesicle fractions in the presence of quaternary ammonium compounds, *Neuropharmacology*, 10, 109, 1971.
74. **Holden, J. T., Rossier, J., Beaujouan, J. C., Guyenet, P., and Glowinski, J.**, Inhibition of high-affinity choline transport in rat striatal synaptosomes by alkyl bisquaternary ammonium compounds, *Mol. Pharmacol.*, 11, 19, 1975.
75. **Simon, J. R., Mittag, T. W., and Kuhar, M. J.**, Inhibition of synaptosomal uptake of choline by various choline analogs, *Biochem. Pharmacol.*, 24, 1139, 1975.
76. **Batzold, F., DeHaven, R., Kuhar, M. J., and Birdsall, N.**, Inhibition of high-affinity choline uptake: structure-activity studies, *Biochem. Pharmacol.*, 29, 2413, 1980.
77. **Kuhar, M. J. and Murrin L. C.**, Sodium-dependent, high-affinity choline uptake, *J. Neurochem.*, 30, 15, 1978.
78. **Jope, R. S.**, High-affinity choline transport and acetyl-CoA production in brain and their roles in the regulation of acetylcholine synthesis, *Brain Res. Rev.*, 1, 313, 1979.
79. **Murrin, L. C.**, High-affinity transport of choline in neuronal tissue, *Pharamcology*, 21, 132, 1980.
80. **Melega, W. P. and Howard, B. D.**, Choline and acetylcholine metabolism in PC12 secretory cells, *Biochemistry*, 20, 4477, 1981.
81. **Yavin, E.**, Regulation of phospholipid metabolism in differentiating cells from rat brain cerebral hemispheres in culture: patterns of acetylcholine, phosphocholine, and choline phosphoglycerides labeling from [methyl-^{14}C]choline, *J. Biol. Chem.*, 251, 1392, 1976.

82. **Meyer, E. M., West, C. M., and Chau, V.**, Antibodies directed against ubiquitin inhibit high-affinity [^{3}H]choline uptake in rat cerebral cortical synaptosomes, *J. Biol. Chem.*, 261, 14365, 1986.
83. **Sandberg, K. and Coyle, J. T.**, Characterization of [^{3}H]hemicholinium-3 binding associated with neuronal choline uptake sites in rat brain membranes, *Brain Res.*, 348, 321, 1985.
84. **Ando, M., Iwata, M., Takahama, K., and Nagata, Y.**, Effects of extracellular choline concentration and K^+ depolarization on choline kinase and choline acetyltransferase activities in superior cervical sympathetic ganglia excised from rats, *J. Neurochem.*, 48, 1448, 1987.
85. **Millington, W. R. and Wurtman, R. J.**, Choline administration elevates brain phosphorylcholine concentrations, *J. Neurochem.*, 38, 1748, 1982.
86. **Reinhardt, R. R. and Wecker, L.**, Evidence for membrane-associated choline kinase activity in rat striatum, *J. Neurochem.*, 41, 623, 1983.
87. **Reinhard, R. R., Wecker, L., and Cook, P. F.**, Kinetic mechanism of choline kinase from rat striata, *J. Biol. Chem.*, 259, 7446, 1984.
88. **Chang, C. C. and Lee, C.**, Studies on the [^{3}H]choline uptake in rat phrenic nerve-diaphragm preparations, *Neuropharmacology,* 9, 223, 1970.
89. **Inaba, M.**, Mechanism of resistance of Yoshida sarcoma to nitrogen mustard. III. Mechanism of supressed transport of nitrogen mustard, *Int. J. Cancer*, 11, 231, 1973.
90. **Goldenberg, G. J., Vanstone, C. L., and Bihler, I.**, Transport of nitrogen mustard on the transport-carrier for choline in L5178Y lymphoblasts, *Science*, 172, 1148, 1971.
91. **Broad, T. E. and Dawson, R. M. C.**, Phospholipid biosynthesis in the anaerobic protozoon *Entodinium caudatum*, *Biochem. J.*, 146, 317, 1975.
92. **Broad, T. E. and Dawson, R. M. C.**, Role of choline in the nutrition of the rumen protozoon *Entodinium caudatum*, *J. Gen. Microbiol.*, 92, 391, 1976.
93. **McCaman, R. E. and Cook, K.**, Intermediatary metabolism of phospholipids in brain tissue. III. Phosphocholine-glyceride transferase, *J. Biol. Chem.*, 241, 3390, 1966.
94. **Kunishita, T., Vaswani, K. K., Morrow, C. R., and Ledeen, R. W.**, Detection of choline kinase in purified rat brain myelin, *Neurochem. Res.*, 12, 351, 1987.
95. **Kunishita, T., Vaswani, K. K., Morrow, C. R., Novak, G. P., and Ledeen, R. W.**, Ethanolamine kinase activity in purified myelin of rat brain, *J. Neurochem.*, 48, 1, 1987.
96. **Upreti, R. K., Sanwal, G. G., and Krishnan, P. S.**, Likely individuality of the enzymes catalyzing the phosphorylation of choline and ethanolamine, *Arch. Biochem. Biophys.*, 174, 658, 1976.
97. **Setty, P. N. and Krishnan, P. S.**, Choline kinase in *Cuscuta reflexa*, *Biochem. J.*, 126, 313, 1972.
98. **McCaman, R. E., Dewhurst, S. A., and Goldberg, A. M.**, Choline kinase: assay and partial purification, *Anal. Biochem.*, 42, 171, 1971.
99. **Burt, A. M. and Brody, S. A.**, The measurement of choline kinase activity in rat brain: the problem of alternate pathways of ATP metabolism, *Anal. Biochem.*, 65, 215, 1975.
100. **Burt, A. M. and Narayanan, C. H.**, Choline acetyltransferase, choline kinase, and acetylcholinesterase activities during the development of the chick ciliary ganglion, *Exp. Neurol.*, 53, 703, 1976.
101. **Dewhurst, S. A.**, Choline phosphokinase activities in the ganglia and neurons of *Aplysia*, *J. Neurochem.*, 19, 2217, 1972.
102. **Browning, E. T.**, Fluorometric enzyme assay for choline and acetylcholine, *Anal. Biochem.*, 46, 624, 1972.
103. **Ansell, G. B. and Spanner, S. G.**, The inhibition of brain choline kinase by hemicholinium-3, *J. Neurochem.*, 22, 1153, 1974.
104. **McCaman, R. E.**, Intermediary metabolism of phospholipids in brain tissue: microdetermination of choline phosphokinase, *J. Biol. Chem.,* 237, 672, 1962.
105. **Weinhold, P. A. and Rethy, V. B.**, The separation, purification and characterization of ethanolamine kinase and choline kinase from rat liver, *Biochemistry*, 13, 5135, 1974.
106. **Ulane, R. E., Stephenson, L. L., and Farrell, P. M.**, A rapid accurate assay for choline kinase, *Anal. Biochem.*, 79, 526, 1977.
107. **Pelech, S. L., Power, E., and Vance, D. E.**, Activities of the phosphatidylcholine biosynthetic enzymes in rat liver during development, *Can. J. Biochem. Cell Biol.*, 61, 1147, 1983.
108. **Nelson, L. D., Brown, N. D., and Wiesmann, W. P.**, Simultaneous assay of choline kinase and choline oxidase in tissue by high-performance cation-exchange chromatography and continuous radioactive detection, *J. Chromatogr.*, 324, 203, 1985.
109. **Ishidate, K., Tsuruoka, M., and Nakazawa, Y.**, Alteration in enzyme activities of *de novo* phosphatidylcholine biosynthesis in rat liver by treatment with typical inducers of microsomal drug-metabolizing system, *Biochim. Biophys. Acta*, 620, 49, 1980.
110. **Ishidate, K., Kihara, M., Tadokoro, K., and Nakazawa, Y.**, Induction of choline kinase by polycyclic aromatic hydrocarbons in rat liver. I. A comparison of choline kinases from normal and 3-methylcholanthrene-induced rat liver cytosol, *Biochim. Biophys. Acta*, 713, 94, 1982.
111. **Ishidate, K., Iida, K., Tadokoro, K., and Nakazawa, Y.**, Evidence for the existence of multiple forms of choline (ethanolamine) kinase in rat tissues, *Biochim. Biophys. Acta*, 833, 1, 1985.

112. **Brostrom, M. A. and Browning, E. T.**, Choline kinase from brewers' yeast: partial purification, properties, and kinetic mechanism, *J. Biol. Chem.*, 248, 2364, 1973.
113. **Haubrich, D. R.**, Partial purification and properties of choline kinase from rabbit brain: measurement of acetylcholine, *J. Neurochem.*, 21, 315, 1973.
114. **Brophy, P. J. and Vance, D. E.**, Copurification of choline kinase and ethanolamine kinase from rat liver by affinity chromatography, *FEBS Lett.*, 62, 123, 1976.
115. **Brophy, P. J., Choy, P. C., Toone, J. R., and Vance, D. E.**, Choline kinase and ethanolamine kinase are separate, soluble enzymes in rat liver, *Eur. J. Biochem.*, 78, 491, 1977.
116. **Ulane, R. E., Stephenson, L. L., and Farrell, P. M.**, Evidence for the existence of a single enzyme catalyzing the phosphorylation of choline and ethanolamine in primate lung, *Biochim. Biophys. Acta*, 531, 295, 1978.
117. **Wharfe, J. and Harwood, J. L.**, Purification of choline kinase from soya bean, in *The Advances in the Biochemistry and Physiology of Plant Lipids*, Appelqvist, L.-A. and Liljenberg, C., Eds., Elsevier/North-Holland, Amsterdam, 1979, 443.
118. **Paddon, H. B., Vigo, C., and Vance, D. E.**, Diethylstilbestrol treatment increases the amount of choline kinase in rooster liver, *Biochim. Biophys. Acta*, 710, 112, 1982.
119. **Ishidate, K., Furusawa, K., and Nakazawa, Y.**, Complete co-purification of choline kinase and ethanolamine kinase from rat kidney and immunological evidence for both kinase activities residing on the same enzyme protein(s) in rat tissues, *Biochim. Biophys. Acta*, 836, 119, 1985.
120. **Kulkarni, G. R. and Murthy, S. K.**, Purification of choline-ethanolamine kinase from chicken liver and its regulation by cAMP, *Indian J. Biochem. Biophys.*, 23, 90, 1986.
121. **Wharfe, J. and Harwood, J. L.**, Lipid metabolism in germinating seeds: purification of ethaolamine kinase from soya bean, *Biochim. Biophys. Acta*, 575, 102, 1979.
122. **Tanaka, K., Tolbert, N. E., and Gohlke, A. F.**, Choline kinase and phosphorylcholine phosphatase in plants, *Plant Physiol.*, 41, 307, 1966.
123. **Ramasarma, T. and Wetter, L. R.**, Choline kinase of rapeseed (*Brassica campestris L.*), *Can. J. Biochem. Physiol.*, 35, 853, 1957.
124. **Kinney, A. J. and Moore, T. S.**, Phosphatidylcholine synthesis in castor bean endosperm: characteristics and reversibility of the choline kinase reaction, *Arch. Biochem. Biophys.*, 260, 102, 1988.
125. **Mellor, R. B., Christensen, T. M. I. E., and Werner, D.**, Choline kinase II is present only in nodules that synthesize stable peribacteroid membranes, *Proc. Natl. Acad. Sci. U.S.A.*, 83, 659, 1986.
126. **Ancelin, M. L. and Vial, H. J.**, Choline kinase activity in *Plasmodium*-infected erythrocytes: characterization and utilization as a parasite specific marker in malarial fractionation studies, *Biochim. Biophys. Acta*, 875, 52, 1986.
127. **Ancelin, M. L. and Vial, H. J.**, Several lines of evidence demonstrating that *Plasmodium falciparum*, a parasitic organism, has distinct enzymes for the phosphorylation of choline and ethanolamine, *FEBS Lett.*, 202, 217, 1986.
128. **Sung, C.-P. and Johnstone, R. M.**, Phosphorylation of choline and ethanolamine in Ehrlich ascites-carcinoma cells, *Biochem. J.*, 105, 497, 1967.
129. **Nishijima, M., Kuge, O., Maeda, M., Nakano, A., and Akamatsu, Y.**, Regulation of phosphatidylcholine metabolism in mammalian cells: isolation and characterization of a Chinese hamster ovary cell pleiotropic mutant defective in both choline kinase and choline-exchange reaction activities, *J. Biol. Chem.*, 259, 7101, 1984.
130. **Sleight, R. and Kent, C.**, Regulation of phosphatidylcholine biosynthesis in cultured chick embryonic muscle treated with phospholipase C, *J. Biol. Chem.*, 255, 10644, 1980.
131. **Cornell, R. B. and Goldfine, H.**, The coordination of sterol and phospholipid synthesis in cultured myogenic cells: effect of cholesterol synthesis inhibition on the synthesis of phosphatidylcholine, *Biochim. Biophys. Acta*, 750, 504, 1983.
132. **Spanner, S. and Ansell, G. B.**, Choline kinase and ethanolamine kinase activity in the cytosol of nerve endings from rat forebrain, *Biochem. J.*, 178, 753, 1979.
133. **Burt, A. M.**, Choline kinase activity in the developing rat spinal cord: differential development of hemicholinium-3-sensitive and -insensitive activity, *J. Neurochem.*, 28, 961, 1977.
134. **Infante, J. P. and Kinsella, J. E.**, Phospholipid synthesis in mammary tissue. Choline and ethanolamine kinases: kinetic evidence for two discrete active sites, *Lipids*, 11, 727, 1976.
135. **Infante, J. P. and Kinsella, J. E.**, Inhibition of phosphatidylcholine synthesis in mammary tissue by 2-chloroethyltrimethylammonium chloride, *Biochem. J.*, 134, 825, 1973.
136. **Stern, P. H. and Vance, D. E.**, Phosphatidylcholine metabolism in neonatal mouse calvaria, *Biochem. J.*, 244, 409, 1987.
137. **Weinhold, P. A., Feldman, D. A., Quade, M. M., Miller, J. C., and Brooks, R. L.**, Evidence for a regulatory role of CTP:choline phosphate cytidylyltransferase in the synthesis of phosphatidylcholine in fetal lung following premature birth, *Biochim. Biophys. Acta*, 665, 134, 1981.

138. **Hamza, M., Lloveras, J., Ribbes, G., Soula, G., and Douste-Blazy, L.**, An *in vitro* study of hemicholinium-3 on phospholipid metabolism of Krebs II ascites cells, *Biochem. Pharmacol.*, 32, 1893, 1983.

139. **Gilad, G. M. and Gilad, V. H.**, Reciprocal regulation of ornithine decarboxylase and choline kinase activities by their respective reaction products in the developing rat cerebellar cortex, *J. Neurochem.*, 43, 1538, 1984.

140. **Schneider, W. J. and Vance, D. E.**, Activities of choline kinase, cholinephosphate cytidylyltransferase and CDP-choline: 1,2-diacylglycerol cholinephosphotransferase in brains from normal and quaking mice, *J. Neurochem.*, 30, 1599, 1978.

141. **Hise, M. K., Harris, R. H., and Mansbach, C. M.**, Regulation of *de novo* phosphatidylcholine biosynthesis during renal growth, *Am. J. Physiol.*, 247, F260, 1984.

142. **Spanner, S. and Ansell, G. B.**, Choline and ethanolamine kinase activity in the cytoplasm of nerve endings from rat forebrain, *Adv. Exp. Med. Biol.*, 101, 237, 1978.

143. **Infante, J. P. and Kinsella, J. E.**, Choline kinase kinetic studies: dual role of Mg^{2+} in the sequential ordered mechanism at low reactant concentrations. Regulatory implications, *Int. J. Biochem.*, 7, 83, 1976.

144. **Oka, T. and Perry, J. W.**, Glucocorticoid stimulation of choline kinase activity during the development of mouse mammary gland, *Dev. Biol.*, 68, 311, 1979.

145. **Infante, J. P., Houghton, G. E., and Kinsella, J. E.**, A novel kinetic mechanism explaining the non-hyperbolic behavior of metal activated enzymes. Case of choline kinase from rat liver, *J. Theor. Biol.*, 86, 177, 1980.

146. **Broad, T. E. and Dawson, R. M. C.**, Distinction between choline and ethanolamine phosphorylation in *Entodinium caudatum*, *Biochem. Soc. Trans.*, 2, 1272, 1974.

147. **Macher, B. A. and Mudd, J. B.**, Partial purification and properties of ethanolamine kinase from spinach leaf, *Arch. Biochem. Biophys.*, 177, 24, 1976.

148. **Ramabrahmam, P. and Subrahamanyam, D.**, Ethanolamine kinase from *Culex pipiens fatigans*, *Arch. Biochem. Biophys.*, 207, 55, 1981.

149. **Shelley, R. M. and Hodgson, E.**, Substrate specificity and inhibition of choline and ethanolamine kinases from the fat-body of *Phormia regina* larvae, *Insect*, 1, 149, 1971.

150. **Clary, G. L., Tsai, C.-F., and Guynn, R. W.**, Substrate specificity of choline kinase, *Arch. Biochem. Biophys.*, 254, 214, 1987.

151. **Fukuyama, H. and Yamashita, S.**, Activation of rat liver choline kinase by polyamines, *FEBS Lett.*, 71, 33, 1976.

152. **Ulane, R. E. and Ulane, M. M.**, The effects of insulin on choline kinase activity in cultured rat liver cells, *Life Sci.*, 26, 2143, 1980.

153. **Tadokoro, K., Ishidate, K., and Nakazawa, Y.**, Evidence for the existence of isozymes of choline kinase and their selective induction in 3-methylcholanthrene- or carbon tetrachloride-treated rat liver, *Biochim. Biophys. Acta*, 835, 501, 1985.

154. **Gilad, G. M. and Gilad, V. H.**, Inhibition of ornithine decarboxylase and glutamic acid decarboxylase activities by phosphorylethanolamine and phosphorylcholine, *Biochem. Biophys. Res. Commun.*, 122, 277, 1984.

155. **Infante, J. P. and Kinsella, J. E.**, Control of phosphatidylcholine synthesis and the regulatory role of choline kinase in rat liver, *Biochem. J.*, 176, 631, 1978.

156. **Tsuge, H., Sato, N., Koshiba, T., Ohashi, Y., Narita, Y., Takahashi, K., and Ohashi, K.**, Change of choline metabolism in rat liver on chronic ethionine-feeding, *Biochim. Biophys. Acta*, 881, 141, 1986.

157. **Ishidate, K., Tsuruoka, M., and Nakazawa, Y.**, Induction of choline kinase by polycyclic aromatic hydrocarbon carcinogens in rat liver, *Biochem. Biophys. Res. Commun.*, 96, 946, 1980.

158. **Ishidate, K., Tsuruoka, M., and Nakazawa, Y.**, Induction of choline kinase by polycyclic aromatic hydrocarbons in rat liver. II. Its relation to net phosphatidylcholine biosynthesis, *Biochim. Biophys. Acta*, 713, 103, 1982.

159. **Ishidate, K., Enosawa, S., and Nakazawa, Y.**, Actinomycin D-sensitive induction of choline kinase by carbon tetrachloride intoxication in rat liver, *Biochem. Biophys. Res. Commun.*, 111, 683, 1983.

160. **Upreti, R. K.**, Influence of choline and ethanolamine administration on choline and ethanolamine phosphorylating activities of mouse liver and kidney, *Can. J. Biochem.*, 57, 981, 1979.

161. **Vigo, C., Paddon, H. B., Millard, F. C., Pritchard, P. H., and Vance, D. E.**, Diethylstilbestrol treatment modulates the enzymatic activities of phosphatidylcholine biosynthesis in rooster liver, *Biochim. Biophys. Acta*, 665, 546, 1981.

162. **Kulkarni, G. R. and Murthy, S. K.**, Induction of choline-ethanolamine kinase in chicken liver by 17-β-estradiol, *Indian J. Biochem. Biophys.*, 23, 254, 1986.

163. **Pelech, S. L., Pritchard, P. H., and Vance, D. E.**, cAMP analogues inhibit phosphatidylcholine biosynthesis in cultured rat hepatocytes, *J. Biol. Chem.*, 256, 8283, 1981.

164. **Bean, G. H. and Lowenstein, L. M.**, Choline pathways during normal and stimulated renal growth in rats, *J. Clin. Invest.*, 61, 1551, 1978.

165. **Gross, I., Smith, G. J. W., Wilson, C. M., Maniscalco, W. M., Ingleson, L. D., Brehier, A., and Rooney, S. A.**, The influence of hormones on the biochemical development of fetal rat lung in organ culture. II. Insulin, *Pediatr. Res.*, 14, 834, 1980.
166. **Gilad, G. M. and Gilad, V. H.**, Increased choline kinase activity in the rat superior cervical ganglion after axonal injury, *Brain Res.*, 220, 420, 1981.
167. **Warden, C. H. and Friedkin, M.**, Regulation of phosphatidylcholine biosynthesis by mitogenic growth factors, *Biochim. Biophys. Acta*, 792, 270, 1984.
168. **Warden, C. H. and Friedkin, M.**, Regulation of choline kinase activity and phosphatidylcholine biosynthesis by mitogenic growth factors in 3T3 fibroblasts, *J. Biol. Chem.*, 260, 6006, 1985.
169. **Ko, K. W. S., Cook, H. W., and Vance, D. E.**, Reduction of phosphatidylcholine turnover in a Nb 2 lymphoma cell line after prolactin treatment: a novel mechanism for control of phosphatidylcholine levels in cells, *J. Biol. Chem.*, 261, 7846, 1986.
170. **Mellor, R. B., Thierfelder, H., Pausch, G., and Werner, D.**, The occurrence of choline kinase II in the cytoplasm of soybean root nodules infected with various strains of *Bradyrhizobium japonicum*, *J. Plant Physiol.*, 128, 169, 1987.
171. **Groener, J. E. M., Klein, W., and Van Golde, L. M. G.**, The effect of fasting and refeeding on the composition and synthesis of triacylglycerols, phosphatidylcholines, and phosphatidylethanolamines in rat liver, *Arch. Biochem. Biophys.*, 198, 287, 1979.
172. **Singh, Y., Chaudhary, V. K., Tyagi, S. R., and Misra, U. K.**, Inhibition of hepatic phosphatidylcholine synthesis by malathion in rats, *Toxicol. Lett.*, 20, 219, 1984.
173. **Bhat, N. R., Rao, A. M., Kulkarni, G. R., and Murthy, S. K.**, *In vitro* inhibition of phospholipid synthesis by cyclic 3′-5′-adenosine monophosphate in chicken liver slices, *Indian J. Biochem. Biophys.*, 16, 288, 1979.
174. **Upreti, R. K. and Shanker, R.**, Effect of oral ethanol administration on choline and ethanolamine phosphorylating activities in liver and brain of mice, *Toxicology*, 11, 297, 1978.
175. **Hosaka, K., Kodaki, T., and Yamshita, S.**, Cloning and characterization of the yeast *CKI* gene and its expression in *Escherichia coli*, (Abstr.) *29th Int. Conf. Biochemistry of Lipids*, Tokyo, September 19 to 22, 1988.
176. **Murray, K. J., El-Maghrabi, M. R., Kountz, P. D., Lukas, T. J., Soderling, T. R., and Pilkis, S. J.**, Amino acid sequence of the phosphorylation site of rat liver 6-phosphofructo-2-kinase/fructose-2,6-bisphosphatase, *J. Biol. Chem.*, 259, 7673, 1984.
177. **Compton, S. K. and Goeringer, G. C.**, Lung development in the chick embryo. II. Choline and ethanolamine kinase activity in the developing lungs of normal and hypophysectomized chick embryos, *Pediatr. Res.*, 16, 561, 1982.
178. **Upreti, R. K.**, Search for sex-dependent and gestation-induced changes in choline and ethanolamine phosphorylating activities, *Experientia*, 34, 166, 1978.
179. **Schneider, W. J. and Vance, D. E.**, Effect of choline deficiency on the enzymes that synthesize phosphatidylcholine and phosphatidylethanolamine in rat liver, *Eur. J. Biochem.*, 85, 181, 1978.
180. **Infante, J. P. and Kinsella, J. E.**, Coordinate regulation of ethanolamine kinase and phosphoethanolamine cytidylyltransferase in the biosynthesis of phosphatidylethanolamine in rat liver: evidence from essential fatty acid-deficient animals, *Biochem. J.*, 179, 723, 1979.
181. **Vigo, C. and Vance, D. E.**, Effect of diethylstilboestrol on phosphatidylcholine biosynthesis and choline metabolism in the liver of roosters, *Biochem. J.*, 200, 321, 1981.
182. **Infante, J. P.**, Rate-limiting steps in the cytidine pathway for the synthesis of phosphatidylcholine and phosphatidylethanolamine, *Biochem. J.*, 167, 847, 1977.
183. **Vietor, K. W., Havsteen, B., Harms, D., Busse, H., and Heyne, K.**, Ethanolaminosis: a newly recognized, generalized storage disease with cardiomegaly, cerebral dysfunction and early death, *Eur. J. Pediatr.*, 126, 61, 1977.
184. **Hosaka, K. and Yamshita, S.**, Isolation and characterization of a yeast mutant defective in cholinephosphotransferase, *Eur. J. Biochem.*, 162, 7, 1987.

Chapter 3

CTP: CHOLINEPHOSPHATE CYTIDYLYLTRANSFERASE

Dennis E. Vance

TABLE OF CONTENTS

I. INTRODUCTION

The reaction catalyzed by CTP:cholinephosphate cytidylyltransferase (CT) was first described by Kennedy and Weiss in 1956.[1] In the same report they demonstrated that the product of this reaction, cytidyl diphosphate (CDP)-choline, was an intermediate in the conversion of cholinephosphate to phosphatidylcholine (PC). The next major development was the report by Schneider in 1963 that the enzyme could be recovered from the cytosol of a rat liver homogenate and the enzyme ativity increased on incubation of the cytosol for several days at 0°C or for several hours at 37°C.[2] Importantly, Schneider also reported CT activity associated with the microsomal fraction. Subsequently, Fiscus and Schneider demonstrated a marked stimulation of CT by phospholipids that had been oxidized and showed that lysophosphatides were particularly effective activators.[3] Another decade elapsed until the next significant step on understanding CT was achieved. In 1977, Choy et al. reported a 963-fold purification of the enzyme.[4] In the same report a high molecular weight aggregated form of the enzyme was described as well as a low molecular weight form ($M_r = 2.0 \times 10^5$, estimated by gel filtration). We now know that the aggregated form of the enzyme from liver is an artifact due to an association of the low molecular weight form with phospholipids present in the cytosol after the homogenization procedure. The partially purified enzyme showed an absolute requirement for phospholipids for activity. It was still another decade before Feldman and Weinhold reported the purification of CT to homogeneity.[5]

Before describing the purification and properties of CT, it is important to note that CT is a distinct enzyme from CTP:ethanolaminephosphate cytidyltransferase, the rate-limiting enzyme involved in phosphatidylethanolamine biosynthesis. The two cytidyltransferases can be separated by Sephadex® G-200 chromatography,[6] and the ethanolaminephosphate cytidyltransferase is absent from partially purified CT.[4] In addition, there is no evidence for translocation of ethanolaminephosphate cytidyltransferase activity between cytosol and membranes as has been described for CT.[7] As noted in Chapter 10, there is genetic evidence that CT and the ethanolaminephosphate cytidyltransferase in yeast are encoded by different genes.

II. PURIFICATION AND PROPERTIES

A. STUDIES ON THE RAT LIVER ENZYME

The purification of CT to homogeneity[5,8] was not reported until 31 years after its initial discovery.[1] Why did it take so long to achieve the purificaton of a cytosolic enzyme from rat liver? I believe the major reason is that we and other laboratories had difficulties because we treated the enzyme as a soluble protein and avoided the use of detergents. The cytosolic enzyme has lipid binding sites, and in the absence of detergents, hydrophobic interactions could promote aggregation of the enzyme or nonspecific binding to columns. Real progress was only achieved when detergents were used, and this led to the reports on purification to homogeneity from Weinhold's laboratory.[5,8] The successful procedure involved incubation of the cytosol with a suspension of PC-oleic acid (1:2), a pH 5.0 precipitation of the lipid-enzyme complex, and extraction of the enzyme from the precipitate by a buffer that contained 20 m*M* octyl glucoside.[8] The redissolved enzyme was purified to homogeneity via a DEAE-Sepharose® column and two passes through a hydroxyapatite column.[5,8] Either octylglucoside or Triton® X-100 was present during the chromatography steps. The CT, purified over 2000-fold from cytosol, was shown to consist of a single subunit with a molecular weight of approximately 45,000 and a specific activity of 47.5 μmol min^{-1} mg^{-1}.[5] The molecular weight based on amino acid composition has been calculated as 49,770.[5] We have repeated the purificaiton of CT from rat liver with a few minor modifications and have confirmed the molecular weight of the subunit and specific activity of the pure preparation.[9] From studies on electrophoresis on gels under nondenaturing conditions, we have evidence that the pure enzyme exists as a dimer with a molecular weight of 90,000 (J. Sanghera and D. E. Vance, unpublished observations). The structure of CT has been studied in both Triton® micelles and bound to membrane vesicles that contained PC, oleic acid, and Triton®.[10] Using cross-linking reagents (e.g., dithiobis [succinimidylpropionate]) evidence was presented that CT, under these conditons, was a dimer of two noncovalently linked subunits.

The pure CT has an apparent K_m for CTP of 0.22 m*M* and for cholinephosphate of 0.24 m*M* when assayed in the presence of saturating amounts of PC:oleic acid (1:1) vesicles.[5] These values are similar to the constants reported earlier for the partially purified enzyme.[4] CT exhibits a broad pH range with an optimum of pH 7.0.[4,8] The enzyme requires certain phospholipids for activity, and this will be described in detail in Section V. The enzyme is inactivated by several sulfhydryl reagents with *p*-chloromercuribenzoate the most potent (80% inactivation at 0.6 μ*M*), and this was not reversed by dithiothreitol.[8] Inactivation by *N*-ethylmaleimide is partially prevented when CTP or cholinephosphate is included in the incubations.[8] Iodoacetamide is one reagent that has no effect on CT activity. CT is usually assayed in the presence of 10 m*M* Mg^{2+}, but other divalent cations were inhibitory at 3 m*M*.[8] Particularly striking is the complete inhibition by Zn^{2+}.[8] The kinetic mechanism for the CT reaction has not been described.

B. CYTIDYLYLTRANSFERASE IN NONHEPATIC TISSUES

CT activity has been found in all animal tissues examined as well as in plants and yeast (Chapter 10). Generally, the activity is recovered in both the cytosol and microsomes, and cytosolic CT is activated by certain phospholipids. The exception to this generalization is the enzyme purified 378-fold from pea stem homogenates to a specific activity of 110 nmol min^{-1} mg^{-1}.[11] The authors claimed that the protein was pure. However, convincing evidence on this point was not presented, and the specific activity was quite low for a pure enzyme, particulary when compared with the specific activity of the pure enzyme from liver (47,500 nmol min^{-1} mg^{-1}). The bulk of the activity of CT in pea stems was recovered from the cytosol, and this activity was not stimulated by phospholipids.[11]

CT from lung has been studied intensely but has not been purified. The properties of this enzyme will be discussed in Chapter 12. The other mammalian enzyme studied in some detail that seems to be different from the liver enzyme is the CT activity found in intestinal mucosa.[12] The cytosolic activity from this tissue was quite high in the absence of exogenous lipid (2.7 nmol min^{-1} mg^{-1}) and was stimulated only severalfold by exogenous lipid. The possibility that the high cytosolic activity might be due to lipid in the cytosol arising from the homogenization was not discussed.

CT has recently been cloned from yeast.[13] The relative molecular mass was calculated to be 49,379, nearly identical to the value (49,770)[5] estimated for the rat liver enzyme. The CT from yeast also appeared to be mostly a hydrophilic protein, even though it is found mainly in the membrane fraction of the cells.[13] The deduced sequence of the enzyme showed local homology with three nucleotidyl-transfer enzymes from *Escherichia coli* and DNA ligase from T_4 phage.[13]

III. SUBCELLULAR LOCATION AND TOPOGRAPHY

It has been clear for some time that CT is an enzyme that is found in both the cytosolic and microsomal compartments of all animal cells examined. The hypothesis that CT in the cytosol is a reservoir of inactive enzyme whereas the microsomal form of the enzyme is the active species[7] will be discussed in Section IV. Since microsomes are a mixture of subcellular membranes, it was not known whether CT was found in other cellular organelles or resided exclusively on the endoplasmic reticulum (ER). We have now shown that Golgi prepared from rat liver homogenates contains CT with a specific activity (0.87 nmol min^{-1} mg^{-1}) slightly higher than recovered in ER (0.57 nmol min^{-1} mg^{-1}).[14] However, because the ER is more abundant than Golgi in liver, the total activity associated with ER was 61 nmol min^{-1} 10 g^{-1} liver compared to 10 nmol min^{-1} 10 g^{-1} liver recovered in the Golgi.[14] There was very low activity associated with plasma membranes or mitochondria, and other subcellular organelles were not examined.[14] In another study CT association with Golgi has also been reported.[15] The function of CT on the Golgi membrane is not known. Since cholinephosphotransferase can also be recovered in Golgi preparations,[14,15] it is possible that Golgi is involved in the synthesis of PC for the export of PC associated with secreted lipoproteins.[14,15] Studies with trypsin treatment of ER and Golgi membranes have shown that CT is rapidly inactivated and therefore is on the cytosolic surface of these membranes.[14,16]

The subcellular localization of CT has been examined in two other studies.[17] In Krebs II cells treated with exogenous phospholipase C, there was translocation of CT from the cytosol to membranes[17] as previously reported from chick embryonic muscle cells.[18] When the membranes were fractionated on Percoll gradients, the CT activity was associated with ER and did not migrate with markers for plasma membranes, *trans* Golgi, lysosomes, or mitochondria.[17] How the generation of diacylglycerol (DG) in the plasma membrane caused an apparently specific translocation of CT to the ER remains to be explained. In another study on choline- and methionine-deficient rats, there was increased activity of CT in the ER but not the Golgi fraction isolated from liver (Yao, Z., Jamil, H., and Vance, D. E., unpublished results). The apparent specific targeting of CT to the ER under these conditions is also not explained.

One other point about subcellular localization should be mentioned. The enzyme used as a marker for Golgi, galactosyltransferase, is localized to *trans* Golgi,[19] and at the present time there is no bona fide enzyme marker for *cis* Golgi. Thus, it is quite feasible that *cis* Golgi are contaminating ER preparations, and we cannot exclude translocation of CT to *cis* Golgi. The definitive experiment on subcellular localization would be immunolocalization of CT detected by electron microscopy of thin sections of a tissue.

IV. THE TRANSLOCATION HYPOTHESIS FOR REGULATION OF CYTIDYLYTRANSFERASE

A. STATEMENT OF THE HYPOTHESIS

In 1984, Pelech and I summarized data from various studies consistent with the idea that the activity of CT in animal cells was enhanced by translocation of inactive enzyme from a cytosolic pool to the ER where it was activated by certain phospholipids.[7] We further proposed that the translocation was regulated by phosphorylation/dephosphorylation reactions, supply of fatty acids, or increased amounts of DG in the ER. This proposal is summarized in Figure 1.

B. EVIDENCE FOR THE TRANSLOCATION MODEL

1. Correlations between Enzyme Activity and Subcellular Location

The first question: how do we know that CT associated with the ER is the active form of the enzyme? This conclusion is based on a striking correlation between the rate of PC biosynthesis in various cellular and whole animal systems and the association of CT with microsomal membranes as summarized in 1984.[7] For example, addition of oleic acid (1 m*M*) to primary hepatocytes in culture caused a twofold increase in the rate of PC biosynthesis which correlated with a 1.8-fold increase of CT activity in the microsomes isolated from these cells.[20] Seven similar correlations between PC biosynthesis and microsomal CT activity were observed.[7] Since then, additional studies in other systems have garnered further support for the proposal.[21-24] Thus, studies from a wide range of experimental systems from many different laboratories support the proposal that the active species of CT is localized on the microsomes. Cytosolic CT appears to be a reservoir of enzyme that can be mobilized at short notice.

2. Fatty Acid-Mediated Translocation and Its Reversibility

Fatty acids have been proposed as regulators of CT binding to cellular membranes (Figure 1).[7] The evidence for this in cultured cells and *in vitro* systems is compelling. Less convincing, and quite unproved, is whether or not fatty acids function as regulators of CT in whole animals.

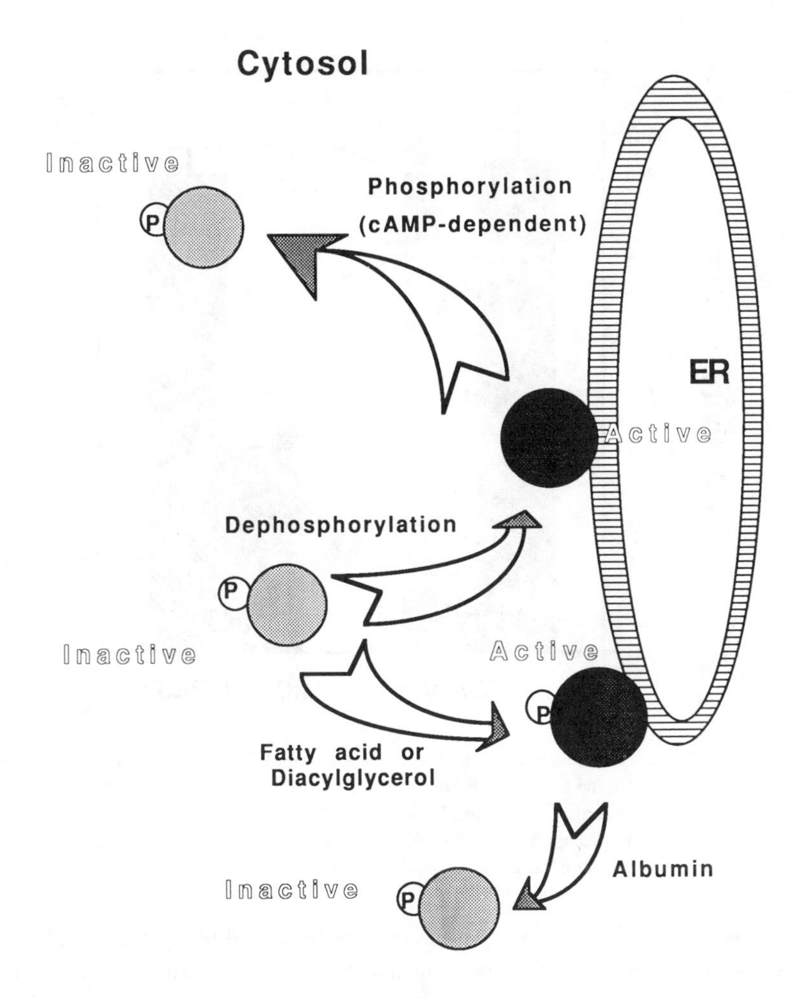

FIGURE 1. Proposed mechanism for translocation of CT between cytosol and ER. It is postulated that CT exists in cytosol in an inactive, phosphorylated form. The enzyme can be translocated to ER by dephosphorylation or the phosphorylated form could be bound to the membrane in the presence of fatty acid or diacylglycerol. Once bound to membranes the lipids in the membrane cause and activation of CT. The association with ER can be reversed by phosphorylation by cAMP–dependent protein kinase or reduction in the levels of diacylglycerol or fatty acid. See text for evidence in support of this model.

The now classic studies of Sundler and Åkesson showed a stimulation of PC biosynthesis in isolated rat hepatocytes by oleic acid.[25] The data at that time suggested that the effect might be by simply supplying more fatty acids for DG synthesis and that DG would be limiting for the cholinephosphotransferase reaction.[25] However, an effect on the CT-catalyzed reaction was not eliminated. Subsequently, addition of albumin-bound fatty acids to cultured hepatocytes demonstrated a two- to threefold stimulation of both the microsomal CT reaction and PC biosynthesis with a concomitant reduction in the cytosolic CT activity.[20] Short-chain fatty acids, such as octanoic acid, were ineffective, but most long-chain fatty acids (saturated and unsaturated) showed a twofold stimulation of PC synthesis.[20] However, oleate itself was unable to activate a partially purified preparation of CT. The data are consistent with the proposal that fatty acids promote the translocation of CT from cytosol to membranes where the enzyme is activiated by certain phospholipids in the membranes.

A difficult problem in studies on regulation of PC biosynthesis is that the changes in activities

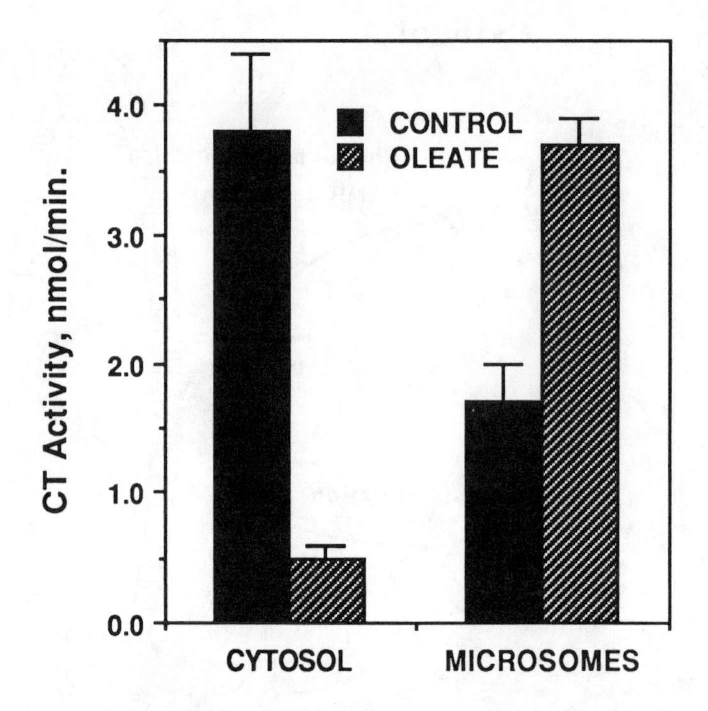

FIGURE 2. *In vitro* translocation of CT from HeLa cell cytosol to microsomes by oleic acid. Cytosol (1.4 mg) isolated from HeLa cells was incubated with 1.4 mg microsomes and no addition or 0.2 m*M* oleate for 10 min at 37°C. Afterward the samples were chilled, centrifuged, and CT assayed in the presence of 2 mg/ml rat liver phospholipid and 0.1 *M* oleate.

are rarely more than twofold. In contrast, the rates of fatty acid biosynthesis and cholesterol biosynthesis can be altered by up to 50-fold. Thus, we were intrigued to discover a five- to tenfold stimulation of PC biosynthesis in HeLa cells incubated with 1 m*M* oleate.[26] If the redistribution of CT in cells were an important regulatory mechanism for PC synthesis, the HeLa cell system should provide a critical test of the hypothesis. The result was unambiguous. In HeLa cells treated with 1 m*M* oleate, there was a dramatic redistribution of CT activity from cytosol to microsomes.[26] Similar results were observed *in vitro* at lower concentrations of oleate (Figure 2).[27]

Artefacts are always a concern in biochemical research, and we worried that the redistribution of activity might have been promoted by the fatty acids during the homogenization of the cells. This possibility was tested by incubating HeLa cells with digitonin which permeabilizes the cell membrane and causes cytosolic enzymes to leak into the medium. After 2 min of incubation at 0° C approximately 60% of the total CT activity was released into the medium of control cells, whereas no CT was released from the HeLa cells treated with 1 m*M* oleate. This result was consistent with the proposal that virtually all of the CT in cells treated with fatty acid was membrane associated. The HeLa cell studies persuaded us that the fatty acid-stimulated increase in PC biosynthesis was mediated by redistribution of CT from cytosol to membranes and was not a homogenization artefact.

The mechanism, specificity, and reversibility of the fatty acid effect on CT binding to membranes of HeLa cells was studied further.[24,27,28] The reversibility of the activation was demonstrated by the addition of bovine serum albumin to HeLa cells that had been pretreated with 0.35 m*M* oleate. The proportion of CT recovered in the membrane fraction in cells treated with albumin was reduced from 100% to approximately 30% (control levels without oleate treatment), and this correlated with extraction of the fatty acid frm the cells by the albumin. Studies with membranes and cytosol *in vitro* yielded similar results.[27] *In vitro* translocation from cytosol to membranes was also promoted by oleoyl alcohol and monoolein.[27] In addition, oleoyl-CoA also promoted redistribution of CT; however, this appeared to be due to hydrolysis of oleoyl-CoA to oleic acid.[27] This was confirmed when oleate, but not oleoyl-CoA, promoted CT binding to liposomes.[28] Numerous related compounds did not promote binding of CT to membranes: lysoPC, lysophosphatidylethanolamine, Triton® X-100, cholate, and calcium palmitate. However, palmitate in the absence of calcium would promote translocation. Apparently, saturated, but not unsaturated, fatty acids pack tightly and form a disoap with calcium (calcium [palmitate]2) and hence, was unavailable for binding to CT.[27] The fatty acid-induced binding of CT to membranes was inhibited by Triton® X-100 but not by 1 *M* KCl. Hence, a variety of compounds with different head groups and hydrocarbon chains promote translocation. Hydrophobic rather than ionic interactions seem to be involved in the binding of CT to membranes.[27,28]

Some convincing evidence that fatty acids play a role in activation of CT *in vivo* came from studies with pre- and postnatal rat lung.[24] Weinhold and co-workers were able to correlate the activity of CT with the fatty acid content in microsomal membranes.[24] In comparisons between fetal rats and rat pups 3-h old, the CT activity in microsomes increased from 1.7 to 6.5 nmol min^{-1} mg^{-1} protein, and the fatty acid content increased from 32 to 62 nmol mg^{-1} protein.[24] The addition of 0.2 m*M* oleate to postmitochondrial supernatants from fetal lung increased microsomal CT activity from 1.3 to 6.8 nmol min^{-1} mg^{-1} protein and the oleic acid content of microsomes increased from a very low level to 68 nmol mg^{-1} protein. Paradoxically, the *in vitro* studies showed a coincident decline in cytosolic CT activity, whereas there was no difference in cytosolic activity between fetal and 3-h-old newborn rats.[24] The *in vitro* translocations to microsomes in the postmitochondrial supernatants were readily reversed by albumin.[24] More recently, fatty acid effects on PC synthesis and CT activity in type II pneumocytes have been reported.[23,29] These cells were isolated from fetal rabbit lung[29] or rat lung[23] and are believed to be the cells in lung responsible for surfactant production. A stimulation of PC synthesis by fatty acids was observed which was attributed to activation of CT by binding to membranes.[23,29]

cis Unsaturated fatty acids stimulated PC synthesis in cultured bovine lymphocytes, whereas saturated and *trans* monounsaturated were ineffective.[30] This result might have been due to calcium binding of the saturated and *trans* fatty acids as discussed above, whereas the *cis* double bonds apparently prevent formation of the dicalcium salt.[27,28] The *cis* fatty acid activation again correlated with increased CT translocation to membranes and the effect was reversed by addition of albumin to the medium.[31] Interestingly, retinoic acid or 5,8,11,14-eicosatetraynoic acid (the triple-bonded analogue of arachidonic acid) activated CT and increased the concentration of CDP-choline in the lymphocytes but stimulated PC biosynthesis marginally.[30] The authors examined the effect of retinoic acid on cholinephosphotransferase *in vitro* and saw only a moderate inhibition of activity.[31] Thus, the mechanism by which retinoic acid stimulates the CT reaction without an effect on PC synthesis remains unexplained.

Studies on PC synthesis in hamster hearts perfused with fatty acids gave different results from those reported above.[22] In this system, CT translocation and PC synthesis were stimulated by stearic acid, but not by oleic acid or arachidonic acid.[22] The reasons for this selectivity are not known.[22]

Thus, there seems little doubt that fatty acids can stimulate CT redistribtuion to microsomal membranes and this translates into increased PC biosynthesis in a variety of *in vitro* and cultured cell systems. The studies in the fetal/newborn lung are the only indications that fatty acid regulation of CT and PC biosynthesis is an important physiological phenomenon. Until more studies are completed, we must reserve judgment on the relevance of fatty acid regulation of CT *in vivo*.

3. Involvement of cAMP and Protein Phosphorylation in Cytidylyltransferase Translocation

cAMP is a secondary messenger that delivers the signal in hepatocytes and other cells and tissues that energy is in short supply. The liver responds by curtailing glycogen synthesis, mobilizing glycogen, and inhibiting fatty acid biosynthesis. Fatty acids are mobilized from adipose tissue and delivered to other tissues as a source of energy. The effect of cAMP on PC biosynthesis in the hepatocytes was not known and was, thus, investigated.[32] Since cAMP is relatively easily degraded, the studies were done with more stable analogues. The results with cultured hepatocytes showed a 40% decrease in PC biosynthesis after a 2-h treatment, and this correlated with a similar reduction in the activity of CT on microsomes isolated from the treated hepatocytes.[32] Longer incubations (12 h or more) of the hepatocytes with cAMP analogues resulted in twofold stimulation of PC biosynthesis and this was accompanied by a 1.4-fold stimulation of microsomal CT activity.[33] This result was in agreement with an earlier report of cAMP analogues causing a stimulation of PC biosynthesis in A549 cells (cells with type 2 pneumocyte characteristics).[34] The activity of CT in these cells was not examined. We also showed that glucagon, a hormone that raises cAMP levels in hepatocytes, inhibited PC biosynthesis by 40% in hepatocytes by inhibition of the CT reaction.[35] Curiously, and still unexplained, we could detect no change in CT activity in microsomes or cytosol isolated from the glucagon-treated hepatocytes.[35] These studies suggest that during the first few hours of treatment an elevation of cAMP in hepatocytes caused an inhibition of PC biosynthesis by release of CT from membranes into the cytosol. A plausible mechanism would be that the cAMP-dependent protein kinase phosphorylates CT, which alters the binding affinity of the enzyme for the membranes.

We wished to gain further evidence in support of the involvement of protein phosphorylation in regulation of CT. Initially, pure enzyme was not available. However, experiment with protein kinase inhibitors and protein phosphatase inhibitors provided additional evidence in support of the model in Figure 1.[36] More recently, with the availability of pure enzyme, we have conducted *in vitro* experiments to show that CT is indeed a substrate for the pure catalytic subunit of cAMP-dependent kinase.[9] In these experiments, we were able to show incorporation of only 0.2 nmol of P_i per nmol of CT subunit.[9] However, this correlated with a 25% release of pure CT bound to PC vesicles. Since CT is isolated from rat liver cytosol, it quite likely exists there in a phosphorylated form, and so it was not very surprising that we observed only a modest effect of the protein kinase on both activity and incorporation of ^{32}P. Fetal pulmonary lung CT (both cytosolic and microsomal forms) has been shown to be inhibited by inclusion of ATP in the incubations. These results were consistent with an inactivation of CT by a phosphorylation reaction.[37] Clearly, the next set of experiments would be to prove phosphorylation of CT *in vivo* as a result of cAMP treatment. However, the essential reagent for these studies is a precipitating antibody to CT. Unfortunately, the antibodies we have prepared recognize CT on immunoblots but will not precipitate the enzyme.

Since fatty acids activated CT and PC synthesis whereas cAMP analogues were inhibitory, we were curious what would happen if hepatocytes were incubated with both compounds. The results clearly showed that fatty acids reversed cAMP-mediated inhibiton of CT and PC biosynthesis and caused a stimulation of CT activity to the same levels observed in the absence of cAMP.[38] Our hypothesis was that fatty acid caused the putative "phosphorylated" form of the enzyme to bind to cellular membranes as shown in Figure 1. Other explanations, such as a fatty acid activation of a protein phosphate phosphatase (that used CT as a substrate and caused binding to membranes), must still be considered.

Presently, there is no evidence in intact animals that cAMP-modulation of CT has an important physiological role. Fasting should cause an increase in cAMP levels in rats, but would at the same time promote the mobilization of fatty acids. From the above studies, we might predict that there would be increased binding of CT to membranes caused by the increased levels of fatty acids in the fasting plasma (0.25 μeq/ml in control plasma vs. 0.8 μeq/ml in fasting plasma).[39] Tijburg et al. have studied the activity and location of CT after a 48-h fast with subsequent refeeding of the rats.[40] They observed over a twofold increase in the microsomal CT in the livers from fasted animals and this was reversed to control levels after refeeding.[40] The mechanism by which the CT was translocated to the membranes was not studied. Two hypotheses would be that there was translocation due to increased levels of fatty acids or DGs (see Section IV) in the livers. The latter seems unlikely since DG levels were measured in an earlier study and the levels in microsomes isolated from 48-h fasted rats were decreased (4.2 vs. 11.8 nmol/mg protein in normally fed rats).[41] Thus, fatty acid levels should be measured in this model system as well as the phosphorylation state of CT when a suitable anitbody is available.

4. Diacylglycerol and PC as Mediators of Cytidylyltransferase Translocation

In vitro experiments clearly demonstrate that DG mediates the translocation of CT from cytosol to membranes. Initial experiments in 1979 showed that DG in cytosol promoted aggregation of the enzyme, whereas other neutral lipids such as triacylglycerol or cholesterol ester were without effect.[42] Similary, incubation of partially purified CT with control cytosol or cytosol isolated from hypercholesterolemic rat livers (enriched more than twofold in DG compared to control livers) resulted in a twofold-enriched binding of CT to the lipid vesicles in hypercholesterolemic compared to control cytosol.[42] In another study, it was demonstrated that CT in HeLa cell cytosol bound to PC vesicles that contained 20 to 30 mol% DG and was fully activated by 29% diolein in egg PC vesicles.[28] Thus, the *in vitro* studies show that DG in membranes will promote CT binding and activation.

In another approach, Kent and co-workers treated cultured cells with exogenous phospholipase C, which degrades PC on plasma membranes to DG and cholinephosphate.[18,21,43-45] The treatment caused translocation of CT to cellular membranes and the process was shown to be reversible and not dependent on protein synthesis. Subsequently, Tercé et al. showed that the phospholipase treatment caused CT translocation mainly to the ER.[17] The conditions of the phospholipase treatment were such that the cells could survive long periods of treatment with the enzyme.[43] A CHO cell mutant defective in CT activity however did not tolerate the phospholipase treatment, indicating the importance of CT to the resynthesis of PC for maintaining the cellular membranes.[43]

A model in which a deficiency of PC in membranes, due to phospholipase treatment, would cause CT translocation was proposed.[44] Whether this hypothesis is true remains to be demonstrated in the phospholipase-treated cells. In two of the cell culture systems examined,[18,45] there was no significant decrease in the PC content of these cells, yet there was at least a twofold increase in CT binding to particulate membranes. In the one system in which DG levels were measured,[18] there was a twofold increase in DG levels in the cells treated with phospholipase C compared to control cells. Thus, an increase in DG, as a result of PC breakdown, would appear to be a more plausible explanation of the phospholipase C-mediated translocation of CT to membranes.

More recent studies with choline-deficient hepatocytes support the level of PC being important in regulating the amount of CT bound to membranes.[52] Hepatocytes prepared from choline-deficient rats, and maintained in a choline- and methionine-deficient medium, showed twofold higher levels of CT bound to the cellular membranes than do hepatocytes supplemented with either choline or methionine.[52] The PC levels in the experimental hepatocytes were reduced from 110 nmol/mg protein in control cells to 75 nmol/mg protein in deficient cells. When supplemented with choline[52] or lyso-PC,[46] the levels of PC increased to control levels after 4-h incubation and the CT was released into the cytosol of the cells. The DG levels were also elevated in the choline- and methionine-deficient hepatocytes and returned to normal after supplementation with choline.[52]

Related to the DG effects on CT, phorbol esters also cause CT to be translocated from cytosol to membranes. This is discussed in Chapter 13.

Although the final answer is not in there is ample evidence to support the hypothesis that DG is a mediator of CT translocation. More recent studies suggest that the level of PC could also be important.

C. IS CYTIDYLYLTRANSFERASE TRANSLOCATION A PHYSIOLOGICALLY IMPORTANT CONTROL MECHANISM?

This is the weakest aspect of the translocation theory because unequivocal data is very difficult to obtain, although there is some evidence in support of this idea. As mentioned in Section IV.B.2, a correlation has been observed between the fatty acid levels in the newborn rat lung and the binding of CT to cell membranes.[24] In another approach, a threefold increase in the rate of PC synthesis was induced in young rats fed a diet enriched in cholesterol and cholate.[47] This correlated with an increase of CT activity in the microsomes isolated from these animals. In another model system, rats that were choline deficient had twice as much CT activity in their liver microsomes than animals fed the same diet supplemented with choline.[52] Since PC biosynthesis was reduced in these animals because of a lack of choline, ostensibly there did not seem to be a correlation of PC biosynthesis with the amount of CT associated with microsomes. However, possibly the liver sensed the requirement for PC synthesis by a yet undefined mechanism and therefore caused CT to bind to the membranes. In this regard, it is noteworthy that PC levels are decreased in the ER from choline-deficient animals. Clearly in the choline-deficient animals the distribution of CT between cytosol and membranes was altered in a whole animal model. However, more data are required to be certain that CT translocation is a significant regulatory mechanism *in vivo*.

V. LIPID REQUIREMENTS FOR ACTIVATION OF CYTIDYLYLTRANSFERASE

It was clear over two decades ago that the activity of CT was stimulated by phospholipids.[3] Since that time several studies have identified the lipid requirements for CT activation.[5,28,48-50] From these studies, the following conclusions were reached. Neutral lipids such as triacylglycerol, cholesterol, and cholesterol ester will not stimulate the enzyme activity. Neutral phospholipids, PC, and PE, were poor activators, whereas, phospholipids carrying a negative charge (phosphatidylserine, phosphatidylinositol) markedly stimulated CT activity. Lyso-PC was not an activator of the partially purified enzyme, whereas lysophosphatidylethanolamine was a very potent activator.[48] In addition, as noted previously, fatty acids did not activate the enzyme in the absence of phospholipids. However, PC vesicles that contained fatty acids were very effective in activating cytosolic CT[28,50] and the pure enzyme.[5] The activity of CT bound to microsomes is not markedly stimulated by addition by exogenous phospholipids. Recently, it has been shown that addition of 10 μM phosphatidylglycerol, diolein, or 1-oleoyl-2-acetylglycerol to the medium of cultured type II pneumocytes promoted activation of the CT reaction in these cells.[51]

Studies with pure CT showed that any of the phospholipids (including PE and PC) would activate the enzyme as well as fatty acids, lyso-PC, and lyso-PE.[53,54]

From these studies it is clear that CT has a requirement for phospholipids, or phospholipid/fatty acid mixtures, in order to have significant catalytic activity. As phospholipids are usually found only in the cellular membranes, this requirement provided additional support for the theory that only the membrane-associated CT is active.

VI. FUTURE DIRECTIONS

Now that CT is available in pure form, albeit in rather small amounts (100 μg from a typical preparation beginning with approximately 200 g of rat liver), future investigations should greatly enhance our understanding of this enzyme and its regulatory role in PC biosynthesis. Cloning of the complementary DNA should yield the primary sequences of the protein. From the cDNA, a probe could be developed to measure the amounts of mRNA for CT under various conditions. There are few, if any, instances where the total CT enzyme activity changes under different perturbations. Thus, the mRNA studies should confirm the tentative conclusion that regulation of the activity of CT at the level of gene expression is not an important control mechanism. The cDNA probe will allow for the identification of the chromosome that contains the gene for CT and permit the cloning and sequencing of the CT gene. Perhaps there will be something of interest for the molecular biologists in the structure of the DNA promoter sequences that code for the expression of an apparently constitutively expressed enzyme. Finally, and perhaps most importantly, the cDNA will allow us to predict the structure of CT. We will then be able to identify sequences in the enzyme that might be important in the reversible binding of CT to membranes, phospholipid vesicles, fatty acids, and DGs. The protein structure will also allow us to begin to identify the sites on CT that are phosphorylated by cAMP-dependent protein kinase[9] and other protein kinases if CT were a substrate.

Pure CT has already allowed us and other laboratories to prepare polyclonal antibodies to the enzyme. This will facilitate numberous studies in intact cells and whole animals. We should be able to determine if phosphorylation/dephosphorylation is an important control mechanism in various model systems. We should be able to determine the half-life of the enzyme. Is the turnover rate of CT in the microsomes the same as in the cytosol, or, is there a precursor-product relationship between CT in the two subcellular locations?

Other structural questions to be asked include whether or not CT is modified by glycosylation or covalently bound fatty acids. What is the stoichiometry of the number of phopsphates on CT? How similar is the structure of CT in the different tissues of an animal and yeast?

Finally, we must continue to look for the unexpected in our studies on CT.

ACKNOWLEDGMENTS

I am most grateful to the Medical Research Council of Canada and the Canadian Heart Foundation for their support of my research on CT for more than a decade. I also acknowledge and thank my many colleagues who have worked so diligently on this enzyme and its regulation. As is usually the case in biochemical research, the daily progress was slow and painstaking. However, when we review the past decade, the progress is visible and real. Finally, I extend my sincere appreciation to Rosemary Cornell, Jean Vance, Richard Kolesnick, Neale Ridgway, and Zemin Yao for their helpful and critical reading of this manuscript.

REFERENCES

1. **Kennedy, E. P. and Weiss, S. B.,** The function of cytidine coenzymes in the biosynthesis of phospholipides, *J. Biol. Chem.*, 222, 193, 1956.
2. **Schneider, W.,** *J. Biol. Chem.*, 238, 3572, 1963.
3. **Fiscus, W. G. and Schneider, W. C.,** The role of phospholipds in stimulating phosphorylcholine cytidyltransferase activity, *J. Biol. Chem.*, 241, 3324, 1966.
4. **Choy, P. C., Lim, P. H., and Vance, D. E.,** Purification and characterization of CTP:cholinephosphate cytidylyltransferase from rat liver cytosol, *J. Biol. Chem.*, 21, 7673, 1977.
5. **Feldman, D. A. and Weinhold, P. A.,** CTP:phosphorylcholine cytidylyltransferase from rat liver; isolation and characterization of the catalytic subunit, *J. Biol. Chem.*, 262, 9075, 1987.
6. **Sundler, R.,** Ethanolaminephosphate cytidylyltransferase; purification and characterization of the enzyme from rat liver, *J. Biol. Chem.*, 250, 8585, 1975.

7. **Vance, D. E. and Pelech, S. L.,** Enzyme translocation in the regulation of phosphatidylcholine biosynthesis, *Trends Biochem. Sci.,* 9, 17, 1984.
8. **Weinhold, P. A., Rounsifer, M. E., and Feldman, D. A.,** The purification and characterization of CTP:phosphorylcholine cytidylyltransferase from rat liver, *J. Biol. Chem.,* 261, 5104, 1986.
9. **Sanghera, J. S. and Vance, D. E.,** CTP:phosphocholine cytidylyltransferase is a substrate for cAMP–dependent protein kinase *in vitro, J. Biol. Chem.,* 264, 1215, 1989.
10. **Cornell, R.,** Chemical cross–linking reveals a dimeric structure for CTP:phosphocholine cytidylyltransferase, *J. Biol. Chem.,* in press.
11. **Price-Jones, M. M. and Harwood, J. L.,** Purification of CTP:cholinephosphate cytidylyltransferase from pea stems, *Phytochemistry,* 24, 2523, 1985.
12. **Mansbach, C. M. and Arnold, A.,** CTP:phosphocholine cytidylyltransferase in intestinal mucosa, *Biochim. Biophys. Acta,* 875, 516, 1986.
13. **Tsukagoshi, Y., Nikawa, J., and Yamashita, S.,** Molecular cloning and characterization of the gene encoding cholinephosphate cytidylyltransferase in *Saccharomyces cerevisiae, Eur. J. Biochem.,* 169, 477, 1987.
14. **Vance, J. E. and Vance, D. E.,** Does rat liver Golgi have the capacity to synthesize phospholipids for lipoprotein secretion?, *J. Biol. Chem.,* 263, 5898, 1988.
15. **Higgins, J. A. and Fieldsend, J. K.,** Phosphatidylcholine synthesis for incorporation into membranes or for secretion as plasma lipoproteins by Golgi membranes of rat liver, *J. Lipid Res.,* 28, 268, 1987.
16. **Vance, D. E., Choy, P. C., Farren, S. B., Lim, P. H., and Schneider, W. J.,** Asymmetry of phospholipid biosynthesis, *Nature,* 270, 268, 1977.
17. **Tercé, F., Record, M., Ribbes, G., Chap, H., and Douste–Blazy, L.,** Intracellular processing of cytidylyltransferase in Krebs II cells during stimulation of phosphtidylcholine synthesis; evidence that a plasma membrane modification promotes enzyme translocation specifically to the endoplasmic reticulum, *J. Biol. Chem.,* 263, 3142, 1988.
18. **Sleight, R.** and **Kent, C.,** Regulation of phosphatidylcholine biosynthesis in cultured chick embryonic muscle treated with phospholipase C, *J. Biol. Chem.,* 255, 10644, 1980.
19. **Farquhar, M. G.,** Progress in unraveling pathways of Golgi traffic, *Annu. Rev. Cell Biol.,* 1, 447, 1985.
20. **Pelech, S. L., Pritchard, P. H., Brindley, D. N., and Vance, D. E.,** Fatty acids promote translocation of CTP:phosphocholine cytidylyltransferase to the endoplasmic reticulum and stimulate rat hepatic phosphatidylcholine synthesis, *J. Biol. Chem.,* 258, 6782, 1983.
21. **Wright, P. S., Morand, J. N., and Kent, C.,** Regulation of phosphatidylcholine biosynthesis in Chinese hamster ovary cells by reversible membrane association of CTP:phosphocholine cytidylyltransferase, *J. Biol. Chem.,* 260, 7919, 1985.
22. **Mock, T., Slater, T. L., Arthur, G., Chan, A. C., and Choy, P. C.,** Effects of fatty acids on phosphatidylcholine biosynthesis in isolated hamster heart, *Biochem. Cell Biol.,* 64, 413, 1986.
23. **Chander A. and Fisher, A. B.,** Cholinephosphate cytidyltransferase activity and phosphatidylcholine synthesis in rat granular pneumocytes are increased with exogenous fatty acids, *Biochim. Biophys. Acta,* 958, 343, 1988.
24. **Weinhold, P. A., Rounsifer, M. E., Williams, S. E., Brubaker, P. G., and Feldman, D. A.,** CTP:phosphorylcholine cytidylyltransferase in rat lung; the effect of free fatty acids on the translocation of activity between microsomes and cytosol, J. *Biol. Chem.,* 259, 10315, 1984.
25. **Sundler, R. and** Åkesson, B., Regulation of phospholipid biosynthesis in isolated rat hepatocytes; effect of different substrates, *J. Biol. Chem.,* 250 3359, 1975.
26. **Pelech, S. L., Cook, H. W., Paddon, H. B., and Vance, D. E.,** Membrane-bound CTP:phosphocholine cytidylyltransferase regulates the rate of phosphatidylcholine synthesis in HeLa cells treated with unsaturated fatty acids, *Biochim. Biophys. Acta,* 795, 433, 1984.
27. **Cornell, R. and Vance, D. E.,** Translocation of CTP:phosphocholine cytidylyltransferase from cytosol to membranes in HeLa cells: stimulation by fatty acid, fatty alcohol, mono- and di-acylglycerol, Bi*ochim. Biophys. Acta,* 919, 26, 1987.
28. **Cornell, R. and Vance, D. E.,** Binding of CTP:phosphocholine cytidylyltransferase to large unilamellar vesicles, *Biochim. Biophys. Acta,* 919, 37, 1987.
29. **Aeberhard, E. E., Barrett, C. T., Kaplan, S. A., and Scott, M. L.,** Stimulation of phosphatidylcholine synthesis by fatty acids in fetal rabbit type II pneumocytes, *Biochim. Biophys. Acta,* 875, 6, 1986.
30. **Anderson, K. E., Whitlon, D. S., and Mueller, G. C.,** Role of fatty acid structure in the reversible activation of phosphatidylcholine synthesis in lymphocytes, *Biochim. Biophys. Acta,* 835, 360, 1985.
31. **Whitlon, D. S., Anderson, K. E., and Mueller, G. C.,** Analysis of the effects of fatty acids and related compounds on the synthesis of phosphatidylcholine in lymphocytes, *Biochim. Biophys. Acta,* 835, 369, 1985.
32. **Pelech, S. L., Pritchard, P. H., and Vance, D. E.,** cAMP analogues inhibit phosphatidylcholine biosynthesis in cultured rat hepatocytes, *J. Biol. Chem.,* 256, 8283, 1981.
33. **Pelech, S. L., Pritchard, P. H., and Vance, D. E.,** Prolonged effects of cyclic AMP analogues of phosphatidylcholine biosynthesis in cultured rat hepatocytes, *Biochim. Biophys. Acta,* 713, 260, 1982.

34. **Niles, R. M. and Makarski, J. S.,** Regulation of phosphatidylcholine metabolism by cyclic AMP in a model alveolar Type 2 cell line, *J. Biol. Chem.*, 254, 4324, 1979.
35. **Pelech, S. L., Pritchard, P. H., Sommerman, E. F., Percival–Smith, A., and Vance, D. E.,** Glucagon inhibits phosphatidylcholine biosynthesis via the CDP-choline and transmethylation pathways in cultured rat hepatocytes, *Can. J. Biochem. Cell Biol.*, 62, 196, 1984.
36. **Pelech, S. L. and Vance, D. E.,** Regulation of rat liver cytosolic CTP:phosphocholine cytidylyltransferase by phosphorylation and dephosphorylation, *J. Biol. Chem.*, 257, 14198, 1982.
37. **Radika, K. and Possmayer, F.,** Inhibition of foetal pulmonary cholinephosphate cytidylyltransferase under conditions favouring protein phosphorylation, *Biochem. J.*, 232, 833, 1985.
38. **Pelech, S. L., Pritchard, P. H., Brindley, D. N., and Vance, D. E.,** Fatty acids reverse the cylcic AMP inhibition of triacylglycerol and phosphatidylcholine synthesis in rat hepatocytes, *Biochem. J.*, 216, 129, 1983.
39. **McGarry, J. D., Meier, J. M., and Foster, D. W.,** The effects of starvation and refeeding on carbohydrate and lipid metabolism *in vivo* and in the perfused rat liver. The relationship between fatty acid oxidation and esterification in the regulation of ketogenesis, *J. Biol. Chem.*, 248, 270, 1974.
40. **Tijburg, L. B. M., Houweling, M., Geelen, M. J. H., and van Golde, L. M. G.,** Effects of dietary conditions on the pool sizes of precursors of phosphatidylcholine and phosphatidylethanolamine synthesis in rat liver, *Biochem. Biophys. Acta*, 959, 1, 1988.
41. **Groener, J. E. M., Klein, W., and van Golde, L. M. G.,** The effect of fasting and refeeding on the composition and synthesis of triacylglycerols, phosphatidylcholines and phosphatidylethanolamines in rat liver, *Arch. Biochem. Biophys.*, 198, 287, 1979.
42. **Choy, P. C., Farren, S. B., and Vance, D. E.,** Lipid requirements for the aggregation of CTP:phosphocholine cytidylyltransferase in rat liver cytosol, *Can J. Biochem.*, 57, 605, 1979.
43. **Sleight, R. and Kent, C.,** Regulation of phosphatidylcholine biosynthesis in mammalian cells; effects of phospholipase C treatment on phosphatidylcholine metabolism in Chinese hamster ovary cells and LM mouse fibroblasts, *J. Biol. Chem.*, 258, 824, 1983.
44. **Sleight, R. and Kent, C.,** Regulation of phosphatidylcholine biosynthesis in mammalian cells; effects of phospholipase C treatment on the activity and subcellular distribution of CTP:phosphocholine cytidylyltransferase in Chinese hamster ovary and LM cell lines, *J. Biol. Chem.*, 258, 831, 1983.
45. **Sleight, R. and Kent, C.,** Regulation of phosphatidylcholine biosynthesis in mammalian cells; effects of alterations in the phospholipid compositions of Chinese hamster ovary and LM cells on the activity and distribution of CTP:phosphocholine cytidylyltransferase, *J. Biol. Chem.*, 258, 836, 1983.
46. **Robinson, B. S., Baisted, D., Yao, Z., and Vance, D. E.,** unpublished results.
47. **Lim, P. H., Pritchard, P. H., Paddon, H. B., and Vance, D. E.,** Stimulation of hepatic phosphatidylcholine biosynthesis in rats fed a high cholesterol and cholate diet correlates with translocation of CTP:phosphocholine cytidylyltransferase from cytosol to microsomes, *Biochim. Biophys. Acta*, 753, 74, 1983.
48. **Choy, P. C. and Vance, D. E.,** Lipid requirements for activation of CTP:phosphocholine cytidylyltransferase from rat liver, J. *Biol. Chem.*, 253, 5163, 1978.
49. **Feldman, D. A., Kovac, C. R., Dranginis, P. L., and Weinhold, P. A.,** The role of phosphatidylglycerol in the activation of CTP:phosphocholine cytidylyltransferase from rat lung, *J. Biol. Chem.*, 253, 4980, 1978.
50. **Feldman, D. A., Rounsifer, M. E.,** and **Weinhold, P. A.,** The stimulation and binding of CTP:phosphorylcholine cytidylyltransferase by phosphatidylcholine-oleic acid vesicles, *Biochim. Biophys. Acta*, 429, 429, 1985.
51. **Rosenberg, I. L., Smart, D. A., Gilfillan, A. M., and Rooney, S. A.,** Effect of 1-oleoyl-2-acetylglycerol and other lipids on phosphatidylcholine synthesis and cholinephosphate cytidylyltransferase activity in cultured type II pneumocytes, *Biochim. Biophys. Acta*, 921, 473, 1987.
52. **Yao, Z., Jamil, H., and Vance, D. E.,** unpublished results.
53. **Sanghera, J.,** unpublished results.
54. **Sleight, R. G. and Dinh, H.–N.,** Activation of CTP:phosphorylcholine cytidylyltransferase by phosphatidylethanolamine containing liposomes, *FASEB J.*, 2, A1333, 1988.

Chapter 4

CHOLINEPHOSPHOTRANSFERASE

Rosemary Cornell

TABLE OF CONTENTS

I. INTRODUCTION

The discovery of cholinephosphotranseferase (CPT) by Kennedy and co-workers[1,2] is described in Chapter 1. The transferase catalyzes the final reaction in the *de novo* synthesis of phosphatidylcholine (PC):

$$\text{CDP–choline} + \text{Diacylglycerol} \rightarrow \text{CMP} + \text{Phosphatidylcholine}$$

The CPT reaction is the dominant means for generating PC in cells of animal[3] or plant origin.[4] Phosphatidylethanolamine (PE) methylation, base exchange, and deacylation/reacylation reactions account for less than 30% of the PC synthesized.[3,5-7] In yeast the relative rates of the cytidine diphosphate (CDP)-choline and methylation pathways depend on the medium supply of precursors; when choline is present the CPT reaction predominates.[8] However, in the bacteria that possess PC, it is synthesized exclusively by the PE methylation pathway.[9] Coleman and Bell[10] have analyzed the tissue distribution of microsomal CPT in rat. Fat cells, liver, and intestine had the highest activity (8 to 9 nmol/min/mg). In other tissues examined CPT activity ranged from 5.8 to 1.1 nmol/min/mg in the order brain > lung > kidney > heart > skeletal muscle.

II. SUBCELLULAR LOCALIZATION

Wilgram and Kennedy[11] first identified enrichment of CPT activity in liver microsomes, along with several other enzymes of lipid synthesis. More extensive membrane fractionation was subsequently carried out by several groups with different results. Van Golde et al.[12] found that the specific activity of CPT in liver was highest in smooth and rough endoplasmic reticulum (ER) (twice as high in smooth), and that any activity in Golgi, mitochondria, plasma, or nuclear membrane could be accounted for by microsomal contamination of these fractions. Other investigations corroborated this result and led to a consensus that the ER is the exclusive site for PC synthesis.[13-16]

On the other hand, a number of more recent studies have indicated high CPT contents in membranes other than the ER. In a thorough investigation of the lipid synthetic capacity of liver membranes, Jelsema and Morré[17] determined that, although 90% of the total cellular CPT activity was in the ER, it was also present in the Golgi and outer membrane of the mitochondria. Their membrane fractions were analyzed for purity not only by characteristic enzyme markers but also by morphometric analysis of electron micrographs. The contamination of Golgi by ER was only 11%, yet the specific activity of CPT was ~40% of that in the ER fraction. This study indicated that Golgi and mitochondria do have the capacity for PC synthesis, although their relative output of PC is small compared to ER (1 and 5% of the total cellular output, respectively). The Golgi apparatus of plant (pea) tissue also has high levels of CPT activity, calculated to account for 25% of the total cellular CPT.[18]

Higgins and Fieldsend,[19] using yet another isolation procedure, confirmed that CPT resides in rat liver Golgi. They separated two Golgi fractions based on density differences and found that the specific activity in the lighter fraction was three times higher than that of the ER. ER contamination of this fraction was only 10%. In this study CPT was assayed using endogenous diacylglycerol (DG) only. The levels of DG endogenous to intracellular membranes are sub-K_m and thus rate limiting. The differences in activities among membrane fractions might reflect in part the variation in DG content rather than CPT content. For example, Possmayer et al.[20] found large distributional differences in CPT activity in the subcellular fractions of brain, depending on whether the enzyme was assayed with endogenous or exogenous DG. The nuclear and mitochondrial fractions appeared richer in CPT when endogenous DG was used as substrate. The problem of differential partitioning of exogenous DG into Golgi, ER, or mitochondrial membranes could be solved by adding a large excess of exogenous DG to ensure saturating membrane levels. DG-phospholipid liposomes are effective in presenting the substrate; the inclusion of detergent which also might have differential effects is not necessary (see Section V.A).

In light of the discrepancies reported for lipid biosynthesis in the Golgi[12-19] and the interest in the role of Golgi in generating PC for lipoprotein assembly, Vance and Vance[21] have compared the activities of CPT in Golgi using three different isolation procedures. Golgi isolated by the procedure of Croze and Morré[22] contained high CPT activity (70% of the ER specific activity), while Golgi isolated by the procedure of Fleischer et al.[12,23] contained none. Sodium dodecylsulfate (SDS) gels showed differences in the protein compositions of these Golgi preparations. It was postulated that the preparations lacking CPT (as well as ethanolaminephosphotransferase [EPT] and PE methyltransferase) contained only the *trans* Golgi compartment.

CPT activity has also been described in neuronal nuclear membranes[24] and in lung mitochondrial preparations.[25,26] However, in the former study[24] ER contamination was not adequately assessed, and in the latter studies[25,26] the activity of the ER marker enzyme, NADPH cytochrome C reductase, was very high in the mitochondrial fraction. Mitochondrial glycerol-3-P acyltransferase and PA phosphohydrolase activities are well documented.[27] Thus, it is possible that the mitochondria could synthesize PC autonomously, via the *de novo* pathway, and would not be totally reliant on the ER as the source of its PC. *In vivo* precursor uptake patterns suggested that the outer membrane of liver mitochondria, along with the ER, is a site for PC synthesis.[28] This may be true generally, but the evidence for transfer between ER and mitochondria is also strong.[29,30]

The interpretation of all the above studies is limited by the ability to isolate pure organelle membranes. It is difficult to eliminate the possibility of contamination of membrane fractions by ER fragments containing CPT but devoid or depleted of the ER marker enzymes used to assess membrane purity.

III. DISTINCTION FROM OTHER PHOSPHOTRANSFERASES

A. ETHANOLAMINEPHOSPHOTRANSFERASE

The enzyme catalyzing the transfer of phosphocholine to DG is distinct from the enzyme catalyzing the transfer of phosphoethanolamine. The two activities, although both predominantly localized in the ER, show different degrees of stimulation by exogenous DG[10,31] and have different acyl chain selectivities.[10,32-35] In yeast EPT activity is more stable to heat and is more sensitive to inhibition by CMP than is CPT.[33] The activities show different sensitivities to trypsin.[10,33] The selectivity for cations differs in some tissues. In liver,[35] fat,[10] heart,[36] and cerebrum[37] CPT requires Mg^{2+} and is inhibited by Mn^{2+}, whereas Mn^{2+} can substitute for Mg^{+2} in the EPT reaction and is the preferred cation for EPT in some tissues.[10,31,35,38,39] EPT is less sensitive to the hyperlipidemic drug, DH-990;[33] the inhibition is competitive, whereas CPT is inhibited noncompetitively. Of the two activities CPT is much more sensitive to delipidation by detergents.[34,40] EPT has been solubilized from rat brain microsomes in Triton® X-100 or octylglucoside.[40] CPT activity was also solubilized in active form by the same procedure,[40] but it disappeared after ion-exchange chromatography (which removes phospholipid), whereas the more stable EPT was purified 37-fold. These studies on their own do not prove that the activities reside on separate enzymes, since it is conceivable that one enzyme could exhibit different properties with respect to different substrates.

More definitive evidence for the distinction between these two enzymes comes from the isolation of mutants defective in either EPT or CPT. Polokoff et al. isolated mutant CHO cells in which EPT activity was six- to tenfold less than the parent cell, yet CPT activity was normal.[41] CPT yeast mutants have also been isolated that have normal EPT activity.[42,43a] Hjelmstad and Bell isolated a gene from a yeast genomic library that corrected the CPT defect, and showed by genetic complementation that it was a structural gene for CPT. Overexpression of this gene in a plasmid resulted in a fivefold elevation of CPT activity with only slightly increased EPT activity.[43a] A separate gene has now been isolated which codes for the yeast EPT.[43b] Overexpression of this gene was accompanied by a 30–fold elevation of EPT activity and, *in addition,* a ~20-fold increase in CPT activity. This suggests that in yeast there are two enzymes which can catalyze the CPT reaction. One, the product of the CPT I gene, is highly selective for phosphocholine transfer, while the other, the product of the EPT I gene, is only slightly selective for phosphoethanolamine transfer. The two transferases differ in their sensitivity to CMP inhibition of phosphocholine transfer.[43b]

B. CDP-CHOLINE:CERAMIDE PHOSPHOCHOLINETRANSFERASE

An enzymatic activity similar to the DG:CDP-choline CPT has been described for sphingomyelin (SM) synthesis in which phosphocholine is transferred from CDP-choline to ceramide.[44] This reaction utilizes short chain *threo* isomers of ceramide (natural SM is *erythro*) and has only been demonstrated *in vitro*. *In vivo* labeling indicates that the final step in the synthesis of SM occurs by transfer of phosphocholine from PC rather than CDP-choline.[45-47] Thus, the terminal phosphocholine transfer step in the synthesis of PC and SM are not catalyzed by the same enzyme.

C. CDP-CHOLINE:1-ALKYL-2-ACETYL GLYCEROL PHOSPHOCHOLINE-TRANSFERASE

Another CPT activity is involved in the synthesis of platelet-activating factor (PAF). There are two alternative routes for PAF synthesis, both of which can operate in the same cell. PAF is synthesized by acetylation of 1-*O*-alkyl-2-lyso glycerol-3-phosphocholine,[48] or by the transfer of phosphocholine from CDP-choline to 1-alkyl-2-acetyl glycerol.[49] The evidence gathered thus far suggests that this latter enzyme is distinct from the CPT involved in the synthesis of long-chain PCs. The reactions differ in responses to thiol-reagents such as dithiothreitol and *N*-ethylmaleimide, in CDP-choline and DG substrate dependencies, and in thermal sensitivity.[50]

IV. MEMBRANE TOPOGRAPHY

CPT is an integral membrane protein. It is resistant to extraction with high ionic strength medium,[51] EDTA,[52] and with low concentrations of detergent.[34,51,52] Its activity can be solubilized only after complete dissolution of the lipid bilayer.[51] Furthermore, its extreme sensitivity to inactivation by numerous detergents (see Section VII.A) suggests that sites critical to its activity are specifically dependent on a phospholipid environment; other amphiphilic molecules are not satisfactory substitutes.

The impermeability of intact microsomal membranes to charged molecules such as CDP-choline[53,54] would lead one to conclude that the active site must face the cytosol. CDP-choline is not permeable to liver microsomal membranes, and CPT activity is not increased by permeabilizing the membrane with detergent.[53] If the active site faced the cytosol, one would expect that at least some of the DG would be localized in the outer leaflet. However, Kanoh and Ohno[55] discovered that DG lipase and DG kinase (soluble enzymes) could not utilize DG generated in microsomes by the back reaction of CPT, although they could utilize DG formed by phospholipase C hydrolysis of the microsomes. The researchers also observed that the DG in cosonicated DG-phospholipid vesicles was not accessible for the reaction. They suggested that DG in the liposomes and the DG generated via CPT was sequestered in the inner leaflet. These observations are not easily reconciled with the rapid rate of transbilayer flip-flop of DG obtained in red cells,[56] fibroblast plasma membranes,[57] or liposomes.[58]

CPT activity is diminished or destroyed by treatment of isolated intact microsomes of liver[53,59,60] or brain[61] with a variety of proteases or with mercury-dextran. A similar finding has been reported for muscle sarcoplasmic reticulum vesicles, although the permeability barrier was not rigorously demonstrated in this study.[52] Inactivation of CPT by membrane-impermeant reagents suggests that part of the enzyme is exposed to the cytosol, but does not prove that the active site is cytosolic. Modification of a protein on one side of the membrane can affect its activity on the other side. For example, hormone binding to receptor on the external side of the plasma membrane can trigger phosphorylation[62] or interaction with a G protein[63] on a cytoplasmic domain. Two groups have tried to identify the leaflet on which PC is synthesized. PC was labeled with CDP-choline and the label in the external leaflet was examined by phospholipase C hydrolysis.[64,65] It was concluded that the site of synthesis is the outer leaflet. However, the results are questionable because their analysis involved treatment with sufficient phospholipase C to degrade 50% of the membrane lipid, and, in order for the membrane to remain intact, transbilayer flip-flop of labeled PC from the inner leaflet would have had to occur.

While the case is strong for a cytosolic domain containing the active site for CPT (and most of the other enzymes of glycerolipid synthesis), the extent to which mammalian CPT penetrates through the bilayer and into the lumen is unknown. Details of its topography await its purification and the generation of specific probes. The sequence of the yeast CPT gene has recently been completed (R. M. Bell, personal communication). It predicts seven transmembrane domains for a protein of molecular weight 46,000 and shares some sequence homology with the nicotinic acetylcholine receptor and several yeast cytidine-binding enzymes.

V. INTERACTION OF SUBSTRATES AND COFACTORS WITH CPT

A. DIACYLGLYCEROL

1. Physical Form of the Substrate

CPT activity is typically measured by monitoring transfer of methyl-labeled phosphocholine from CDP-choline to a DG acceptor. Endogenous microsomal DG or exogenously added DG have been used as acceptors. Addition of exogenous DG to microsomes generally stimulates the reaction 10- to 30-fold,[10,35,36,66,67] but rate enhancements of 2.5-fold[31] and 300-fold[68] have been reported. Because DG is insoluble a variety of schemes for introducing the exogenous DG have been tried. Tween® emulsions have been the most common method for adding DG.[2] Other preparations include dispersion in ethanol,[10,32,43,66] sonication in Tween®-deoxycholate mixtures,[34,35] sonication in Tween®-phospholipid mixtures,[51,68] or sonication with phospholipid.[68,69]

The variation in the extent of stimulation is influenced by the different physical forms of the substrate that are presented to the enzyme. Miller and Weinhold[68] compared the stimulatory effects of DG prepared by different methods on the activation of lung microsomal CPT. Dipalmitoylglycerol stimulated activity 20-fold when sonicated with Tween®; however, it stimulated activity 300-fold when cosonicated with phosphatidylglycerol (PG) and Tween®. The highest specific activities were obtained with diolein:phosphatidylglycerol:Tween® in a molar ratio of 1/0.8/0.15. The V_{max} obtained, 43 nmol/min/mg, is the highest reported activity for microsomal CPT in any tissue. We have obtained similar specific activities for the liver enzyme using cosonicated diolein:asolectin:Tween® (R. Cornell, unpublished). It is likely that only the exogenous DG that is transferred into the microsomal membrane will bind to the transferase and stimulate activity. The extent of transfer from the added droplet, micelle, or vesicle has not been quantified. The temperature at which disaturated DG/PG mixtures were sonicated affected the stimulation of CPT.[36,68] Temperatures exceeding the phase transition for the DGs were required, presumably for the formation of mixed liposomes. Numerous studies comparing the preference of CPT toward saturated and unsaturated DGs have failed to take into account the effects of the physical state on the ability of the DGs to partition into the microsomal membrane.

The CPT reaction has also been studied using DG generated in the membrane by treatment of microsomes with phospholipase C,[16,70] stimulation of the back reaction with CMP,[71-73] or stimulation of its *de novo* synthesis from glycerol-3-P and acyl CoA via phosphatidic acid.[74-76] In a few of these studies the DG content of the membrane has been measured chemically, permitting calculation of K_m values. Values of 28.5 nmol/mg microsomal protein using DG generated via *de novo* synthesis[75] and 12.8 nmol/mg using DG generated by promoting the reverse reaction[74] have been reported for the rat lung enzyme. The DG content in lung microsomes is only 8 nmol/mg.[75] Thus, the CPT reaction could be subject to regulation by membrane DG levels *in vivo*.

2. Utilization of Newly Synthesized vs. Endogenous DG — Is There a Preferred Pool?

CPT activity has also been measured by monitoring the transfer of label from DG in the presence of cold CDP-choline. The label has been incorporated into DG by incubation with radiolabeled glycerol-3-P prior to addition of CDP-choline.[75,78-80] This reaction thus measures the utilization of DG generated in the membrane by *de novo* synthesis. Binaglia et al.[78] and Rustow and Kunze[79] compared the kinetics of utilization of DG labeled from [^{3}H]- glycerol-3-P vs. the total endogenous membrane DG, which was monitored by [^{14}C] CDP-choline incorporation. The incorporation of label from glycerol-3-P was rapid and showed saturation within 10 min, whereas the label from CDP-choline was incorporated at a slower linear rate for 20 min[79] or 2 h.[78] The two distinct kinetic curves were interpreted by both groups to mean that there are two different pools of DG in the membrane, designated the *de novo* and endogenous pools. Both are accessible to CPT, but the *de novo* pool is preferentially utilized. If there were complete rapid mixing of the two pools, the kinetics of incorporation of the two labels should have been the same.

A closer analysis of the initial kinetics of the incorporation of [^{14}C] CDP-choline into PC revealed a hyperbolic rate of incorporation for the first 2 min followed by linear incorporation for 20 min.[80] The magnitude of the nonlinear part of the curve increased proportionally with the concentration of glycerol-3-P used in the preincubation. The size of the *de novo* pool of DG, estimated from the area under the curve of nanomoles PC formed vs. time, could be increased from less than 5 to 30% of the total endogenous DG pool by incubation with increasing concentrations of glycerol-3-P. The initial rates of PC formation increased proportionately. A K_m for the *de novo* DG pool was calculated to be tenfold higher than the size of the *de novo* pool in freshly isolated microsomes. Hence, the availability of *de novo*-formed DG could limit the rate of PC synthesis.[80] Ide and Weinhold[75] have also increased the *de novo* pool of DG to 80% of the total membrane pool by incubation of lung microsomes with glycerol-3-P. In this case the initial rates of [^{14}C]-CDP-choline and [^{3}H]-glycerol-3-P incorporation were identical. They suggested that the DG pool generated from glycerol-3-P was the *only* pool utilized by CPT.

To account for the preferential utilization of the newly synthesized DG, it was proposed[78-80] that the enzymes involved in PC synthesis from glycerol-3-P are organized in a multienzyme complex, wherein the metabolic intermediates are channeled directly from one enzyme active site to the next without mixing with the bulk membrane (endogenous) lipid. Compartmentalization of membrane DG pools has also been postulated in other studies.[81,82]

Since these interpretations propose novel ideas for the regulation of PC synthesis, it is important to take note of the methodology used. In all these studies PC was synthesized in three separate steps. Initially, glycerol-3-P was incorporated into phosphatidic acid in a 30-min reaction at 37°C; the microsomes were sedimented by a 1-h centrifugation and resuspended in a new medium. Next, dephosphorylation of phosphatidate was carried out in a second 30-min incubation at 37°C. The final step was initiated by adding CDP-choline. In order to maintain a separate *de novo* pool, the DG generated must have been unable to diffuse out of the putative enzyme complex where it was formed during the incubations at 37°C and during the time elapsed during centrifugation. Measured lateral diffusion rates for lipids in membranes(10^{-8} cm^2/s)[83]

suggest that complete mixing should have occurred within seconds. The proposal thus implicates a very tight binding between DG and the enzyme complex.

Another problem relates to the stoichiometry between the complex of enzymes and the newly synthesized DG pool. A model in which there is direct channeling of DG from PA phosphatase to CPT would predict a small number of DG molecules associated per complex, especially since in these *in vitro* reactions the rate of DG production via PA phosphatase is slow compared to the rate of DG utilization via CPT.[78] The estimated size of the *de novo* DG pool was up to 8 nmol/mg microsomal protein.[80] The molar content of the total microsomal protein is only 18 nmol/mg using an average molecular weight of 55,000. A high estimate of 5% of the total protein for the CPT + PA phosphatase content would still give only 0.9 nmol/mg for these enzymes. Thus, the nature of the interaction of the intermediates with the putative enzyme complex is a problem that must be addressed.

This provocative hypothesis invites further testing. It is possible to generate PC from glycerol-3-P, an acyl-CoA generating system, CDP-choline, and microsomes in a single incubation.[5,84] This approach would greatly shorten the time in which diffusion of newly synthesized intermediates and their metabolism to products other than PC could occur.

Infante[85] has recently criticized the DG compartmentalization model and has offered an alternative explanation for the different rates of precursor incorporation based on a newly proposed *de novo* pathway for PC synthesis.[85] He has postulated that there are two separate pathways for CDP-choline incorporation into PC: (1) via condensation with DG and (2) via condensation with glycerol-3-P to form glycerol-phosphocholine (GPC) followed by two acyl transfers to produce PC. The evidence for this latter pathway is that GPC was formed by incubation of liver homogenates with labeled CDP-choline or glycerol-3-P, and that unlabeled GPC competed in the formation of PC (but not PE) from [^{14}C] glycerol-3-P, suggesting an involvement as a metabolic intermediate.[86] If in the double–label experiments described above a portion of the CDP-choline label was entering PC via a different route than that taken by the glycerol-3-P-labeled DG, the kinetics of uptake would likely be different.

B. INTERACTION WITH CDP-CHOLINE

1. K_m Measurements

Reported values for the K_m for CDP-choline range from 10 μ*M* in castor bean endosperm[5] to 104 μ*M* in HeLa cell microsomes.[87] Inclusion of exogenous DG in the assay influenced the apparent K_m values for CDP-choline. In HeLa cell microsomes the K_m was 27 μ*M* when the enzyme was assayed with endogenous DG, but increased to 104 μ*M* when exogenous DG was added.[87] The K_m for CDP-choline increased seven- to tenfold in rabbit neuronal tissue when exogenous DG was added.[24] In rat liver microsomes the K_m for CDP-choline was 50 μ*M* when assayed with endogenous DG and 200 μ*M* when assayed with exogenous DG (R. Cornell, unpublished). One likely reason for the DG effect is that endogenous DG levels are subsaturating. The affinity for CDP-choline may also have been influenced by the Tween®-20 which was added with the DG, rather than the DG itself.

2. Substrate Specificity

Work has begun toward establishing the importance of the functional groups on CDP-choline which mediate its interaction with CPT. Early work established that the requirement for a 2′ hydroxyl is flexible.[88] Pontoni et al.[89] synthesized a series of CDP-choline analogues containing modifications in the ribose, cytidine base, pyrophosphate linkage, and choline. None of the analogues with the exception of dCDP-choline showed significant activity as substrates. Deoxy-CDP-choline, ara-CDP-choline, UDP-choline, 5-iodo CDP-choline, and CMP-choline were weak inhibitors. The type of inhibition, reported only for dCDP-choline, araCDP-choline, and CMP-choline, was uncompetitive.[89] Although several of the modifications involved substitution with bulky groups which could have prevented binding due to steric effects rather than a missing function, potential interaction sites were identified including the 2′ and 3′ positions on ribose and the C-4 amino group on the pyrimidine ring. The two phosphate groups and choline appeared to be required for substrate binding. In short, it appears that the interaction between the enzyme and CDP-choline requires multiple binding sites.

Considering the synthetic effort undertaken and the potential usefulness of such a study, the inhibitor studies were rather weak on numbers. For most of the compounds the effects at only one or two concentrations were shown. Although no inhibition by CDP was seen in this study using rat liver microsomes, competitive inhibition by CDP has been observed for CPT from rat adipocytes[10] and yeast.[33] We have synthesized 5-iodo CDP-choline and dCDP-choline by the procedures of Pontoni et al.;[89] however, in our hands no inhibition was seen for the iodo derivative below 3 m*M*, and dCDP-choline was a competitive inhibitor of CDP-choline incorporation into PC with a K_i of 0.6 μ*M* (R. Cornell and A. Biro, unpublished).

Several studies have shown that there is selectivity toward the degree of methylation of the head group. CDP-ethanolamine inhibits competitively with a K_i of 227 μ*M* (K_m CDP-choline = 24 μ*M*) in adipocytes[10] and a K_i of 350 μ*M* (K_m for CDP-choline = 36 μ*M*) in solubilized and partially purified liver CPT.[34] The *in vivo* incorporation of monomethyl– and dimethyl-ethanolamine into phospholipid[6] indicated that these CDP-bases can compete with CDP-choline or CDP-ethanolamine. Monomethylethanolamine can compete with ethanolamine for EPT binding and dimethylethanolamine can compete with choline for CPT binding.[6]

Studies of this type may prove helpful in the design of affinity gels or photoaffinity labels (see Section VII.C). For example, CPT did not bind to a CDP-choline agarose gel (R. Cornell, unpublished) in which the spacer arm, adipic acid, was intercalated into the ribose ring between the 2′ and 3′ hydroxyls. This same type of coupling to CDP-DG provided the main step in the purification of PS and PG synthases.[90,91]

3. Reaction Mechanism

Pontoni et al.[89] also studied the catalytic mechanism of rat liver microsomal CPT by kinetic analysis. They obtained intersecting double reciprocal plots for both substrates, indicative of a sequential mechanism. They proposed a predictable base-catalyzed deprotonation of the DG hydroxyl which would lead to a nucleophilic attack on the β-phosphorus of CDP-choline.

C. METAL ION REQUIREMENTS

As mentioned in Section III.A, the preference for Mg^{2+} vs. Mn^{2+} varies from tissue to tissue. Since the substrate remains the same, the enzymes from various sources must discriminate between the geometry of Mn^{2+}- vs. Mg^{2+}-complexed phosphate groups. Solubilized rat liver CPT has been partially separated into Mn^{2+}- and Mg^{2+}-requiring components by density gradient centrifugation,[34] which suggests that two isomers of CPT might exist in liver.

Ca^{2+} inhibits both CPT and EPT in every tissue tested. The sensitivity of CPT to Ca^{2+} is at least two orders of magnitude greater than the sensitivity of cytidylyltransferase,[92] PA phosphatase,[93] or glycerol-P acyltransferase.[94] Taniguchi et al.[95] have investigated the nature of this inhibition in platelet membranes. In these membranes the order of stimulation of CPT activity was $Mn^{2+} > Mg^{2+} > Co^{2+}$. Ca^{2+} and La^{2+} were the most potent inhibitors of a series of divalent cations. The inhibition was competitive toward Mg^{2+} and Mn^{2+} and was reversed by EGTA. Higher concentrations of Ca^{2+} were required to inhibit the Mg^{2+}-stimulated activity than the Mn^{2+}-stimulated activity (600 vs. 20 μ*M*). The mechanism of the cofactor inhibition was not resolved; however, a reasonable suggestion is that the inhibiting ion competes for complexation with CDP-choline, and the resulting complex is unfavorable as a substrate.

The possibility that intracellular calcium levels could regulate PC synthesis has been considered.[96] Treatment of rat hepatocytes with the calcium ionophore A23187 resulted in a transient inhibition of choline incorporation into PC, as did vasopressin and angiotensin, two hormones which elevate cytosolic calcium. The effect was attributed to an inhibition of CPT since it was inhibited by micromolar Ca^{2+}, although other Ca^{2+}-sensitive sites were not ruled out. Glycerol-P acyltransferase, PA phosphatase, and cytidylyltransferase have rate-limiting roles in the supply of PA, DG, and CDP-choline, respectively, and are sensitive to millimolar calcium.[92-94] Tijburg et al.[97] subsequently examined the effects of vasopressin on choline flux through choline metabolites and into PC. An effect on cytidylyltransferase was clearly indicated by a reduced rate of phosphocholine turnover which paralleled the inhibition of incorporation into PC. The effects of vasopressin on PC synthesis may not be mediated by Ca^{2+}, since hormone binding generates multiple cellular signals. Whether changes in calcium concentrations in response to hormones provide a mechanism for adjusting the rates of PC synthesis is a question that needs further study.

VI. REGULATION

A. REGULATION OF THE ACYL COMPOSITIONS OF PCS

Conditions wherein CPT regulates the rate of PC synthesis have not been discovered, whereas there is much evidence suggesting a regulatory role for cytidylyltransferase (see Chapter 3). There appears to be an excess of CPT in the ER; the rate of the reaction is controlled by substrate rather than enzyme levels (see Section VI.B). CPT may however have a role in determining the acyl composition of newly synthesized PCs. The selectivity of CPT toward DGs containing a selection of acyl or alkyl chains has been studied ad nauseam. Part of the keen interest stems from the unusual situation in lung where a distinct PC species, dipalmitoyl PC, is synthesized for secretion as the major component of surfactant. Whether this PC is generated by *de novo* synthesis via CPT or by a two-step remodeling reaction in which the *sn*-2 unsaturated chain is replaced with palmitate is discussed in Chapter 6. The consensus is that both pathways play a role.

The process of selecting the acyl chain composition of PC begins with the acyltransferases. A saturated fatty acid is usually esterified at C-1 and an unsaturated fatty acid at C-2 of glycerol-3-P.[27] The species composition of PC is not the same as that of DG, TG, PA, or other phospholipids in tissues such as liver,[98] brain,[99] and heart.[36] Either CPT is selective in its utilization of DG species or it utilizes the whole DG pool nonselectively, and the acyl chains are subsequently modified by deacylation and reacylation. To answer this question many approaches have been used. The most informative method is to monitor the transfer of label between glycerol- or fatty acid-labeled DG and PC. This method avoids the problems of insolubility and differential membrane partitioning associated with adding exogenous DGs. The results gained from this approach suggest the following: in lung CPT is nonselective toward DG species including dipalmitin;[73,75,100–102] in liver CPT shows a selectivity for 16:0 and discriminates against 18:0 in position 1, but otherwise is nondiscriminating;[69,70] and in peas and soybeans CPT prefers 16:0, 18:2 species but is otherwise nonselective.[103] Using the solubilized, partially purified liver enzyme, Morimoto and Kanoh[104] found that even 1-unsaturated, 2-saturated species were used effectively, demonstrating that CPT does not really discriminate for or against chains with double bonds. However, the solubilized enzyme again discriminated against chains longer than 17 carbons in the C-1 position. It may be that a pocket in the enzyme which binds DG cannot accommodate longer chains.

Recently, CPT has been shown to be involved in the synthesis of plasmenylcholine, the major phospholipid of myocardial membranes.[105] At physiological CDP-choline concentrations CPT utilized 1-*O*-alk-1-enyl glycerol derived from plasmenyl PE. Although the diacylglycerols are 20 times more abundant, the flux of CDP-choline into PC and plasmenylcholine were similar, indicating a preference in the CPT reaction for the vinyl ether-linked substrate.[105]

B. SUBSTRATE REGULATION OF CPT

The evidence for and mechanism by which the CDP-choline supply regulates the CPT reaction has been discussed in Chapter 1. That the supply of DG can also be rate limiting is more debatable. The concentrations of both microsomal DG and cytosolic CDP-choline are sub-K_m. Thus, either or both *could* control the reaction, but whether the rate of the reaction *is* ever controlled *in vivo* by changes in the DG concentration is not known. Comparison of the intracellular concentrations of DG and CDP-choline is not useful in assessing which of the two is rate-limiting, since DG is compartmentalized into membranes and CDP-choline is soluble.

Sundler and Akesson[6] found that glycerol incorporation into PC of intact hepatocytes could be stimulated by oleic acid (30%) and by choline (48%), and almost threefold by a combination of the two. The DG content rose approximately threefold in the presence of 1 m*M* oleic acid, which raised the possibility that the availability of DG could affect the rate of PC synthesis. Pelech et al.[106] later determined that oleic acid stimulated the rate of conversion of phosphocholine to CDP-choline in hepatocytes; thus, the stimulation of PC synthesis might have been due to an increased CDP-choline supply. There is evidence that the enzymes catalyzing the synthesis of TG, PE, and PC from DG compete for a common DG pool but that flux into TG is more sensitive to variations in the DG pool size.[107] A lower affinity for DG by the DG acyltransferase compared to EPT or CPT would explain this finding. The K_m values for all three solubilized enzymes are comparable: 60 μ*M* for DG acyltransferase,[108,109] 81 μ*M* for CPT,[34] and 63 μ*M* for EPT.[34] Thus, the affinities of the solubilized enzymes do not appear very different, but the properties of detergent-solubilized and native enzymes may not be the same.

In several different experimental systems there was a lack of correlation between the rate of PC synthesis and the size of the DG pool.[97,110,111] However, the CPT reaction was stimulated 2 to 2.5-fold by addition of acyl CoA to permeabilized HeLa cells, which resulted in a 2.5-fold enrichment in membrane DG.[87] This stimulation was observed even at a physiological, sub-K_m concentration of CDP-choline, which implied that altered DG levels could well influence the rate of the reaction *in vivo*. More experimentation of this type is needed before drawing a firm conclusion concerning the regulation of CPT by DG levels. Changing DG levels *in vivo* without affecting CDP-choline supply may prove difficult, since altered membrane DG content triggers the reversible translocation and activation of cytidylyltransferase, and thereby regulates the CDP-choline supply.[112–115]

For example, inhibition of PA phosphatase by treating intact hepatocytes with chlorpromazine decreased the incorporation of glycerol or oleate into DG by 75%.[116] The incorporation into TG was inhibited to a similar extent. Incorporation into PC was also reduced but was less sensitive to the degree of change in the supply of DG. Since cytidylyltransferase is also inhibited by chlorpromazine,[117] the reduced PC synthesis could have been caused by a decreased supply of CDP-choline. Leli and Hauser[118] observed that chlorpromazine inhibited glycerol incorporation into DG by 50% and had a much larger effect on incorporation into PC. The inhibition of PC synthesis could be partially reversed by adding 1-oleoyl,2-acetyl glycerol (OAG) to the hepatocytes. However, addition of OAG to cells stimulates the cytidylyltransferase reaction.[114] Again, the difficulty in independently altering the two substrates, CDP-choline and DG, is illustrated.

C. REVERSIBILITY OF THE CPT REACTION

Bjornstad and Bremer injected labeled choline or methionine into rats and measured the specific radioactivities of PC, CDP-choline, and phosphocholine.[119] Within 20 min after the methionine injection the specific radioactivities of CDP-choline and PC were the same. When choline was injected the specific radioactivity of CDP-choline was much lower than that for phosphocholine after 20 min. Both these results can be explained by the operation of CPT in the reverse direction.

SAM* CMP

1. 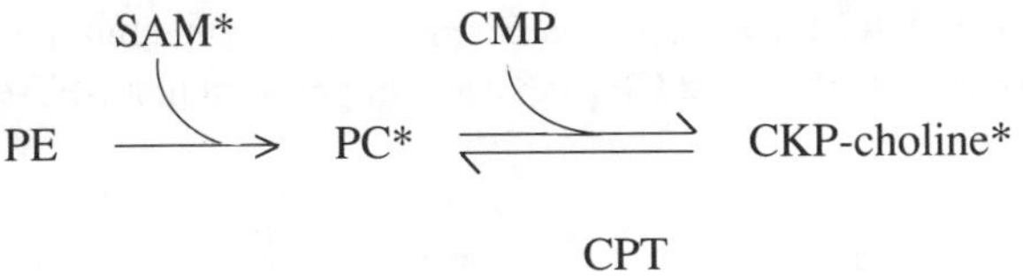 PE ⟶ PC* ⇌ CKP-choline*

CPT

CMP

2. 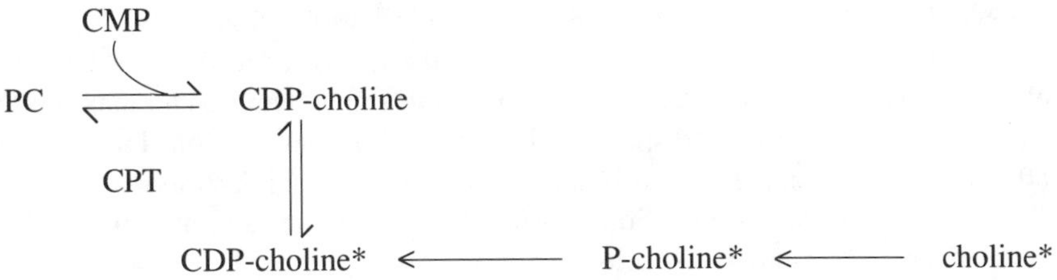 PC ⇌ CDP-choline

CPT ⇅

CDP-choline* ⟵ P-choline* ⟵ choline*

Sundler and Akesson[120] showed that in the first 5 min after injection with [^{3}H] choline the specific radioactivity of CDP-choline and phosphocholine were very similar, but thereafter the specific radioactivity of the CDP-choline decreased relative to that of phosphocholine. This suggested that the rate of the forward reaction exceeded that of the reverse. The flux for the forward direction was calculated to be 0.5 to 0.7 μmol/min, compared to 0.3 to 0.4 μmol/min for the reverse direction.[120] This study[120] and others[107,121] provide strong evidence for the reversibility of CPT *in vivo*.

The back reaction of CPT has been promoted by incubating isolated microsomes with CMP to enlarge the DG pool size[81,82] or to study the acyl chain selectivity of CPT.[71,72] The K_m for CMP determined in this way was 0.18 m*M* for rat liver[71] and 0.35 m*M* for rat brain.[122] These values are 5 and 11 times higher than the K_m values for CDP-choline in the same tissues.[34,24] The CMP concentration in liver is <10 μ*M*[123] and CDP-choline is ~10 μ*M*.[7,124] Thus, changes in the concentration of CMP or CDP-choline would influence the direction of the flux.

There has been speculation that the back reaction might be of significance in the generation of DG and fatty acid from PC in response to a cellular stimulus. CPT was implicated in the release of CDP-choline[125,126] as well as fatty acids[122] from brain microsomal phospholipids in experimental ischemia. A role for the CPT reverse reaction has also been hypothesized in the release of arachidonic acid from PC for prostaglandin synthesis in human platelets.[127] Platelets have a reversible (CMP-sensitive) CPT activity,[127] but there is as yet no evidence for this role. In summary, there is good evidence that the CPT catalyzed reaction can operate in the reverse as well as the forward direction *in vivo*. However, a specific role for the generation of CDP-choline and DG from PC has not been established.

VII. SOLUBILIZATION, RECONSTITUTION, AND PURIFICATION ATTEMPTS

Our knowledge of CPT is very limited compared to many other membrane enzymes which have been studied for 30 years. Many of the unsolved issues raised in this review, such as the membrane topology and subcellular location, the reaction mechanism, its sequestration in an enzyme complex, or the role of its reverse reaction in generating fatty acids in response to a cell stimulus, will not be completely resolved until the enzyme is isolated and specific probes are generated. Genetic approaches might also shed light on the question of enzyme complexes or the significance of the back reaction, for example, by analysis of the phenotype of a mutant in the structural gene and reconstitution of function by transfection of the gene into the mutant cell.

These possibilities are a long way down the road for the mammalian enzyme, but much nearer at hand for yeast CPT. The isolation of the yeast CPT gene and expression in a CPT-defective mutant cell[43] are described in Chapter 10.

A. DETERGENT EFFECTS ON CPT

As a prelude to isolation of CPT, the effects of detergents on CPT activity have been examined. Low subsolubilizing concentrations of detergents stimulated enzyme activity. This was observed with Tween®-20,[2,128] Triton® X-100,[51,31,128] lysolecithin,[129] and deoxycholate.[34,31] This stimulation was likely due to the ability of the detergent to disperse DG. There is a correlation between the concentration of a particular detergent required for maximum stimulation of activity and its ability to disperse and maintain DG in suspension. Tween® and Triton® are effective in dispersing DG; the bile salts are not.[16] The molar ratio of Tween®:DG or Triton®:DG for maximum stimulation was 0.5;[128] whereas the ratio of deoxycholate:DG or taurocholate:DG was 2 and 5, respectively.[34,35]

Solubilizing concentrations of detergents are inactivating. Cornell and MacLennan compared the detergent concentration curves for inactivation and solubilization of sarcoplasmic reticulum membranes containing CPT.[51] The curves coincided for cholate, deoxycholate, Triton® X-100, and sodium dodecylsulfate (SDS). The enzyme was inhibited by concentrations of Tween® and octylglucoside at concentrations well below the range required for solubilization. In every case, membrane solubilization inactivated the enzyme >90%. Rat liver CPT is also severely inhibited by Zwittergent® and the zwitterionic cholate derivative CHAPS (R. Cornell, unpublished). By contrast, EPT activity is stable in 30 m*M* octylglucoside[69,40] or 1% (15 m*M*) Triton®,[40] PE methyltransferase is stable in 0.5 m*M* Triton® when supplemented with 20 m*M* phosphate and 20% glycerol,[130] and DG acyltransferase is stable in 9 m*M* CHAPS.[108] CPT is unusually detergent sensitive, even for a lipid–metabolizing membrane enzyme. Nevertheless, this stringent requirement for an unperturbed lipid environment suggests that there are intriguing CPT-lipid interactions to be studied when the enzyme is finally isolated.

Kanoh and Ohno purified the solubilized liver enzyme approximately fourfold by a two-step sonication procedure.[34] The enzyme remained in the 100,000 × *g* pellet after a brief sonication in 4 m*M* deoxycholate, 20% glycerol at pH 7.5, but was extracted after a second sonication in 5 m*M* deoxycholate, 20% glycerol at pH 8.5. This supernatant was dialyzed to reduce the detergent concentration before measuring the activity. The deoxycholate-solubilized enzyme eluted in the void volume of a Sepharose® 4B column, indicating that it was part of a large aggregate. Further treatment with Triton® led to inactivation.[34]

B. RECONSTITUTION OF DETERGENT-INACTIVATED CPT

Cornell and MacLennan[51] reconstituted CPT activity after solubilization and inactivation of the sarcoplasmic reticulum enzyme with cholate, deoxycholate, Triton®, or octylglucoside at variable detergent:membrane ratios. Soybean phospholipid (asolectin) was added to the solubilized preparations, and the detergent was removed by dialysis, gel filtration, or by adsorption onto Biobeads SM-2. Activity recoveries were 60 to 100% when the initial detergent:protein ratios in the solubilized preparations were <2. At higher concentrations (initial detergent:protein >10) recovery was very poor unless stabilizers were present. DG and glycerol were found to be effective stabilizers, protecting the enzyme from irreversible inactivation if they were included in the solubilization medium. Reconstitution of cholate-dissolved membranes having an initial detergent:protein ratio of 20 resulted in recovery of only 3% of the presolubilized activity. When DG (2%) and glycerol (40%) were included, the recovery was 99%. The DG may have protected the portion of the enzyme containing the active site from denaturation at high levels of detergent.[51] Stabilizing effects of glycerol and DG have recently been described for EPT;[40] by including DG in the detergent media a 37-fold purification was achieved. Unfortunately, CPT activity was lost during the purification scheme. No attempt was made to reconstitute membrane vesicles before assaying activity.[40]

C. PURIFICATION IDEAS

Classic brute force — This approach would involve the usual series of ion-exchange columns, etc. and the inclusion of DG dissolved in the column-washing buffers. This requirement would preclude the use of bile salt detergents which are unable to maintain DG in solution for extended periods.[16] Each fraction would have to be reconstituted into lipid vesicles by a convenient method. This approach would be laborious but might work with luck and perseverance.

Photoaffinity labels — A CPT-specific photoaffinity label modeled after one of its substrates might be a useful means for identifying CPT on an SDS gel. An antibody could be prepared against the eluted protein. A photoreactive analogue of DG (1-palmitoyl-2-[*m*-diazirinophenoxy] glycerol) was tried unsuccessfully as a specific label for CPT (R. Cornell and D. E. Vance, unpublished). A specific label modeled after CDP-choline holds more promise. Examples of successful photoaffinity labeling with azido derivatives of nucleotide bases are numerous.[131] However, the K_m values for the nucleotides are usually submicromolar, whereas the K_m for CDP-choline is ~50 μ*M*. A further complication is that addition of a photoreactive group would likely reduce the affinity. 2′-Azido CDP-choline is a competitive inhibitor (R. Cornell, unpublished), but placement of the azido on a carbon bonded to hydrogen would lead to an internal rearrangement and generation of an unreactive imine.[131] Synthetic procedures for azido derivatives on nucleotide bases which would generate a reactive aryl azide have been published for adenine, guanine, uracil, and thymine,[132-133] but not for cytosine.

In the past few years direct photoactivation with underivatized nucleoside phosphates has also proved a successful method for labeling a number of enzymes.[134-138] Very recently we have photolabeled purified cytidylyltransferase with [^{32}P] CTP (T. Cutforth and R. Cornell, unpublished). The labeling was dependent on irradiation, specific for the 45-kDa subunit, and competed out with cold CTP. It is hoped that photoirradiation of microsomes with [^{32}P] CDP-choline will lead to the identification and the eventual isolation of CPT. Given the disappointing progress on the purification of mammalian CPT, it might be wiser to first pursue the isolation and sequencing of the cDNA. From the derived protein sequence peptide fragments could be synthesized to generate antibodies which could be used to dissect the topography of the enzyme in the membrane, and to isolate the solubilized protein by immunoaffinity chromatography. In *yeast* screening libraries by genetic complementation of mutant cells has proved to be a successful approach for cloning the genes for CPT[43a] and EPT.[43b] Although isolation of *mammalian* cell mutants and monitoring the phenotype switching is considerably more difficult, CHO mutants have been isolated at several points in PC biosynthesis including the EPT[41] and CT[139] genes.

REFERENCES

1. **Kennedy, E. P. and Weiss, S. B.,** The function of cytidine coenzymes in the biosynthesis of phospholipids, *J. Biol. Chem.,* 222, 193, 1956.
2. **Weiss, S. B., Smith, S. W., and Kennedy, E. P.,** The enzymatic formation of lecithin from cytidine diphosphate choline and *D*-1,2-diglyceride, *J. Biol. Chem.,* 231, 53, 1958.
3. **Ansell, G. B. and Spanner, S.,** Phophatidylserine, phosphatidylethanolamine, and phosphatidylcholine, in *Phospholipids,* Vol. 4, Hawthorne, J. N. and Ansell, G. B., Eds., Elsevier, Amsterdam, 1982, 1.
4. **Moore, T. S.,** Phosphatidylcholine synthesis in castor bean endosperm, *Plant Physiol.,* 57, 383, 1976.
5. **Fox, P. L. and Zilversmit, D. N.,** High *de novo* synthesis of glycerolipids compared to deacylation-reacylation in rat liver microsomes, *Biochim. Biophys. Acta,* 712, 605, 1982.
6. **Sundler, R. and Akesson, B.,** Regulation of phospholipid biosynthesis in isolated rat hepatocytes, *J. Biol. Chem.,* 250, 3359, 1975.
7. **Sundler, R., Arvidson, G., and Akesson, B.,** Pathways for the incorporation of choline into rat liver phosphatidylcholine *in vivo, Biochim. Biophys. Acta,* 280, 559, 1972.

8. **Waechter, C. J. and Lester, R. L.,** Differential regulation of the *N*-methyl transferase responsible for phosphatidylcholine synthesis in *Saccharomyces cerevisiae, Arch. Biochem. Biophys.,* 158, 401, 1973.
9. **Goldfine, H.,** Comparative aspects of bacterial lipids, in *Advances in Microbial Physiology,* Vol. 8, Rose, A. H. and Tempest, D. W., Eds., Academic Press, New York, 1972, 1.
10. **Coleman, R. and Bell, R. M.,** Phospholipid synthesis in isolated fat cells, *J. Biol. Chem.,* 252, 3050, 1977.
11. **Wilgram, G. R. and Kennedy, E. P.,** Intracellular distribution of some enzymes catalyzing reactions in the biosynthesis of complex lipids, *J. Biol. Chem.,* 238, 2615, 1963.
12. **Van Golde, L. M. G., Fleischer, B., and Fleischer, S.,** Some studies on the metabolism of phospholipids in Golgi complex from bovine and rat liver in comparison to other subcellular fractions, *Biochim. Biophys. Acta,* 249, 318, 1971.
13. **Van Golde, L. M. G., Rabin, J., Batenburg, J. J., Fleischer, B., Zambrano, F., and Fleischer, S.,** Biosynthesis of lipids in Golgi complex and other subcellular fractions from rat liver, *Biochim. Biophys. Acta,* 360, 179, 1974.
14. **McMurray, W. C. and Dawson, R. M. C.,** Phospholipid exchange reactions within the liver cell, *Biochem. J.,* 112, 91, 1969.
15. **McMurray, W. C.,** Lecithin biosynthesis in liver mitochondrial fractions, *Biochem. Biophys. Res. Commun.,* 58, 467, 1974.
16. **McMurray, W. C.,** Biosynthesis of mitochondrial phospholipids using endogenously generated diglycerides, *Can. J. Biochem.,* 53, 784, 1975.
17. **Jelsema, C. L. and Morré, D. J.,** Distribution of phospholipid biosynthetic enzymes among cell components of rat liver, *J. Biol. Chem.,* 253, 7960, 1978.
18. **Montague, M. J. and Ray, P. M.,** Phospholipid-synthesizing enzymes associated with Golgi dictyosomes from pea tissue, *Plant Physiol.,* 59, 225, 1977.
19. **Higgins, J. A. and Fieldsend, J. K.,** Phosphatidylcholine synthesis for incorporation into membranes or for secretion as plasma lipoproteins by Golgi membranes of rat liver, *J. Lipid Res.,* 28, 268, 1987.
20. **Possmayer, F., Kleine, L., Duwe, G., Stewart-De Hann, P. J., Wong, T., MacPherson, C., and Harding, P. G. R.,** Differences in the subcellular and subsynaptosomal distribution of the putative endoplasmic reticulum markers NADPH-cytochrome c reductase, estrone sulfate sulfohydrolase, and CDP-choline:diacylglycerol cholinephosphotransferase in rat brain, *J. Neurochem.,* 32, 889, 1979.
21. **Vance, J. E. and Vance, D. E.,** Does rat liver Golgi have the capacity to synthesize phospholipids for lipoprotein secretion?, *J. Biol. Chem.,* 263, 5898, 1988.
22. **Croze, E. M. and Morré, D. J.,** Isolation of plasma membrane, Golgi apparatus, and endoplasmic reticulum fractions from single homogenates of mouse liver, *J. Cell. Physiol.,* 119, 46, 1984.
23. **Fleischer, S. and Kervina, M.,** Subcellular fractionation of rat liver, *Methods Enzymol.,* 31, 6, 1974.
24. **Baker, R. R. and Chang, H. Y.,** Cholinephosphotranferase activities in microsomes and neuronal nuclei isolated from immature rabbit cerebral cortex: the use of endogenously generated diacylglycerols as substrate, *Can. J. Biochem.,* 60, 724, 1982.
25. **Sikpi, M. D. and Das, S. K.,** The localization of cholinephosphotransferase in the outer membrane of guinea-pig lung mitochondria, *Biochim. Biophys. Acta,* 899, 35, 1987.
26. **Harding, P. G. R., Chan, F., Casola, P. G., Fellows, G. F., Wong, T., and Possmayer, F.,** Subcellular distribution of the enzymes related to phospholipid synthesis in developing rat lung, *Biochim. Biophys. Acta,* 750, 373, 1983.
27. **Brindley, D. N. and Sturton, R. G.,** Phosphatidate metabolism and its relation to triacylglycerol biosynthesis, in *Phospholipids,* Vol. 4, Hawthorne, J. N. and Ansell, G. B., Eds., Elsevier, Amsterdam, 1982, 179.
28. **Bygrave, F.,** Studies on the biosynthesis and turnover of the phospholipid components of the inner and outer membranes of rat liver mitochondria, *J. Biol. Chem.,* 244, 4768, 1969.
29. **Jungalwala, F. B. and Dawson, R. M. C.,** Phospholipid synthesis and exchange in isolated liver cells, *Biochem. J.,* 117, 481, 1970.
30. **Eggens, I., Valtersson, C., Dallner, G., and Ernster, L.,** Transfer of phospholipid between the endoplasmic reticulum and mitochondria in rat hepatocytes *in vivo, Biochim. Biophys. Res. Commun.,* 91, 709, 1979.
31. **Vial, H. J., Thuet, M., and Philippot, J. R.,** Cholinephosphotransferase and ethanolaminephosphotransferase activities in *Plasmodium knowlesi*-infected erythrocytes, *Biochim. Biophys. Acta,* 795, 372, 1984.
32. **Smith, J. D.,** Differential selectivity of cholinephosphotransferase and ethanolaminephosphotransferase of Tetrahymena for diacylglycerol and alkylacylglycerol, *J. Biol. Chem.,* 260, 2064, 1985.
33. **Percy, A. K., Carson, M. F., and Waechter, C. J.,** Control of phosphatidylethanolamine metabolism in yeast: diacylglycerol ethanolaminephosphotransferase and diacylglycerol cholinephosphotransferase are separate enzymes, *Arch. Biochem. Biophys.,* 230, 69, 1984.
34. **Kanoh, H. and Ohno, K.,** Solubilization and purification of rat liver microsomal 1,2-diacylglycerol: CDP-choline cholinephosphotransferase and 1,2-diacylglycerol: CDP-ethanolamine ethnolaminephospho-transferase, *Eur. J. Biochem.,* 66, 201, 1976.
35. **Kanoh, H. and Ohno, K.,** 1,2-Diacylglycerol: CDP choline cholinephosphotransferase, *Methods Enzymol.,* 71, 536, 1981.

36. **Arthur, G. and Choy, P. C.,** Acyl specificity of hamster heart CDP-choline: 1,2-diacylglycerol phosphocholinetransferase in phosphatidylcholine biosynthesis, *Biochim. Biophys. Acta,* 795, 221, 1984.
37. **Dorman, R. V., Bischoff, S., and Terrian, D. M.,** Choline and ethanolamine phosphotransferase activities in glomerular particles isolated from bovine cerebral cortex, *Neurochem. Res.,* 11, 1167, 1986.
38. **Ansell, G. B. and Metcalfe, R. F.,** Studies on CDP-ethanolamine: 1,2-diglyceride ethanolaminephosphotransferase of rat brain, *J. Neurochem.,* 18, 647, 1971.
39. **Liteplo, R. G. and Sribney, M.,** Inhibition of rat liver CDPethanolamine: 1,2-diacylglycerol ethanolaminephosphotransferase activity by ATP and pantothenic acid derivatives, *Can. J. Biochem.,* 55, 1049, 1977.
40. **Vecchini, A., Roberti, R., Freyz, L., and Binaglia, L.,** Partial purification of ethanolaminephosphotransferase from rat brain microsomes, *Biochim. Biophys. Acta,* 918, 40, 1987.
41. **Polokoff, M., Wing, D. C., and Raetz, C. H. R.,** Isolation of somatic cell mutants defective in the biosynthesis of phosphatidylethanolamine, *J. Biol. Chem.,* 256, 7687, 1981.
42. **Hosaka, K. and Yamashita, S.,** Isolation and characterization of a yeast mutant defective in cholinephosphotransferase, *Eur. J. Biochem.,* 162, 7, 1987.

43a. **Hjelmstad, R. H. and Bell, R. M.,** Mutants of *Saccharomyces cerevisiae* defective in *sn*-1,2-diacylglycerol cholinephosphotransferase, *J. Biol. Chem.,* 262, 3909, 1987.

43b. **Hjelmstad, R. H. and Bell, R. M.,** The sn-1,2-diacylglycerol ethanolamine-phosphotransferase activity of *S. cerevisiae, J. Biol. Chem.,* 263, 19748, 1988.

44. **Sribney, M. and Kennedy, E. P.,** The enzymatic synthesis of sphingomyelin, *J. Biol. Chem.,* 233, 1315, 1958.
45. **Marggraf, H. and Anderer, F. A.,** Alternative pathways in the biosynthesis of sphingomyelin and the role of phosphatidylcholine, CDP-choline and phosphorylcholine as precursors, *Hoppe-Seyler's Z. Physiol. Chem.,* 355, 803, 1974.
46. **Ullman, M. D. and Radin, N. S.,** The enzymatic formation of sphingomyelin from ceramide and lecithin in mouse liver, *J. Biol. Chem.,* 249, 1509, 1974.
47. **Voelker, D. R. and Kennedy, E. P.,** Cellular and enzymatic synthesis of sphingomyelin, *Biochemistry,* 21, 2753, 1982.
48. **Mueller, H. W., O'Flaherty, J. J., and Wykle, R. L.,** Biosynthesis of platelet activating factor in rabbit polymorphonuclear neutrophils, *J. Biol. Chem.,* 258, 6213, 1983.
49. **Renooij, W. and Snyder, F.,** Biosynthesis of 1-alkyl-2-acetyl-*sn*-glycero-3-phosphocholine (platelet activating factor and a hypotensive lipid) by cholinephosphotransferase in various rat tissues, *Biochim. Biophys. Acta,* 663, 545, 1981.
50. **Woodward, D. S., Lee, T. C., and Snyder, F.,** The final step in the *de novo* biosynthesis of platelet activating factor, *J. Biol. Chem.,* 262, 2520, 1987.
51. **Cornell, R. and MacLennan, D. H.,** Solubilization and reconstitution of cholinephosphotransferase from sarcoplasmic reticulum: stabilization of solubilized enzyme by diacylglycerol and glycerol, *Biochim. Biophys. Acta,* 821, 97, 1985.
52. **Pikula, S., Szymanska, G., and Sarzala, G.,** Transbilayer distribution of phospholipid synthesizing enzymes in the sarcoplasmic reticulum membrane, *Int. J. Biochem.,* 18, 1023, 1986.
53. **Ballas, L. M. and Bell, R. M.,** Topography of phosphatidylcholine, phosphatidylethanolamine, and triacylglycerol biosynthetic enzymes in rat liver microsomes, *Biochim. Biophys. Acta,* 602, 578, 1980.
54. **Coleman, R. and Bell, R. M.,** Topography of membrane-bound enzymes, in *The Enzymes,* Vol. 16, Boyer, P., Ed., Academic Press, Orlando, FL, 1983, 607.
55. **Kanoh, K. and Ohno, T.,** Utilization of diacylglycerol in phospholipid bilayers by pig brain diacylglycerol kinase and Rhizopus arrhizus lipase, *J. Biol. Chem.,* 259, 11197, 1984.
56. **Allan, D., Thomas, P., and Michell, R. H.,** Rapid transbilayer diffusion of 1,2-diacylglycerol and its relevance to control of membrane curvature, *Nature (London),* 276, 189, 1978.
57. **Pagano, R. E. and Longmuir, K. J.,** Phosphorylation, transbilayer movement, and facilitated intracellular transport of diacylglycerol are involved in the uptake of a flourescent analog of phophatidic acid by cultured fibroblasts, *J. Biol. Chem.,* 260, 1909, 1985.
58. **Ganong, B. R. and Bell, R. M.,** Transmembrane movement of phosphatidylglycerol and diacylglycerol sulfhydryl analogues, *Biochemistry,* 23, 4977, 1984.
59. **Vance, D., Choy, P. C., Farren, S. B., Lim, P., and Schneider, W. J.,** Asymmetry of phospholipid biosynthesis, *Nature (London)* 270, 268, 1977.
60. **Coleman, R. and Bell, R. M.,** Evidence that biosynthesis of phosphatidylethanolamine, phosphatidylcholine, and triacylglycerol occurs on the cytoplasmic side of microsomal vesicles, *J. Cell Biol.,* 76, 245, 1978.
61. **Arienti, G., Corazzi, L., Freyz, L., Binaglia, L., Roberti, R., and Porcellati, G.,** Sidedness of phosphatidylcholine-synthesizing enzymes in rat brain microsomal vesicles, *J. Neurochem.,* 44, 38, 1985.
62. **Rosen, O.,** After insulin binds, *Science,* 237, 1452, 1987.
63. **Gilman, A. G.,** G proteins: transducers of receptor-generated signals, *Annu. Rev. Biochem.,* 56, 615, 1987.
64. **Freyz, L., Binaglia, L., Dreyfus, H., Massarelli, R., Golly, F., and Porcellati, G.,** Topological biosynthesis of phosphatidylcholine in brain microsomes, *J. Neurochem.,* 45, 57, 1985.

65. **Higgins, J. A.,** Asymmetry of the site of choline incorporation into phosphatidylcholine of rat liver microsomes, *Biochim. Biophys. Acta,* 558, 48, 1979.
66. **Kameyama, Y., Yoshioka, S., Hasegawa, I., and Nozawa, Y.,** Studies of diacylglycerol cholinephosphotransferase and diacylglycerol ethanolamine-phosphotransferase activities in tetrahymena microsomes, *Biochim. Biophys. Acta,* 665, 195, 1981.
67. **Cornell, R. and MacLennan, D. H.,** The capacity of the sarcoplasmic reticulum for phospholipid synthesis: a developmental study, *Biochim. Biophys. Acta,* 835, 567, 1985.
68. **Miller, J. C. and Weinhold, P. A.,** Cholinephosphotransferase in rat lung, *J. Biol. Chem.,* 256, 12662, 1981.
69. **Radominska-Pyrek, A., Pilarska, M., and Zimniak, P.,** Solubilization of microsomal phosphoethanolaminetransferase by octylglucoside, *Biochem. Biophys. Res. Commun.,* 85, 1074, 1978.
70. **Sarzala, M. G. and Van Golde, L. M. G.,** Selective utilization of unsaturated phosphatidylcholines and diacylglycerols by cholinephosphotransferase in lung microsomes, *Biochim. Biophys. Acta,* 441, 423, 1976.
71. **Kanoh, H. and Ohno, K.,** Studies on 1,2-diglycerides formed from endogenous lecithins by the back-reaction of rat liver microsomal CDPcholine: 1,2-diacylglycerol cholinephosphotransferase, *Biochim. Biophys. Acta,* 326, 17, 1973.
72. **Kanoh, H. and Ohno, K.,** Subtrate-selectivity of rat liver microsomal 1,2-diacylglycerol:CDPcholine (ethanolamine) phosphotransferase in utilizing endogenous substrates, *Biochim. Biophys. Acta,* 380, 199, 1975.
73. **Van Heusden, G. P. H. and Van den Bosch, H.,** Utilization of disaturated and unsaturated phosphatidylcholine and diacylglycerols by cholinephosphotransferase in rat lung, *Biochim. Biophys. Acta,* 711, 361, 1982.
74. **Rustow, B. and Kunze, D.,** Synthesis of phosphatidylcholine and phosphatidylethanolamine in relation to the concentration of membrane-bound diacylglycerols of rat lung microsomes, *Biochim. Biophys. Acta,* 793, 372, 1984.
75. **Ide, H. and Weinhold, P. A.,** Cholinephosphotransferase in rat lung, *J. Biol. Chem.,* 257, 14926, 1982.
76. **Fallon, H. J., Barwick, J., Lamb, R. G., and Van den Bosch, H.,** Studies of rat liver microsomal diglyceride acyltransferase and cholinephosphotransferase using microsomal-bound substrate, *J. Lipid Res.,* 16, 107, 1975.
77. **Rustow, B. and Kunze, D.,** The availability of endogenous and exogenous disaturated diacylglycerol for the diacylglycerol-consuming reactions in lung microsomes, *Biochim. Biophys. Acta,* 796, 359, 1984.
78. **Binaglia, L., Roberti, R., Vecchini, A., and Porcellati, G.,** Evidence for a compartmentation of brain microsomal diacylglycerol, *J. Lipid Res.,* 23, 955, 1982.
79. **Rustow, B. and Kunze, D.,** Diacylglycerol synthesis *in vitro* from sn-glycerol 3-phophate and endogenous diacylglycerol are different substrate pools for the biosynthesis of phosphatidylcholine in rat lung microsomes, *Biochim. Biophys. Acta,* 835, 273, 1985.
80. **Rustow, B. and Kunze, D.,** Further evidence for the existence of different diacylglycerol pools of the phosphatidylcholine synthesis in microsomes, *Biochim. Biophys. Acta,* 921, 552, 1987.
81. **Baker, R. R. and Chang, H.-Y.,** An increased incorporation of fatty acid into triacylglycerols of neuronal nuclei *in vitro* in the presence of CMP and EGTA, *Can. J. Biochem. Cell Biol.,* 62, 379, 1984.
82. **Tsao, F. H. C.,** Reversibility of cholinephosphotransferase in lung microsomes, *Lipids,* 21, 498, 1986.
83. **Jacobson, K., Hou, Y., Derzko, Z., Wojcieszyn, J., and Organisciak, D.,** Lipid lateral diffusion in the surface membrane of cells and in multibilayers formed from plasma membrane lipids, *Biochemistry,* 20, 5268, 1981.
84. **Walton, P. A. and Possmayer, F.,** The role of Mg^{2+}-dependent phosphatidate phosphohydrolase in pulmonary glycerolipid biosynthesis, *Biochim. Biophys. Acta,* 796, 364, 1984.
85. **Infante, J. P.,** Biosynthesis of acyl-specific glycerophospholipids in mammalian tissues: postulation of new pathways, *FEBS Lett.,* 170, 1, 1984.
86. **Infante, J. P.,** *De novo* CDP-choline-dependent glycerophosphorylcholine synthesis and its involvement as an intermediate in phosphatidylcholine synthesis, *FEBS Lett.,* 214, 149, 1987.
87. **Lim, P., Cornell, R., and Vance, D. E.,** The supply of both CDP-choline and diacylglycerol can regulate the rate of phosphatidylcholine synthesis in Hela cells, *Biochem. Cell Biol.,* 64, 692, 1986.
88. **Kennedy, E. P., Borkenhagen, L. F., and Smith, S. W.,** Possible metabolic function of deoxycytidine diphosphate choline and deoxycytidine diphosphate ethanolamine, *J. Biol. Chem.,* 234, 1998, 1959.
89. **Pontoni, G., Manna, C., Salluzzo, A., Del Piano, L., Galletti, P., De Rosa, M., and Zappia, V.,** Studies on enzyme-substrate interactions of cholinephosphotranferase from rat liver, *Biochim. Biophys. Acta,* 836, 222, 1985.
90. **Hirabayashi, T., Larson, T. J., and Dowhan, W.,** Membrane-associated phosphatidylglycerophosphate synthetase from Escherichia coli: purification by substrate affinity chromatography on CDP-diacylglycerol sepharose, *Biochemistry,* 15, 5205, 1976.
91. **Bae-Lee, M. S. and Carman, G. M.,** Phosphatidylserine synthesis in *S. cerevisiae, J. Biol. Chem.,* 259, 10857, 1984.

92. **Weinhold, P. A., Rounsifer, M. E., and Feldman, D. A.,** The purification and characterization of CTP: phosphocholine cytidylyltransferase from rat liver, *J. Biol. Chem.*, 261, 5104, 1986.
93. **Sturton, R. G. and Brindley, D. N.,** Factors controlling the metabolism of phosphatidate by phosphohydrolase and phospholipase A activities, *Biochim. Biophys. Acta*, 619, 494, 1980.
94. **Soler-Argilaga, C., Russell, R. L., and Heimberg, M.,** Reciprocal relationship between uptake of Ca^{2+} and biosynthesis of glycerolipids from sn-glycerol-3-phosphate by rat liver microsomes, *Biochem. Biophys. Res. Commun.*, 78, 1053, 1977.
95. **Taniguchi, S., Morikawa, S., Hayashi, H., Fujii, K., Mori, H., Fujiwara, M., and Fugiwara, M.,** Effects of Ca^{2+} on ethanolaminephosphotransferase and cholinephosphotransferase in rabbit platelets, *J. Biochem.*, 100, 485, 1986.
96. **Alemany, S., Varela, I., and Mato, H.,** Inhibition of phosphatidylcholine synthesis by vasopressin and angiotensin in rat hepatocytes, *Biochem. J.*, 208, 453, 1982.
97. **Tijburg, L., Schuurmans, E., Geelen, M., and Van Golde, L. M. G.,** Effects of vasopressin on the synthesis of phosphatidylethanolamines and phosphatidylcholines by isolated rat hepatocytes, *Biochim. Biophys. Acta*, 919, 49, 1987.
98. **Holub, B. J. and Kuksis, A.,** Metabolism of molecular species of diacylglycerophospholipids, in *Advances Lipid Research*, Vol. 16, Paoletti, R. and Kritchevsky, D., Eds., Academic Press, New York, 1978, 1.
99. **Roberti, R., Binaglia, L., and Porcellati, G.,** Synthesis of molecular species of glycerophospholipids from diglyceride-labeled brain microsomes, *J. Lipid Res.*, 21, 449, 1980.
100. **Ishidate, K. and Weinhold, P. A.,** The content of diacylglycerol, triacylglycerol, and monoacylglycerol and a comparison of the structural and metabolic heterogeneity of diacylglycerols and phosphatidylcholine during rat lung development, *Biochim. Biophys. Acta*, 664, 133, 1981.
101. **Van Heusden, G. P. H., Rustow, B., Van der Mast, M. A., and Van den Bosch, H.,** Synthesis of disaturated phosphatidylcholine by cholinephosphotransferase in rat lung microsomes, *Biochim. Biophys. Acta*, 666, 313, 1981.
102. **Rustow, B., Kunze, D., Rabe, H. and Reichman, G.,** The molecular species of phosphatidic acid, diacylglycerol, and phosphatidylcholine synthesis from sn-glycerol 3-phosphate in rat lung microsomes, *Biochim. Biophys. Acta*, 835, 465, 1985.
103. **Justin, A. M., Demandee, C., and Mazliak, P.,** Choline and ethanolamine-phosphotransferase from pea leaf and soya beans discriminate 1-palmitoyl-2-linoleoyl diacylglycerol as a preferred substrate, *Biochim. Biophys. Acta*, 992, 364, 1987.
104. **Morimoto, K. and Kanoh, H.,** Acyl chain length dependency of diacylglycerol cholinephosphotransferase and diacylglycerol ethanolaminephosphotransferase, *J. Biol. Chem.*, 253, 5056, 1978.
105. **Ford, D. A. and Gross, R. W.,** Identification of endogenous 1-*O*-alk-1′-enyl-2-acyl-sn-glycerol in myocardium and its effective utilization by cholinephosphotransferase, *J. Biol. Chem.*, 263, 2644, 1988.
106. **Pelech, S. L., Pritchard, P. H., Brindley, D. N., and Vance, D. E.,** Fatty acids promote translocation of CTP:phosphocholine cytidylyltransferase to the endoplasmic reticulum and stimulate rat hepatic phosphatidylcholine synthesis, *J. Biol. Chem.*, 258, 6782, 1983.
107. **Sundler, R. and Akkeson, B.,** Factors controlling the biosynthesis of individual phosphoglycerides in liver, *Biochem. Soc. Trans.*, 5, 43, 1977.
108. **Kwanyuen, P. and Wilson, R. F.,** Isolation and purification of diacylglycerol acyltransferase from germinating soybean cotyledons, *Biochim. Biophys. Acta*, 876, 238, 1986.
109. **Polokoff, M. A. and Bell, R. M.,** Solubilization, partial purification and characterization of rat liver microsomal diacylglycerol acyltransferase, *Biochim. Biophys. Acta*, 618, 129, 1980.
110. **Vance, D. E., Trip, E., and Paddon, H.,** Poliovirus increases phosphatidylcholine biosynthesis in Hela cells by stimulation of the rate-limiting reaction catalyzed by CTP phosphocholine cytidylyltransferase, *J. Biol. Chem.*, 255, 1064, 1980.
111. **Pritchard, P. H. and Brindley, D. N.,** Studies on the ethanol-induced changes in glycerolipid synthesis in rats and their partial reversal by benfluorex, *J. Pharm. Pharmacol.*, 29, 343, 1977.
112. **Pelech, S. L. and Vance, D. E.,** Regulation of phosphatidylcholine biosynthesis, *Biochim. Biophys. Acta*, 779, 217, 1984.
113. **Cornell, R. and Vance, D. E.,** Translocation of CTP: phosphocholine cytidylyltransferase from cytosol to membrance in Hela cells: stimulation by fatty acid, fatty alcohol, and mono- and di-acylglycerol, *Biochim. Biophys. Acta*, 919, 26, 1987.
114. **Rosenberg, I. L., Smart, D. A., Gilfillan, A. M., and Rooney, S. A.,** Effect of 1-oleoyl,2-acetylglycerol and other lipids on phosphatidylcholine synthesis and cholinephosphate cytidylyltransferase activity in cultured type II pneumocytes, *Biochim. Biophys. Acta*, 876, 581, 1987.
115. **Tercé, F., Record, M., Ribbes, G., Chap, H., and Douste-Blazy, L.,** Intracellular processing of cytidylyltransferase in Krebs II cells during stimulation of phosphatidylcholine synthesis: evidence that a plasma membrane modification promotes enzyme translocation specifically to the endoplasmic reticulum, *J. Biol. Chem.*, 263, 3142, 1988.

116. **Martin, A., Hopewell, R., Martin-Sanz, P., Morgan, J., and Brindley, D. N.**, Relationship between the displacement of phosphatidate phosphohydrolase from the membrane-associated compartment by chlorpromazine and the inhibition of the synthesis of triacylglycerol and phosphatidylcholine in rat hepatocytes, *Biochim. Biophys. Acta,* 876, 581, 1986.
117. **Pelech, S. L., Jetha, F., and Vance, D. E.**, Trifluoperazine and other anaesthetics inhibit rat liver CTP: phosphocholine cytidylyltransferase, *FEBS Lett.,* 158, 89, 1983.
118. **Leli, U. and Hauser, G.**, Modification of phospholipid metabolism induced by cholorpromazine, desmethylimipramine, and propranolol in C_6 glioma cells, *Biochem. Pharmacol.,* 36, 31, 1987.
119. **Bjornstad, P. and Bremer, J.**, *In vivo* studies on pathways for the biosynthesis of lecithin in the rat, *J. Lipid Res.,* 7, 38, 1966.
120. **Sundler, R. and Akesson, B.**, Biosynthesis of phosphatidylethanolamines and phosphatidylcholines from ethanolamine and choline in rat liver, *J. Lipid Res.,* 146, 309, 1975.
121. **Slack, C. R., Campbell, L. C., Browse, J. A., and Roughan, P. G.**, Some evidence for the reversibility of the cholinephosphotransferase-catalysed reaction in developing linseed cotyledons *in vivo, Biochim. Biophys. Acta,* 754, 10, 1983.
122. **Goracci, G., Francescangeli, E., Horrocks, L. A., and Porcellati, G.**, The reverse reaction of cholinephosphotransferase in rat brain microsomes: a new pathway for degradation of phosphatidylcholine, *Biochim. Biophys. Acta,* 664, 373, 1981.
123. **Reiss, P. D., Zuurendonk, P. K., and Veech, R. L.**, Measurement of tissue purine, pyrimidine and other nucleotides by radial compression high performance liquid choromatography, *Anal. Biochem.,* 140, 162, 1984.
124. **Wilgram, G., Holoway, C. F., and Kennedy, E. P.**, The content of cytidine diphosphate choline in the liver of normal and choline-deficient rats, *J. Biol. Chem.,* 235, 37, 1960.
125. **Porcellati, G., De Medio, G., Fini, C., Florida, A., Goracci, G., Horrocks, L., Lazarewicz, J., Palmeini, C., Strosznajder, J., and Trovarelli, G.**, Phospholipid and its metabolism in ischemia, in *Proc. Eur. Soc. Neurochem.,* Vol. 1, Neuhoff, V., Ed., Verlag Chemie, 1978, 285.
126. **Goracci, G., Francescangeli, E., Mozzi, R., Porcellati, S., and Porcellati, G.**, Regulation of phospholipid metabolism by nucleotides in brain and transport of CDPcholine into brain, in *Novel Biochem., Pharmacol. and Clinical Aspects of CDP-Choline,* Zappia, V., Kennedy, E. P., Nilson, B., and Galleti, P., Elsevier Science, 1985, 105.
127. **Goracci, G., Gresele, P., Arienti, G., Porrovecchio, P., Nenci, G., and Porcellati, G.**, Cholinephosphotransferase activity in human platelets, *Lipids,* 18, 179, 1983.
128. **Arthur, G., Tam, S. W., and Choy, P. C.**, The effects of detergents on CDP-choline: 1,2-diacylglycerol phosphocholinetransferase from hamster heart, *Can. J. Biochem. Cell Biol.,* 62, 1059, 1984.
129. **Parthasarathy, S. and Baumann, W.**, Lysolecithin as a regulator of *de novo* lecithin synthesis in rat liver microsomes, *Biochem. Biophys. Res. Commun.,* 91, 637, 1979.
130. **Ridgeway, N. D. and Vance, D. E.**, Purification of phosphatidylethanolamine *N*-methyltransferase from rat liver, *J. Biol. Chem.,* 262, 17231, 1987.
131. **Bayley, H.**, Photogenerated reagents in biochemistry, in *Laboratory Techniques in Biochemistry and Molecular Biology,* Vol. 12, Work, T. S., and Burdow, R. H., Elsevier, Amsterdam, 1983, 25.
132. **Czarnecki, J., Geahlen, R., and Haley, B.**, Synthesis and use of azido photoaffinity analogs of adenine and guanine nucleotides, *Methods Enzymol.,* 56, 642, 1979.
133. **Evans, R. and Haley, B.**, Synthesis and biological properties of 5-azido-2′-deoxyuridine-5′-triphosphate, a photoactive nucleotide suitable for making light-sensitive DNA, *Biochemistry,* 26, 269, 1987.
134. **Caras, I. W. and Martin, D. W.**, Direct photoaffinity labeling of an allosteric site of subunit protein M1 of mouse ribonucleotide reductase by dATP, *J. Biol. Chem.,* 257, 9508, 1982.
135. **Caras, I. W., Jones, T., Eriksson, S., and Martin, D. W.**, Direct photoaffinity labeling of the catalytic site of mouse ribonucleotide reductase by CDP, *J. Biol. Chem.,* 258, 3064, 1983.
136. **Lee, R. W. H., Suchanek, C., and Huttner, W. B.**, Direct photoaffinity labeling of proteins with adenosine 3′-(^{32}P) phosphate 5′ phosphosulfate, *J. Biol. Chem.,* 259, 11153, 1984.
137. **Biswas, S. B. and Kornberg, A.**, Nucleoside triphosphate binding to DNA polymerase III holoenzyme of *Escherichia coli, J. Biol. Chem.,* 259, 7990, 1984.
138. **Moriyama, Y. and Nelson, N.**, Nucleotide binding sites and chemical modification of the chromaffin granule proton ATPase, *J. Biol. Chem.,* 262, 14723, 1987.
139. **Esko, J. D., Nishijima, M., and Raetz, C. H. R.**, Animal cells dependent on exogenous phosphatidylcholine for membrane biogenesis, *Proc. Natl. Acad. Sci. U.S.A.,* 79, 1698, 1982.

Chapter 5

PHOSPHOLIPASE D AND THE BASE EXCHANGE ENZYME

Julian N. Kanfer

TABLE OF CONTENTS

I. INTRODUCTION

These independent enzyme activities are occasionally considered together because they react with the same moiety of the phospholipid molecule. This is at the phosphodiester bond linking the component characterizing each particular phospholipid, as ethanolamine being the determinant for the classification as a phosphatidylethanolamine, and the diglyceride backbone. The base exchange enzymes catalyze the substitution of free ethanolamine, serine, and choline for a similar substitutent on the preexisting phospholipids. There is an absence of unequivocal evidence indicating that the base exchange enzymes are capable of operating as hydrolases.

Phospholipase D (PLD), however, is a hydrolytic enzyme which liberates the amino alcohol components from intact phospholipids. Under special *in vitro* circumstances PLD can catalyze the incorporation of a variety of substances containing alcohol groups into their corresponding phospholipid. This is referred to as the "transphosphatidylation" activity of PLD. The base exchange reactions and PLD activity are represented as follows.

1. Phospholipase D

(a) Hydrolysis

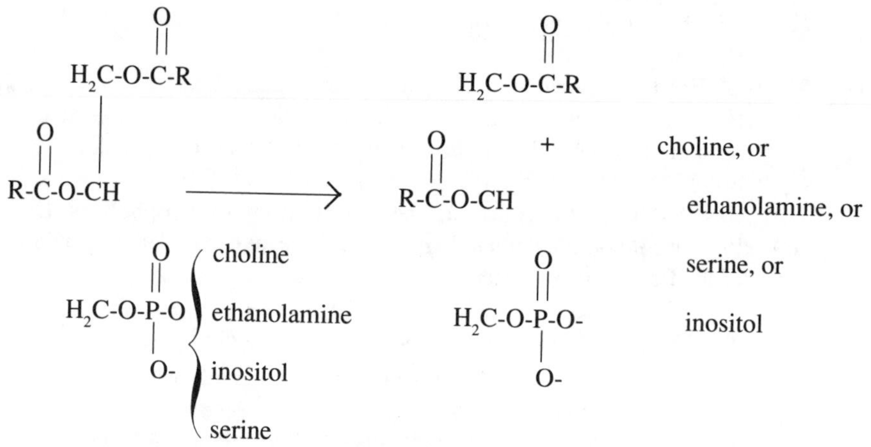

```
        O                                O
        ||                               ||
     H2C-O-C-R                        H2C-O-C-R
  O     |
  ||    |                           O       +        choline, or
R-C-O-CH        ------------>       ||
                                  R-C-O-CH            ethanolamine, or
        O  / choline
        || |                             O            serine, or
     H2C-O-P-O | ethanolamine            ||
        |  |                          H2C-O-P-O-      inositol
        O- | inositol                    |
           |                             O-
           \ serine
```

Phosphatidyl—serine, ethanolamine, inositol, choline → Phosphatidic acid

(b) "Transphosphatidylation"

```
        O                                 O
        ||                                ||
     H2C-O-C-R                         H2C-O-C-R
                      HOY ---->
R-C-O-CH        +   <---- HOY      R-C-O-CH   +   HOX

        O                                 O
        ||                                ||
     H2C-O-P-OX                        H2C-O-P-OY
        |                                 |
        O-                                O-
```

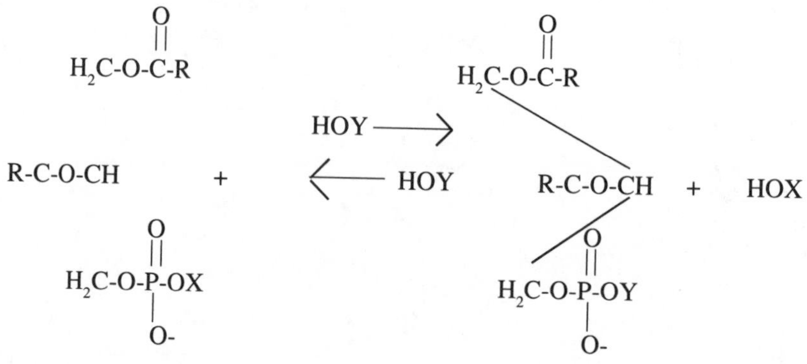

Phosphatidyl X → Phosphatidyl Y

These two activities are postulated to occur by the release of a "phosphatidyl" group from an intermediate which is speculated to be enzyme bound.

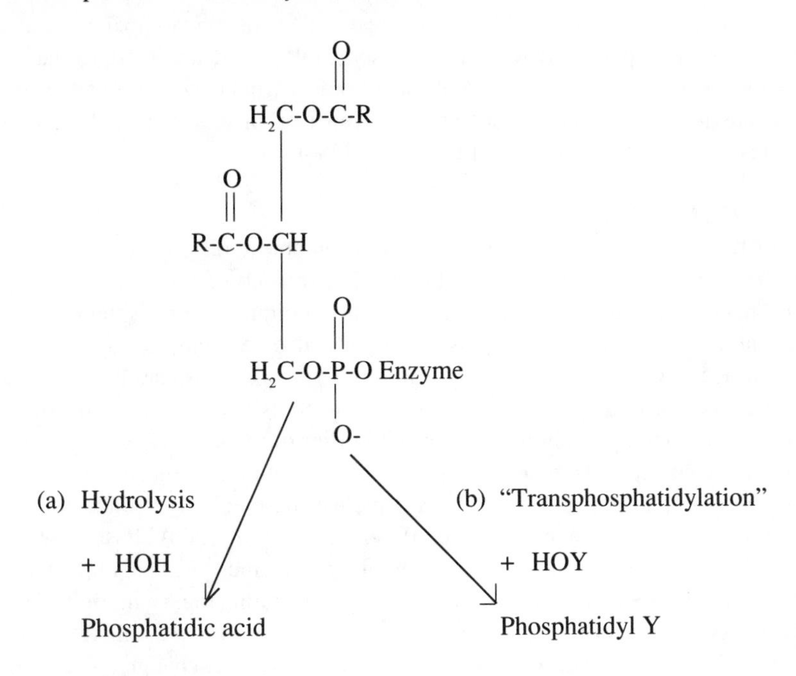

2. The Base Exchange Enzymes

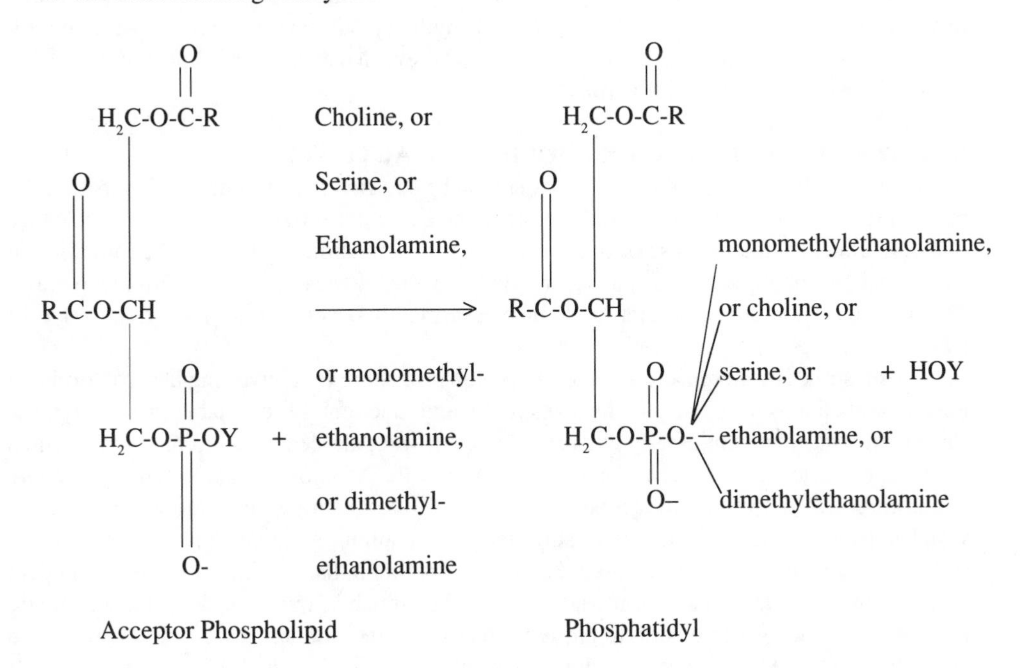

Reviews have appeared about PLD and the base exchange enzymes of mammalian tissues[1,2] and PLD of various sources.[3,125]

II. PHOSPHOLIPASE D

There has been considerably less interest in this enzyme compared to interest in the more fashionable phospholipase A_2, which is responsible for release of prostaglandin precursors,[4] or the phosphatidylinositol specific phospholipase C, believed responsible for producing two separate intracellular second messengers.[5] However, there appears to be a recent awakening of interest in a potential function of PLD in cell biology.

A. UTILIZATION

PLD has been employed as one of the available tools for investigating the asymmetric distribution of membrane phospholipids.[6] The transphosphatidylation reaction catalyzed by PLD has been exploited to produce unusual phospholipid analogues such as phosphatidylmethanol and phosphatidylethanol.[7] This activity has also been employed to prepare dimethylethanolamine plasmalogen,[8] phosphatidylhomoserine,[9] and phosphatidylnucleosides.[10]

PLD has also been utilized as one of the tools for demonstrating the contribution of membrane-bound phospholipids to the properties of membranes. Exposure of rat liver microsomes to cabbage PLD reduces the phosphatidylcholine content and results in the inactivation of glucose-6-phosphatase and NADH dehydrogenase activity.[11] Phospholipase D-treated rat kidney mitochondria have reduced phospholipid, protein, and ATPase activity, suggesting that phospholipids have a role in the function of this enzyme.[12] PLD disrupts long-range arrays of *Neurospora* mitochondria membrane channels, indicating that zwitterionic lipids may stabilize the arrays of channel-forming proteins.[13] Treatment of sarcoplasmic reticulumn vesicles with PLD reduces Ca^{+2} binding[14] and stimulates a Na^{+}-Ca^{+2} exchange.[15] The surface charge and electrostatic potential of mitochondrial membranes were increased by PLD treatment and resulted in a tenfold increase of the K_m for D-β-hydroxy butyrate dehydrogenase activity.[16] Phospholipase D modification of human erythrocytes revealed that separate lipid domains exist, possessing specific physical properties.[17]

B. DETERMINATION OF PHOSPHOLIPASE D ACTIVITY

In principle the presence of PLD activity should be detected after appropriate incubations by measuring (1) the reduction of phosphatidylcholine content, (2) the increase of/or appearance of free choline, (3) the increase of/or appearance of phosphatidic acid, or (4) the formation of an unusual phospholipid such as phosphatidylethanol during the course of *in vitro* incubations. The most common substrate employed in measuring PLD activity is either radioactive or nonradioactive phosphatidylcholine.

This substrate is water insoluble, and a common procedure for converting it into a form that is most aesthetically acceptable to the investigator and functionally acceptable to the enzyme is sonication. This treatment usually will provide a clear-to-opalescent aqueous medium in which the phosphatidylcholine is present as some type of micelle. A major advantage of this procedure is the ability to deliver reproducible identical substrate quantities into individual reaction vesicles. However, this usually is not sufficient for obtaining good enzyme activity, and the addition of activators is often required. Early workers found that a volume of ether equivalent to the volume of aqueous solution was effective. Presumably, this is because the enzymatic reaction occurs at a water-ether interphase.[18] The use of ether for activation of PLD appears to have been replaced by the use of sodium dodecylsulfate (SDS), which is most effective at a lecithin:SDS ratio of about 1 to 2 or 3.[18,19] Ca^{+2} is usually required with the highly purified, commercially available PLD preparations from plant tissues.

C. OCCURRENCE OF PHOSPHOLIPASE D

The overwhelming body of information concerning this enzymatic activity has become available from studies of the preparations from the plant kingdom. One of the earliest

TABLE 1
Some Properties of Purified Phospholipase D of Plant Sources

Molecular weight	Peanut	Cabbage	Citron
By equilibrium centrifugation	22,000	116,600	—
By gel filtration	200,000	—	90,500
By SDS electrophoresis	48,500	112,500	90—94,000
K_m	3.38 m*M*	2.66 m*M*[a]	—
V_{MAX}	—	399 μmol/mg/min[a]	—
pH optimum	5.6	7.256.5	
Ca^{2+}	—	0.21[b]	50 m*M*
N–terminal amino acid	Glycine	——	
pI	4.5—4.6	—5.0	

[a] Dihexanoyl phosphatidylcholine was used as substrate.
[b] K_m value for Ca^{2+}.

suggestions predicting PLD existence was obtained with extracts of carrots.[20] A phospholipid fraction was prepared from carrot slices directly or from carrot slices previously exposed to steam for 5 min. Analysis of the phospholipid fraction showed that the amount of lipid phosphorus, which is a measure of the quantity of phospholipid, and the amount of fatty acid recovered were identical from both sources. However, there was a very low choline or lipid nitrogen content in the material obtained from the sample that was directly extracted as compared to the steam-heated sample.[20] This observation prompted these individuals to search for an enzyme that might be responsible for this difference in phospholipid composition. They were successful in characterizing some of the properties of a PLD in studies using the decrease of ether-soluble nitrogen as the enzyme assay procedure.[21] It should be indicated that the early literature refers to this enzyme activity either as lecithinase C or phospholipase C.

There will not be an attempt to provide a catalog of all the reports of the presence of PLD in plants, microorganisms, and mammalian tissues since this information is available.[3,4]

Apparently homogeneous preparations of PLD have been obtained from peanuts,[22] savoy cabbage,[23] and cultured citrus seeds.[24] Selected properties of these three enzymes is shown as Table 1.

Dihexanoyl phosphatidylcholine was utilized as the substrate in the purification of the cabbage enzyme because it is a water-soluble lecithin analogue. This enzyme preparation was relatively stable when stored in a 10% inositol-containing solution and was activated by dithiothreitol, glycerol, and at pHs above 7.4. The enzyme was less stable at –20°C under these conditions. However, storage at 4°C did not result in significant losses of activity, even in the absence of glycerol.[22] The instability at –20°C might be explained by disaggregation of the high molecular weight active enzyme to less active monomers. Both of the hydrolytic and the transphosphatidylation activities of the 1000-fold enriched PLD of peanut were copurified, and the ratio of both activities at each stage of purification was approximately 1:1.2.[26] PLD activity is present in many plant tissues.

Escherichia coli[27] and *H. parainfluenzae*[28] have a PLD activity that hydrolyzes cardiolipin to phosphatidylglycerol and phosphatidic acid. A PLD capable of hydrolyzing sphingomyelin to ceramide phosphate and free choline has been found in toxins of *C. ovis*[29] and *V. damsela.*[30]

Although PLD was postulated to be present in mammalian tissues, its occurrence was only unequivocally demonstrated when it was detected in solubilized preparations of rat brain microsomes[31] and subsequently detected in all solid tissues of the rat.[32] PLD activity has been detected in human erythrocytes,[33,34] human[35] and rat eosinophils,[36] rat heart,[37] and in plasma in a form that may be specific for the hydrolysis of the phosphatidylinositol anchor of membrane proteins.[38]

D. PROPERTIES OF MAMMALIAN PHOSPHOLIPASE D

The enzyme activity from rat brain microsomal membranes was solubilized and partially purified approximately 240-fold. Solubilization was achieved with an uncommon detergent, Miranol H2M, and enzyme enrichment through the use of gel exclusion and DEAE cellulose column chromatography.[39] The enzyme had an apparent molecular weight of about 200,000, and the pH optimum was 6.0. Both phosphatidylethanolamine and phosphatidylcholine served as substrates with K_ms of 0.75 and 0.91 m*M*, respectively; however, lysolecithin was not hydrolyzed. The activity of this partially purified PLD was not dependent upon added cation, but was stimulated 65 and 80% by Ca^{+2} and Fe^{+2}, respectively, and was unaffected by the presence of ether. Para chloromercurobenzoate completely abolished this enzyme activity. The base exchange enzyme activities were undetectable in this preparation, indicating that mammalian PLD is incapable of catalyzing this reaction. A less purified solubilized rat-brain enzyme preparation was shown to catalyze the transphosphatidylation reaction in the presence of glycerol, producing phosphatidylglycerol.[40] The product was a racemic mixture believed to be 3 *sn*-diacylglycerol-phospho-1′,3′-*sn* glycerol. The transphosphatidylation activity was optimum at pH 6.0, and required detergent, with 8 m*M* taurodeoxycholate being the most effective of several tested. The K_m value for glycerol was 0.2 *M* and for phosphatidylcholine was 3.5 m*M*. Other properties were consistent with a single enzyme being responsible for both the hydrolytic and the transphosphatidylation reactions.

This enzyme *in situ* is embedded in membranes, and its activity was undetectable with exogenous substrate in the absence of added detergent.[22] Therefore, it is conceivable that the detergent was merely required to facilitate the presentation of the substrate in a form acceptable to the membrane-bound enzyme. This possibility was tested by exploiting the transphosphatidylation activity of the membrane-bound enzyme, since the substrate would be endogenous phospholipid located in the same membranes. It was apparent that both the hydrolytic and transphosphatidylation activities of the rat-brain PLD were latent. The most effective activators were the monunsaturated fatty acids, oleic and palmitoleic acids, which stimulated this activity 166-fold. Saturated fatty acids and a variety of detergents were inert, and diunsaturated were less effective than monounsaturated.[41] PLD activity is demonstrable in an axolemma-enriched fraction from rat central nervous system, with a specific activity equivalent to that of the brain microsomal fraction.[42]

Further support for the observed latency of membrane-bound PLD activity was obtained from studies utilizing prelabeled rat brain microsomes. Animals were administered with either [^{3}H]oleic acid or [C^3H_3]choline intracerebrally to label total lipids and lecithin, respectively. These labeled particles were incubated *in vitro* and either the lipid-soluble or the aqueous-soluble products measured. There was a dependence on the presence of oleic acid, 5 m*M* being optimum, to demonstrate the formation of either phosphatidic acid or choline from endogenous prelabeled substrate.[43] Considerable quantities of diglyceride were also obtained, presumably as a result of phosphatidic acid phosphatase hydrolysis of phosphatidic acid. This interpretation is consistent with the formation of free choline, but not choline phosphate as the aqueous soluble product, and can be explained as a result of coupled PLD and phosphatidic acid phosphatase activities as:

$$\text{(a) Phosphatidylcholine} \xrightarrow[(C_{18:1})]{\text{PL-D}} \text{choline + phosphatidic acid}$$

$$\text{(b) Phosphatidic acid} \xrightarrow[\text{PTase}]{\text{P.A.}} \text{Diglyceride} + H_3\,PO_4$$

Alternately, for phospholipase C activity the products would be

$$\text{(c) Phosphatidylcholine} \xrightarrow{\text{PL-C}} \text{Diglyceride + choline phosphate}$$

The coupling of reactions (a) and (b) are favored because (1) choline and not choline phosphate was liberated, (2) KF which partially inhibits phosphatidic acid phosphatase increased the amount of phosphatidic acid recovered, and (3) the formation of phosphatidic acid and choline were both oleate dependent.

Attempts were made to demonstrate that phospholipase A_2 activity, also present in these membranes and which liberates free fatty acids from endogenous substrates, could activate PLD in these same membranes. Mellitin which activates phospholipase A_2 was added to the prelabeled membranes and caused the release of free choline, suggesting that the oleate liberated from endogenous phospholipids was capable of activating PLD. The sequence can be formulated:

(d) Phospholipid $\xrightarrow[\text{(Mellitin)}]{PLA_2}$ lysophospholipid + oleate

(e) Phosphatidylcholine $\xrightarrow[\text{(oleate)}]{\text{PL-D}}$ Phosphatidic acid + choline

(f) Phosphatidic acid $\xrightarrow[\text{(KF inhibits)}]{\text{P.A.-PTase}}$ Diglyceride + H_3PO_4

This illustrates a coordinated system contained in a biological membrane where the product of one enzymatic reaction becomes the substrate for another enzyme.

A human eosinophil PLD was partially purified from sonicates using a colorimetric choline assay for quantitation. The enzyme had a pH optimum from 4.5 to 6, a pI of 5.8 to 6.2, and a molecular weight of about 60,000.[35] Phosphatidic acid was the lipid produced in the reaction as seen by thin-layer chromatography. This is an unexpected observation, since the *in vitro* incubations usually contained 25% ethanol. Under these conditions the expected product would have been phosphatidylethanol due to the transphosphatidylation activity of the enzyme.

Phosphatidyl-*N*-palmitoyl-ethanolamine was hydrolyzed by the mitochondria and the microsomes of canine heart tissue. Diglyceride was the principal lipid soluble product in the absence of added KF, but in the presence of KF it was phosphatidic acid.[37] This reinforces the proposal that the ubiquitous phosphatidic acid phosphatase is capable of effectively utilizing phosphatidic acid formed by PLD as substrate. Of several detergents tested Triton-X-100 incresed the hydrolytic activity, and cations were not required. The activity was assayed at pH 5.5 and 7.0; however, there did not appear to be a discrete optimum.

E. LYSOPHOSPHOLIPASE D

The hydrolysis of hexadecyl-*sn*-glycero-3-phosphorylethanolamine or the corresponding choline derivative was demonstrated with rat brain microsomes, and the product obtained was hexadecylglycerol. However, if NaF was included in these incubations, hexadecylglycerol-phosphate was produced. This suggests that a lysophosphatidic acid phosphatase was present which cleaved the hexadecylglycerolphosphate formed by the lysophospholipase D. The lysophospholipase D activity was stimulated by Mg^{2+}, but not by Ca^{2+}, Zn^{2+}, or Mn^{2+} and was most active at pH 7.2. Acylation of the free hydroxyl group at C-2 of the substrate reduced the hydrolysis of these ether-containing lyso phospholipids.[44] Enzyme activity with essentially identical characteristics was enriched in the microsomal fraction from rat kidney, intestine, lung, tests, and liver.[45] The enzyme present in rat liver microsomes does not appear to hydrolyze acyl-containing substrates, preferring those with ether or alkyl linkages. The 1-alkyl-2-acetyl-*sn*-glycero-3-phosphorylethanolamine or the corresponding phosphorylcholine-containing compound were not substrates.[46] Incubation of plasma from several mammalian species results in the formation of a vasoactive substance suggested to be a lysophospholipid, perhaps lysophosphatidic acid. This compound appears to be produced by an endogenous lysophospholipase

D present in serum. Intact platelet activating factor (PAF), which is a 1-alkyl-2-acetyl-*sn*-3-glycerylphosphorylcholine, is not a substrate.[47]

F. TRANSPHOSPHATIDYLATION

This reaction has been best studied for the enzymes isolated from plant sources. Purified enzymes from cabbage were used by two separate groups who reported similar results nearly simultaneously.[48,49] Progressive increases in phosphatidylglycerol formation in concert with progressive decreases in phosphatidic acid formation were observed with increasing glycerol concentration present in the *in vitro* incubations. The concentrations of the alcohol acceptors routinely employed were in the range of 0.2 to 0.8 *M*. It was found that glycerol, methanol, ethanol, *n*-propanol, propanediol, ethylene glycol, choline, and ethanolamine were converted to substances presumed to be their corresponding phospholipid. The only product well characterized was phosphatidylglycerol. A variety of alcohols were not active acceptors, including inositol, serine, glucose, glycerol-1-phosphate, isopropanol, sucrose, galactose, malate, citrate, lactate, or glycolate.

An enzyme from cauliflower florets converts phosphatidylinositol to bis (phosphatidyl) inositol rather than the principal anticipated product, phosphatidic acid. This unusual inositide may be formed by the transphosphatidylation between two molecules of phosphatidylinositol.[50] This peculiarity of the enzyme had earlier been described for commercial cabbage PLD. Cardiolipin was produced in incubations containing phosphatidylglycerol as substrate.[51]

Membranes prepared from *Cl. butyricium* incubated with labeled phosphatidylserine, phosphatidylglycerol, or phosphatidylethanolamine in the presence of Triton X-100 yielded three products. These appeared to correspond to cardiolipin, phosphatidic acid, and phosphatidyltriton, presumably due to the transphosphatidylation activity of PLD.[52]

G. PHYSIOLOGICAL FUNCTION OF PHOSPHOLIPASE D

The simplest function of this enzyme could be to remove "old" phospholipid molecules. However, this task is generally assigned to the enzyme repertoire of the "lysosomal" compartment of mammalian tissues.[53] PLD, however, has not been detected in these organelles.

An acidic phospholipid not normally present in mammalian tissues was detected in selected organs of rats exposed to ethanol.[54] This unusual compound was eventually identified as phosphatidylethanol and, in analogy to the plant enzyme was speculated to arise through the transphosphatidylation activity of PLD.[55] Preliminary results suggested that PLD of rat brain membranes could produce phosphatidylethanol *in vitro*.[56] This postulated involvement of PLD in phosphatidylethanol formation was independently confirmed simultaneously by two laboratories.[57-58] An oleate dependence for radioactive ethanol incorporation into phosphatidylethanol and the exclusion of a base exchange-type reaction was demonstrated with microsomes.[57] Phosphatidylethanol formation by synaptosomes was dependent upon linoleic, linolenic, oleic, or arachidonic acids, but not a variety of other fatty acids and detergents. The reaction was inhibited by para chloromercuro-benzoate, phosphatidylcholine was the preferred substrate, and the activity was present to varying degrees in the microsomal fraction of several rat organs.[58] The potential contribution of this abnormal component in the synaptosomal membrane to the signs and symptoms of the fetal alcohol syndrome[59] is a matter of speculation.

Phosphatidylethanol appearance has facilitated the demonstration of PLD involvement in the biological responses of cell cultures exposed to several membrane perturbators. The tumor-promoting phorbol ester, 12-0-tetradecanoylphorbol acetate (TPA), elicits various responses depending upon the particular cell type, presumably particular cell type, presumably through interaction with a TPA receptor and a Ca^{+}-dependent protein kinase C. The addition of TPA to cultures of bovine lymphocytes[60] and HL-60 cells[61] containing ethanol in the growth medium resulted in the appearance of phosphtidylethanol. A similar observation was obtained when the HL-60 cells were exposed to a chemotactic peptide, *n*-formyl-met-leu-phe.[62] Vasopressin, angiotensin II, and epinephrine, Ca^{2+}-mobilizing hormones, when added to culture medium of hepatocytes result in phosphatidylethanol appearance. GTPγS stimulated phosphatidylethanol formation by rat liver plasma membranes in the presence of ethanol.[63] These independent observations provide experimental evidence, exploiting transphosphatidylation activity as a tool, that PLD is in some way linked to receptor-mediated events of biological membranes. The responsiveness to external signals may also reflect the latency of this membrane-bound enzyme activity which could be of advantage to the cell from the perspective of regulation and cellular homeostasis.

Increased mass of phosphatidic acid and decreased mass of phosphatidylcholine were found with isolated hepatocyte plasma membranes exposed to GTPγS. There were increased phosphatidic acid levels of intact hepatocytes in response to vasopressin, epidermal growth factor, epinephrine, angiotensin, a Ca^{2+} ionophore A23187, but not to cholera toxin. The changes were accompanied by increases in choline and phosphorylcholine.[64] These observations support a probable functional role of membrane-bound PLD in receptor- and ligand-induced cellular reactions.

Perfusion of isolated chicken hearts and rat cortex *in vitro* with muscarinic agonists or cholinesterase inhibitors caused the release of free choline.[65] This choline release most likely was at the expense of tissue phosphatidylcholine, presumably catalyzed by PLD, since efflux was enhanced in the presence of oleic acid.[66]

The source of choline for acetylcholine formation has been of interest for years, and it was conjectured that lecithin was a potential candidate.[67,68] Unfortunately experimental evidence had been lacking in support of this hypothesis. Rat brain subfractions were prepared, and it was found that purified synaptic membranes contained a PLD with high specific activity. It was also demonstrated that the choline released by this PLD could be coupled to acetylcholine production with rat brain synaptosomes. This was the first experimental evidence supporting the postulate.[69] The formation of acetylcholine from acetyl-CoA by intact synaptosomes was dependent upon the presence of sodium oleate. Synaptosomal PLD is latent and its activity barely detectable in the absence of oleate.[70] Therefore, it appears reasonable to propose a functional contribution of PLD in acetylcholine homeostasis of cholinergic neurons.

Animal carcinogenesis is thought to be a multistage process that may evolve over a considerable lifespan of the organism. A useful tool employed experimentally to understand this process is the tumor-promoting agents such as TPA. TPA has the capability of either enhancing or interfering with differentiation of a variety of different cell systems. Interest has focused on the effects of TPA upon cell membrane phospholipids. It is generally observed that there is an increase in diglyceride formation presumably at the expense of lecithin, although not from the phosphoinositides, perhaps by phospholipase C or D activation.[71]

H. TPA EFFECT: PHOSPHOLIPASE D OR PHOSPHOLIPASE C?

It is possible to locate publications that examine the nature of water-soluble and lipid-soluble products formed in cells exposed to TPA. Radioactive precursors (fatty acids, glycerol, or choline) prelabel the cellular phospholipids with general lipid precursors, or lecithin and sphingomyelin with choline. The products are usually determined chromatographically, employing appropriate techniques. The following selected examples are for illustration of the phenomenon rather than attempting to be encyclopedic.

HeLa cells were prelabeled with [^{3}H]choline, and after 24 h no free choline was detectable in these cells. TPA caused a release of both phosphocholine and choline into the medium, which reflected an origin from the cellular phospholipid rather than the cellular phosphocholine pool. There was no information provided regarding the nature of the lipid products. Even though it was concluded that choline and not phosphocholine was released from phospholipid as a consequence of TPA, the enzyme postulated to be involved was phospholipase C (PLC), not D.[72] The fatty acid composition of phosphatidic and diglyceride produced by chemoattractant stimulation of human neutrophils did not resemble that of cellular phosphatidylinositols.[73] Unfortunately, similar data for phosphatidylcholine was not provided; however, the fatty acid composition of the diglyceride as described in this publication is similar to the lecithin of most tissues. The authors postulated the existence of a unique pool of phosphoinositides which were cleaved by a phospholipase C.[73] There was no consideration of a lecithin precursor. Mandin-Darby canine kidney cells were prelabeled with [^{3}H]choline or 1-0[^{3}H]hexadecyl-2-acyl-*sn*-glycero-3-phosphocholine. TPA stimulation caused the release of labeled choline and phosphocholine into the medium, but the relative proportions of these two compounds or the time course of release was not provided. TPA stimulation also resulted in labeled diglyceride, triglyceride, and phosphatidic acid formation. The earliest time point was 1 h, at which time 2.5 and 7.5% of the total [^{3}H] disintegrations per minute were in phosphatidic acid and diglyceride, respectively. If it is assumed that the cellular phosphatidic acid content is appreciably lower than the diglyceride content, then the specific activity of the former would be greater than the latter. This would suggest that phosphatidic acid was a precursor of the diglyceride. These authors concluded that a PLC activation was responsible, without considering a PLD involvement.[74] The preadipocytic cell line 3T3-L1 or HL-60 cells were exposed to either [^{3}H]choline, [^{14}C]oleate, or [^{32}P]H_3PO_4 to label cellular phosphatidylcholine and other phospholipids.[75] The phosphatidate produced by TPA exposure was reported to have a fatty acid composition like phosphatidylcholine. The kinetics of diglyceride appearance were presented, but similar data for phosphatidate were not presented. Phosphocholine was reported to be the principal intracellular product. The extracellular water-soluble products were not identified. The sum of the intra- and the extracellular water-soluble products was about 3 nmol total, which represented about 13% of the cellular phosphatidylcholine. The endogenous phosphorylcholine pool was 16-fold greater than the choline pool in those cells.[75] These results could alternatively be explained by a PLD stimulation, since intracellular choline is rapidly converted to its phosphorylated derivative.

The diacylglycerol content was increased from 0.7 to 2 nmol/100 nmol of phospholipid in Swiss 3T3 fibroblasts exposed to TPA. There were no detectable changes in any of the cellular phospholipids. Water-soluble choline metabolites appeared in the medium; however, these were not identified. The authors conclude that PLC hydrolysis of lecithin was responsible for these results.[76] Cultured human umbilical endothelial cells were prelabeled with fatty acids and exposed to either thrombin or the calcium ionophore A23187.[77] The appearance of labeled phosphatidic acid preceded the appearance of labeled diacylglycerol, presumably at the expense of phosphatidylcholine rather than of the phosphoinositides. Despite this result these authors suggest that PLC may be involved.[77] Phospholipids of Swiss 3T3 cells were prelabeled with [^{32}P]H_3PO_4, [^{3}H]glycerol, or [^{3}H]choline and incubated with either bombesin or TPA.[78] There was an approximate twofold increase in labeled diacylglycerol and the release of both phosphocholine and choline. The changes in phosphatidic acid labeling were not provided.[78] A neuroblastoma glioma hybrid cell line, NG108-15, was prelabeled with [^{3}H]choline and exposed to TPA, and the product released into the medium was choline.[79] Similarly with ethanolamine prelabeled cells, ethanolamine was released into the medium. PLD involvement was suggested to explain these results.[79]

A rat embryo cell line, REF 52, was prelabeled with either [^{3}H]choline or [^{3}H]glycerol. The presence of TPA or vasopressin resulted in an increase of cellular diacylglycerol content and release of free choline into the culture medium.[80] The hydrolysis of phosphorylcholine did not occur during *in vitro* incubations. The release of choline was accompanied by the appearance of phosphatidic acid, which preceded diglyceride appearance. These results suggest that PLD rather than PLC is involved in this ligand-mediated hydrolysis of lecithin.[81]

The appearance of diglyceride and/or phosphocholine with intact cells can no longer be regarded as unequivocal evidence for PLC hydrolysis of lecithin. It is equally plausible that PLD produced phosphatidic acid, which can be rapidly cleaved to diglyceride by phosphatidic acid phosphatase. This possibility has been described in this chapter.

There is an extensive literature demonstrating that choline is very rapidly and efficiently phosphorylated by choline kinase in intact cells (Chapter 2). As described in this chapter there is very little choline in cells, compared to phosphocholine. These pathways are

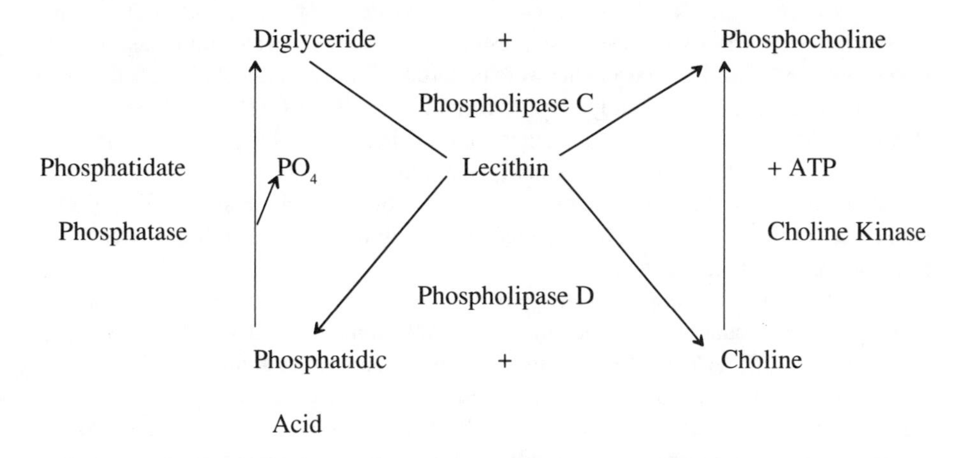

It is, therefore, necessary that caution be exercised in the interpretation of information upon the appearance of diglyceride, phosphocholine, choline, and labeling of phosphatidic acid in cell biological experiments. Perhaps the principal role of the phosphatidylinositol-specific PLC is to liberate only the various water-soluble phosphorylated inositols. The principal role of PLD may be to produce diglyceride for the particular protein kinase C and phosphatidic acid for phospholipid biosynthesis.

III. THE BASE EXCHANGE ENZYMES

The base-exchange enzyme reactions are shown as Reaction 2. A comprehensive review of the literature relating to these reactions was published in 1980,[1] therefore, this early material will not be covered in detail.

Operationally, the base exchange reactions can be defined as those which catalyze the incorporation of the amino alcohols, normally found in phospholipids, into their corresponding phospholipids. The base-exchange enzyme reactions are not responsible for net snythesis of any phospholipid, but rather for the remodeling of preexisting membrane phospholipids. The compounds classically considered as substrates for these enzymes are choline, serine, ethanolamine, and, more recently added to this list, monomethylethanolamine and dimethylethanolamine. Alcohols such as ethanol and methanol are unreactive. Characteristically, these incorporations are energy independent and stimulated by or dependent upon Ca^{2+}. They possess an alkaline pH optimum and, when membrane bound, prefer endogenous phospholipids as acceptors. They are detectable in all plants, animals, and a number of unicellular organisms, but undetectable in "true" bacteria. The properties of the base-exchange enzyme activities of brain tissue have been the most extensively studied.

A. HISTORICAL SURVEY

An understanding of the early work from the laboratory of Hübscher and Dils on these reactions will assist in placing in perspective the misconception that the base exchange reactions merely represent the reversal of PLD activity. These workers intended to determine if phosphatidylserine was biosynthesized through the *de novo* pathway responsible for the formation of phosphatidylcholine and phosphatidylethanolamine. They reported that rat liver mitochondria were capable of producing phosphatidylserine from L serine-^{14}C. The reaction appeared to require the presence of Mg^{2+} and was stimulated by ATP, CMP, and CoA. Ca^{2+} and

CTP interfered with the serine incorporation in the presence of the former cofactors.[82] The researchers also showed that phosphorylserine was not involved in the reactions. Although Ca^{2+} inhibited at 10^{-2} *M*, it stimulated at 4×10^{-3} *M*, the pH optimum was 7.5, and the optimum CMP concentration was 3.5×10^{-4} *M*. These characteristics eliminated a *de novo* pathway for phosphatidylserine biosynthesis and suggested that a reversal of a phospholipase activity might be responsible.[83] Similar experiments were undertaken substituting labeled choline for labeled serine to investigate phosphatidylcholine formation. ATP or CTP did not have any effects. However, CMP alone or in the presence of ATP increased phosphatidylcholine formation sixfold, and Ca^{+} at 3.7×10^{-3} *M* in the absence of other additions resulted in a 20-fold increase. These authors again suggested the possible reversal of a phospholipase activity. They considered PLD as a likely candidate, but rejected this possibility because of the absence of reports of its presence in mammalian tissues.[84]

It was subsequently shown that these particular enzyme activities were associated with rat liver microsomes rather than mitochondria. The reaction with microsomes had a pH optimum of about 8.5, was linear for about 30 min. Ca^{+} was optimum at about 3×10^{-3} *M*. The K_m for choline was 8×10^{-3} *M*, and both the D and L isomers of serine were competitive inhibitors with K_is of 5×10^{-3} and 7.7×10^{-4} *M*. The researchers were unsuccessful in demonstrating choline release from the microsomal phosphatidylcholine pool. Using glycerophosphate produced by saponification of total extracted phospholipids as an index of phosphatidic acid formation, there appeared to be a Ca^{2+}- dependent increase of this phospholipid. This probably represents one of the earliest suggestions that PLD was present in mammalian liver. Therefore, the authors suggested that this incorporation of choline was a result of PLD reversibility.[85] Similar observations were then made with serine incorporation. It was proposed that separate enzymes are responsible for the L-serine and choline incorporations by these exchange reactions.[86] These pioneers in the studies of the base reaction, therefore, postulated that the reversible reaction of a PLD could be responsible for their observations because of a lack of other models at that time. Unfortunately, they did not offer alternative possibilites. As a result, considerable effort has been required to show that this hypothesis was incorrect.

B. MEMBRANE-BOUND BASE EXCHANGE ENZYME: PROPERTIES

The major portion of the subsequent information was derived from investigations on the properties of the enzyme activities of rodent brain. These studies have contributed to a better understanding of the relationship between the base exchange activities and PLD.

Perhaps the earliest characterization of these reactions occurring in brain tissue was with chick brain microsomes. Selected properties are contained in Table 2. Ethanolamine and serine were reciprocal competitive inhibitors of one another's incorporation into lipid.[87] These authors suggested that there might be an exchange of free bases for that present in endogenous phospholipids rather than a reversal of PLD, and that perhaps a single enzyme was responsible.

Selected properties of the rat-brain microsomal, base-exchange enzymes are provided in Table 2. The three substrates utilized — choline, serine, and ethanolamine — were mutual inhibitors of each other's incorporations, and Ca^{2+} was required for these incorporations. The membrane phospholipids were labeled by the base exchange incorporation of serine, choline, or ethanolamine. Nonradioactive serine and ethanolamine increased the release of labeled serine and ethanolamine, but not choline, from these prelabeled particles. It was postulated that these three compounds are substrates for separate enzymes, based upon differences in inhibitions by structural analogues, responses to phospholipase treatments, and stabilities.[88] A reciprocal relationship between optimum Ca^{2+} concentration and the pH of the *in vitro* incubations was described for rat brain microsomes. The optimum pH for all three bases at 0.25 m*M* Ca^{2+} was between 8.2 and 9.5. In contrast, at 25 m*M* Ca^{2+} the optimum pH decreased to 7.5 or 8.0. These data are provided in Table 2. The possible existence of separate enzymes responsible for the incorporation of each of these three bases was suggested, based upon these kinetic differences.[89]

TABLE 2
Selected Properties of the Base Exchange Enzymes from Various Sources

	Ethanolamine	Serine	Choline
	Chicken brain microsomes[87]		
K_m	0.3 mM	0.33 mM	—
V_{MAX}	—	90	—
pH optimum	8 to 9	8 to 9	—
Ca^{2+} optimum	25 to 30 mM	25 to 30 mM	—
	Rat brain microsomes[88]		
K_m	0.015 mM	0.04 mM	0.14 mM
K_I ethanolamine	—	0.057 mM	0.57 mM
K_I serine	1 mM	—	1 mM
K_I choline	—	0.04 mM	—
pH optimum	9.0	8.5	9.5
	Rat brain microsomes[89]		
K_m	0.08 to 0.16 mM	0.04 mM	0.0588 mM
pH optimum	8.1	8.1	8.1
Ca^{2+} optimum	2 to 10 mM	10 mM	2 mM
	Rat liver microsomes[90]		
K_m	0.095, 0.024 mM	0.062 mM	0.023 mM
V_{MAX}	11.4, 17	14.7	2
pH optimum	8.8	8.8	8.8
Ca^{2+} optimum	0.16 mM	0.16 mM	0.16 mM
K_I ethanolamine	—	0.11 mM	0.09 mM
K_I serine	0.072 mM	—	0.07 mM
K_I choline	0.29 mM	0.06 mM	—
	"Solubilized" rat brain microsomes[31]		
K_m	0.013 mM	0.043 mM	0.675 mM
V_{MAX}	16	22	14.6
pH optimum	7.2	7.2	7.2
Ca^{2+}	8 to 10 mM	25 mM	8 to 10 mM
	Partially purified serine base exchange enzyme[93]		
K_m	—	0.4 mM	—
V_{MAX}	—	57.1	—
pH optimum	—	8	—
Ca^{2+} optimum	—	10 mM	—
	Partially purified serine–ethanolamine base exchange enzyme[94]		
K_m	0.02 mM	0.11 mM	—
V_{MAX}	40	330	—
pH optimum	7.0	7.0	—
Ca^{2+} optimum	8 mM	10 mM	—
K_I ethanolamine	—	0.025 mM	—
K_I serine	0.123 mM	—	—

TABLE 2 (continued)
Selected Properties of the Base Exchange Enzymes from Various Sources

	Ethanolamine	Serine	Choline
	Canine cardiac microsomes		
K_m	0.13 m*M*	0.06 m*M*	0.49 m*M*
V_{MAX}	26	130	27
Ca^{2+} optimum	0.36 m*M*	0.22 m*M*	0.39 m*M*
K_I ethanolamine	—	0.11 m*M*	0.11 m*M*
K_I serine	0.11 m*M*	—	0.11 m*M*
K_I choline	6.9 m*M*	0.64 m*M*	—
	Rat heart[96]		
Sarcolemma K_m	0.014 m*M*	0.025 m*M*	—
Sarcolemma V_{MAX}	102	34	—
Sarcoplasmic reticulum K_m	0.01 m*M*	0.04 m*M*	—
Sarcoplasmic reticulum V_{MAX}	77	30	—
	Bovine retina microsomes[97]		
K_m	0.063 m*M*	0.14 m*M*	0.62 m*M*
V_{MAX}	32	24	18

[a] All V_{MAX} values expressed as nmol/mg/h.

The inverse relationship between optimum Ca^{2+} concentration and the pH of the incubations had been observed for rat liver microsomes, and these data are in Table 2. It was concluded that a single enzyme is not responsible for the Ca^{2+}-stimulated incorporation of these three bases.[90]

C. SOLUBILIZED BASE EXCHANGE ENZYMES

The suspicions relating the base-exchange enzyme activities and PLD as well as the possible existence of separate base exchange enzymes responsible for the incorporation of each substrate became resolved when these enzyme activities were successfully solubilized.[31] The enzyme and the phospholipid substrate are both firmly bound to the intact membranes *in situ* so that exogenous phospholipids have no appreciable effect on the reaction. This difficulty was overcome upon solubilization of the rat brain system with a detergent, Miranol H2M. There was almost no incorporation in the absence of added phospholipid acceptor by this solubilized extract. Selected properties of the solubilized enzymes are in Table 2. The previous suggestions of the existence of separate base exchange enzymes was reinforced because of differential responses to several inhibitors and differences in heat stability of the individual enzyme activities.[91] The participation of PLD could not be excluded, since it was also present in these extracts.[31] However, as indicated in the previous section, partially purified rat brain PLD was found devoid of base-exchange enzyme activity.[39] Unequivocal evidence for the existence of separate enzymes for the incorporation of ethanolamine, choline, and serine became available when they were successfully separated chromatographically.[92]

A serine base exchange enzyme was partially purified from solubilized extracts of rat brain membranes, and some of its characteristics are presented in Table 2. There was no inhibition of L-serine incorporation by choline or ethanolamine. Phosphatidylethanolmaine was the most effective cosubstrate of several phospholipids tested with a K_m value of about 0.25 m*M*. There was no detectable PLD activity present in the most purified enzyme sample.[93]

Serine and ethanolamine base-exchange enzyme activities were copurified from solubilized rat brain membranes to a stage of apparent homogeniety based upon polyacrylamide gel electrophoresis. It had an apparent mass of 100 kDa, and some of its properties are in Table 2. Both of these substrates were mutually competitive inhibitors of each other's incorporation. Phosphatidylethanolmine was the only phospholipid of those tested capable of being the cosubstrate for this enzyme.[94]

D. BASE EXCHANGE ENZYMES AND THEIR RELATIONSHIP TO PHOSPHOLIPASE D

It is reasonable to conclude that these activities are present in separate proteins, at least in mammalian tissues. This conclusion is based upon the following observations:

1. Membrane bound PLD is latent and requires an activator such as oleate. There is no activator required to demonstrate base-exchange enzyme activity.
2. Membrane-bound PLD, when activated, will utilize both endogenous and exogenous phospholipids as substrates. Membrane-bound base exchange enzymes will not utilize exogenous phospholipids as substrates.
3. Phospholipase D will utilize a spectrum of alcohols for transphosphatidylation activity. Base exchange enzymes will only utilize amino alcohols as substrates and will not utilize ethanol or methanol.
4. The K_ms for the the alcohols acting as substrates for the transphosphatidylation activity of PLD are in the tenths-molar concentration range. The K_ms for the base exchange reactions are 10^3 or 10^4 lower, being in the tenths- or hundredths-millimolar range.
5. Membrane-bound PLD is unresponsive to added Ca^{2+}. There is a very appreciable stimulation of the membrane-bound base exchange by added Ca^{2+}.
6. Partially purified mammalian PLD does not possess base exchange enzyme activity, and partially purified base exchange enzymes do not have PLD activity.

E. PHARMACOLOGY, PHENOMENOLOGY, OCCURRENCE, AND POSTULATED FUNCTIONS OF THE BASE EXCHANGE REACTIONS

The literature about aspects of the base exchange reactions prior to 1980 was adequately covered previously,[1] and only post-1980 information will be included.

1. Occurrence

Canine cardiac microsomes possess the base-exchange enzyme activities, and some of their properties are listed in Table 2. The incorporations were Ca^{2+} stimulatable and inhibited by La^{3+}.[95]

Sarcolemma and sarcoplasmic reticulum of rat heart were isolated and their base-exchange enzyme activity investigated. The kinetic data for serine and ethanolamine are included in Table 2. The data for choline base exchange enzyme are not included since its activity is extremely low. The optimum Ca^{2+} concentrations were 2 to 4 m*M* for these two substrates, with a pH optimum of 7.5 to 8. Interestingly, both activities of the sarcolemma were stimulated by the addition of Azolectin, a commercial mixture of soybean phospholipids.[96] Bovine retinal microsomes possess the base-exchange enzyme activities, and these data are in Table 2. There was no discrete pH optimum, and maximum incorporation required between 10 and 15 mM Ca^{2+}.[97]

Although there was no enrichment over the initial homogenates, the ethanolamine and serine base-exchange enzyme activities appeared associated with an axolemma-enriched fraction isolated by zonal centrifugation.[98] Rat brain microsomes were fractionated by zonal centrifugation and the distribution of various enzymes located. The base exchange enzymes, cyclic nucleotide phosphodiesterase, NADPH cytochrome C reductase, and cholinephosphotransferase, were recovered in the least dense portion of the gradient in the region usually associated with myelin. This region had low activity of Na^+, K^+-ATPase, acetylcholinesterase, and 5′

nucleotidase.[99] Rat brain synaptosomes appear to have the ability to catalyze the base exchange reactions. However, there is no enrichment of the activities in these membranes, and the specific activities are lower than the starting homogenates.[70,100]

2. Properties of Membrane-Bound Enzyme Activity

The major portion of rat brain microsomal phosphatidylethanolamine is present in the interior, with smaller amounts on the external or cytosolic surface of these vesicles. Rat brain microsomal phosphatidylethanolamine was labeled by the base exchange reaction for 20 min and the location of the product probed by reacting with trinitrobenzoensulfonic acid (TNBS). Nearly all of the labeled molecules, but only 24% of the nonradioactive molecules, reacted with this reagent.[101] Microsomes were incubated for various times with labeled ethanolamine under appropriate conditions for incorporation by the base exchange enzyme and the localization of labeled phosphatidylethanolamine determined with TNBS. The specific activity of the phosphatidylethanolamine on the cytosolic leaflet was three times greater than the internal leaflet after 2 min of incubation. At 30 min of incubation these were equal.[102] Treatment of these membrane vesicles with trypsin,[102] pronase,[101] or mercury-dextran[101] reduced the ethanolamine base-exchange enzyme activity. These observations suggest a cytosolic surface localization of this enzyme activity in these membranes.

The molecular species of phosphatidylserine in rat liver microsomes was compared to the molecular species of phosphatidylserine produced by the base exchange reaction. The biosynthetic product was not identical to that of the endogenous molecular species. It was suggested that modifications through deacylation-reacylation reactions could account for these fatty acid differences.[103] In addition to labeled phosphatidylserine, labeled lecithin and sphingomyelin were produced, but not phosphatidylethanolamine.

Rat brain microsomes were treated with 25 to 40 m*M* octylglucopyranoside, resulting in 60 to 80% solubilization of protein and lipid. These conditions inactivated the ethanolamine base-exchange enzyme activity. Dialysis of the solubilized mixture restored 85% of this activity.[104] The presence of 5 m*M* Ca^{2+} and a pH of 6.0 in the dialysis fluid was required for maximal recovery of the base-exchange enzyme activity and phospholipid. Phosphatidylethanolamine, but not phosphatidylcholine supplementation to the detergent-solubilized mixture, resulted in reduced base-exchange enzyme activity in the postdialysis reaggregates.[105]

Base exchange enzyme catalyzed incorporation of monomethylethanolamine, and dimethylethanolamine was demonstrated to occur with rat brain microsomes.[106] Maximal incorporations were obtained at 1 to 4 m*M* Ca^{2+}, at pH 7.5 to 8.5. The K_m and V_{MAX} values were 0.97 m*M* and 9.6 nmol/mg/h for dimethylethanolamine and 0.5 m*M* and 6.25 nmol/mg/h for monomethylethanolamine. Choline, ethanolamine, serine, and TRIS were competitive inhibitors. These activities were detected in the microsomal fraction of several tissues of the rat.[106]

3. Effects of Various Compounds on the Base Exchange Activities

The base-exchange enzyme activities of rabbit platelet membranes were stimulated by Ca^{2+}, and the optimum concentration required was reduced by the calmodulin antagonists chloropromazine, trifluoperazine, and *N*(6-aminohexyl)-5-chloro-1-napthalene sulfonamide. Calmodulin additions to the incubation mixtures reduced the choline base exchange, but had no effect on the serine or ethanolamine base-exchange enzyme activities.[107] These antagonists reduced the endogenous level of serine, increased the level of choline and ethanolamine, but had no effect on the levels of free amino acids.[108] The antagonists had no effect on the K_m values and only increased the V_{MAX} values of the three base-exchange enzyme activities. They also decreased the binding of Ca^{2+} to the membranes, suggesting perhaps that these are unrelated phenomena.[109]

Unidentified products present in conditioned growth medium decrease the incorporation of labeled serine into phosphatidylserine by HL60 cells, and TPA, the tumor promotor, interferes with this effect.[110] Halothane inhibited the incorporation of serine into phosphatidylserine by longitudinal muscle strips of guinea pig ilium, even at clinical concentrations of this anesthetic.[111]

The base-exchange enzyme activity of human neutrophil or lymphocyte membranes was increased by calmodulin antagonists, zymosan, and Concanavalin A. There were no effects on the activity of either the choline- or ethanolaminephosphotransferase activities involved in the *de novo* pathway of phospholipid biosynthesis.[112] Elevations of the ethanolamine base-exchange enzyme activity were seen with lymphocytes and neutrophils from humans with active Behcet's disease, active systemic lupus erythematosus, and severe bacterial infections. Serine and choline base exchange activities, and ethanolamine- and cholinephosphotransferase activities were unaffected.[113] These workers speculated that the phosphatidylethanolamine produced by the base exchange enzyme is a specific pool supporting the phospholipid-*N*-methylation pathway.

4. Cell Biology of the Base Exchange Enzymes

The labeling of phosphatidylethanolamine or phosphatidylcholine by the addition of ethanolamine or choline to cell cultures in most cases does not reflect base exchange activities, but rather the *de novo* pathway. L-Serine labeling of phospholipids, in contrast, reflects an initial formation of phosphatidylserine by the serine base exchange enzyme.

A mouse neuroblastoma cell line, C1300, and a human glioma cell line, 138MG, were harvested after growth in complete and serum-free medium. The ethanolamine and serine base exchange activities of homogenates of the cells from serum-free medium were increased between 23 to 76%, depending upon the substrate and cell line.[114]

Serine was demonstrated to maintain the cellular requirements for ethanolamine of rat brain cells in culture. This would occur by the base exchange formation of phosphatidylserine, which could be decarboxylated to provide phosphatidylethanolamine.[115] Myelinating explants of neonatal rat cerebellum maintained in culture were capable of labeling both serine-containing and ethanolamine-containing phospholipids from [3-^{14}C]serine, but [1-^{14}C]DL serine only labeled phosphatidylserine. This demonstrates the contribution of phosphatidylserine decarboxylase to the cellular phosphatidylethanolamine pool.[116] This contribution of l serine to phosphatidylethanolamine formation is not restricted to nervous tissue and has been observed with isolated hepatocytes.[117] Rats received an intracranial injection containing both l[U-^{14}C]serine and [1-^{3}H]ethanolamine, and the ratios of radioactivity were determined in microsomal and mitochondrial phosphatidylethanolamine. A minimum of 7% was estimated for the contribution of phosphatidylserine to phosphatidylethanolamine formation *in vivo*.[118]

Rat mammary 64-24 cells responsive to ethanolamine and unresponsive 22-1 cells were grown in the presence of radioactive serine and the labeling of phospholipids examined.[119] The labeling of both serine and ethanolamine containing phosphoglycerides were reduced in the 62-24 cells as compared to the 22-1 cells. This could reflect reduced serine base-exchange enzyme activity. However, this activity was identical in homogenates prepared from both cell lines. Reduced serine transport does not explain this phenomenon since acid-soluble pools and protein labeling was the same in both cell lines.[119] Mutant Chinese hamster ovary (CHO) cells have been isolated which appear to possess reduced serine base exchange activity.[120,122] The enzyme activities for the *de novo* pathway of phosphatidylethanolamine biosynthesis and the ethanolamine base exchange were determined during the growth of rat astrocytes in culture. These activities had different developmental patterns, and this base-exchange enzyme activity appeared associated with a mitochondria-enriched fraction.[123]

Incorporation of ethanolamine and monomethylethanolamine into their corresponding phospholipid by their base exchange enzymes activated a Na^+-K^+-ATPase of rat brain microsomes. The serine and dimethylethanolamine base-exchange enzyme-catalyzed incorporations resulted in the inhibition of this particular ATPase.[124]

ACKNOWLEDGMENTS

This research was supported by grants from the Medical Research Council of Canada and the Alzheimer's Disease and Related Disorders Association, Inc., Chicago.

REFERENCES

1. **Kanfer, J. N.,** The base exchange enzymes and phospholipase D of mammalian tissues, *Can. J. Biochem.,* 12, 1370, 1980.
2. **Hattori, H., Witter, B., Chalifour, R., and Kanfer, J. N.,** Metabolic alteration of phospholipid polar head groups by rat brain particulates, in *Phospholipids in the Nervous System,* Vol. 2, Horrocks, L. A., Kanfer, J. N., and Porcellati, G., Eds., Raven Press, New York, 1985, 107.
3. **Waite, M.,** *The Phospholipases,* Plenum Press, New York, 1987.
4. **Bailey, J. M.,** *Prostaglandins, Leukotrienes and Lipoxins: Biochemistry, Mechanisms of Action and Clinical Applications,* Plenum Press, New York, 1985.
5. **Berridge, M. J.,** Inositol trisphosphate and diacylglycerol: two interating second messengers, *Annu. Rev. Biochem.,* 56, 159, 1987.
6. **Op den Kamp, J. A. F.,** Lipid asymmetry in membranes, *Annu. Rev. Biochem.,* 48, 47, 1979.
7. **Eibl, H. and Kovatchev, S.,** Preparation of phospholipids and their analogs by phospholipase D, *Methods Enzymol.,* 72, 632, 1981.
8. **Achterberg, V., Freike, H., and Gercken, G.,** Conversion of radiolabelled ethanolamine plasmalogen into the dimethylethanolamine and choline analogue via transphosphatidylation by phospholipase D from cabbage, *Chem. Phys. Lipids,* 41, 349, 1986.
9. **Shuto, S., Imamura, S., Fukukawa, K., Sakakibara, H., and Murese, J.,** A facile one-step synthesis of phosphatidylhomoserine by phospholipase D catalyzed transphosphatidylation, *Chem. Pharm. Bull.,* 35, 447, 1987.
10. **Shuto, S., Ueda, S., Itoh, H., Endo, E., Fukukawa, F., Imamura, S., Tsujino, M., Matsuda, A., and Ueda, T.,** Synthesis of 5′ phosphatidylnucleosides by phospholipase D catalyzed transphosphatidylation, *Nucleic Acids Symp. Ser.,* 17, 73, 1986.
11. **Lumper, L., Zubrzycki, Z., and Staudinger, H.,** The inactivation of microsomal enzymes by phospholipase D, *Hoppe-Seyler's Z. Physiol. Chem.,* 350, 163, 1969.
12. **Zaidi, S. N. A., Shipstone, A. C., and Garg, N. K.,** Effect of phospholipase D on rat kidney mitochondria, *J. Biosci.,* 1, 75, 1979.
13. **Mannella, C. A.,** Effects of phospholipases C and D on ordering of channel proteins in the mitochondrial outer membranes, *Biochim. Biophys. Acta,* 861, 67, 1986.
14. **Fiehn, W.,** The effect of phospholipase D on the function of fragmented sarcoplasmic reticulum, *Lipids,* 13, 264, 1978.
15. **Philipson, K. D. and Nishimoto, A. Y.,** Stimulation of Na^+-Ca^{+2} exchange in cardiac sarcoplasmal vesicles by phospholipase D, *J. Biol. Chem.,* 259, 16, 1984.
16. **Clancy, R. M., Wissenberg, A. R., and Glaser, M.,** Use of phospholipase D to alter the surface charge of membranes and its effect on enzymatic activity of D-β hydroxybutyrate dehydrogenase, *Biochemistry,* 21, 6060, 1981.
17. **Witt, W. and Gercken, G.,** Modification of phospholipids in erythrocyte membranes by phospholipase D. A fluorescence and ESR spectroscopic study, *Biochim. Biophys. Acta,* 862, 100, 1986.
18. **Heller, M. and Arad, R.,** Properties of phospholipase D from peanut seeds, *Biochim. Biophys. Acta,* 210, 276, 1970.
19. **Dawson, R. M. C. and Hemmington, N.,** Some properties of purified phospholipase D and especially the effect of amphipathic substances, *Biochem. J.,* 102, 76, 1967.
20. **Hanahan, D. J. and Chaikoff, I. L.,** The phosphorus-containing lipids of the carrot, *J. Biol. Chem.,* 168, 233, 1946.
21. **Hanahan, D. J. and Chaikoff, I. L.,** A new phospholipid splitting enzyme specific for the ester linkage between the nitrogenous base and the phosphoric acid group, *J. Biol. Chem.,* 169, 699, 1947.
22. **Heller, M., Mozes, N., Peri, I., and Maes, E.,** Phospholipase D from peanut seeds. IV. Final purification and some properties of the enzyme, *Biochim. Biophys. Acta,* 369, 397, 1974.
23. **Allgyer, T. T. and Wells, A. M.,** Phospholipase D from savoy cabbage: purification and preliminary kinetic characterization, *Biochemistry,* 24, 5348, 1979.

24. **Witt, W., Yelenosky, G., and Mayer, R. T.,** Purification of phospholipase D from citrus callus tissue, *Arch. Biochem Biophys.*, 259, 164, 1987.
25. **Heller, M., Greenzaid, P., and Lichtenberg, D.,** The activity of phospholipase D on aggregates of phosphatidylcholine, dodecylsulfate and Ca^{2+}, in *Enzymes of Lipid Metabolism,* Gatt, S., Freysz, L., and Mandel, P., Eds., Plenum Press, New York, 1977, 213.
26. **Tzur, R. and Shapiro, B.,** Purification of phospholipase D from peanuts, *Biochim. Biophys. Acta,* 280, 290, 1972.
27. **Cole, R., Benns, G., and Proulx, P.,** Cardiolipin specific phospholipase D activity in *E. coli* extracts, *Biochim. Biophys. Acta,* 337, 325, 1974.
28. **Ono, Y. and White, D. C.,** Cardiolipin-specific phospholipase D of *H. parinfluenzae*. II. Characteristics and possible significance, *J. Bacteriol.*, 104, 712, 1970.
29. **Saucek, A., Michalec, C., and Souckova, A.,** Identification and characterization of a new enzyme of the group "Phospholipase D" isolated from *Corynebacterium ovis, Biochim Biophys. Acta,* 227, 116, 1971.
30. **Kreger, A. S., Bernheimer, A. W., Etkin, L. A., and Daniel, L. W.,** Phospholipase D activity of *Vibrio damsela* cytolysin and its interaction with sheep erythrocytes, *Infect. Immun.*, 55, 3209, 1987.
31. **Saito, M. and Kanfer, J. N.,** Solubilization and properties of a membrane-bound enzyme from rat brain catalyzing a base-exchange reaction, *Biochem. Biophys. Res. Commun.*, 53, 391, 1973.
32. **Chalifour, R. J. and Kanfer, J. N.,** Microsomal phospholipase D of rat brain and lung tissue, *Biochem. Biophys. Res. Commun.*, 96, 742, 1980.
33. **Gercken, G. and Witt, W.,** Transphosphatidylierungs — Reacktionen mit Phospholipase D in Erythrozytenmembranen, *Acta Biol. Med. Ger.*, 36, 837, 1977.
34. **Chapman, B. E. and Beilharz, G. R.,** Endogenous phospholipase and choline release in human erythrocytes: a study using H NMR spectroscopy, *Biochem. Biophys. Res. Commun.*, 105, 1280, 1982.
35. **Kater, L. A., Goetzl, E. J., and Austen, K. F.,** Isolation of human eosinophil phospholipase D, *J. Clin. Invest.*, 57, 1173, 1976.
36. **Lempereur, C., Capron, M., and Capron, A.,** Identification and measurement of rat eosinophil phospholipase D. Its activity on schistosomula phospholipids, *J. Immunol. Methods,* 33, 249, 1980.
37. **Schmid, P. C., Reddy, P. V., Natarajan, V., and Schmid, H. O.,** Metabolism of *N*-acyl-ethanolamine phospholipids by a mammalian phosphodiesterase of a phospholipase D type, *J. Biol. Chem.*, 258, 9302, 1983.
38. **Low, M. G. and Prasad, A. R. S.,** A phospholipase D specific for the phosphatidylinositol anchor of cell surface proteins is abundant in plasma, *Proc. Natl. Acad. Sci. U.S.A.*, 85, 980, 1988.
39. **Taki, T. and Kanfer, J. N.,** Partial purification and properties of a rat brain phospholipase D, *J. Biol. Chem.*, 254, 9761, 1979.
40. **Chalifour, R. J., Taki, T., and Kanfer, J. N.,** Phosphatidylglycerol formation via transphosphatidylation by rat brain extracts, *Can. J. Biochem.*, 58, 1189, 1980.
41. **Chalifour, R. J. and Kanfer, J. N.,** Fatty acid activation and temperature perturbation of rat rain microsomal phospholipase D, *J. Neurochem.*, 39, 199, 1982.
42. **DeVries, G. H., Chalifour, R. J., and Kanfer, J. N.,** The presence of phospholipase D in rat central nervous system axolemma, *J. Neurochem.*, 40, 1189, 1983.
43. **Witter, B. and Kanfer, J. N.,** Hydrolysis of endogenous phospholipids by rat brain microsomes, *J. Neurochem.*, 44, 155, 1985.
44. **Wykle, R. L. and Schremmer, J. C.,** A lysophospholipase D pathway in the metabolism of ether-linked lipids in brain microsomes, *J. Biol. Chem.*, 249, 1742, 1974.
45. **Wykle, R. L., Kraemer, W. F., and Schremmer, J. M.,** Studies of phospholipase D of rat liver and other tissues, *Arch. Biochem. Biophys.*, 184, 149, 1977.
46. **Wykle, R. L., Kraemer, W. F., and Schremmer, J. M.,** Specificity of lysophospholipase D, *Biochim. Biophys. Acta,* 619, 58, 1980.
47. **Tokumura, A., Harada, K., Fukuzawa, K., and Tsukatoni, H.,** Involvement of lysophospholipase D in the production of lysophosphatidic acid in rat plasma, *Biochim. Biophys. Acta,* 875, 31, 1986.
48. **Yang, S. F., Freer, S., and Benson, A. A.,** Transphosphatidylation by phospholipase D, *J. Biol. Chem.*, 242, 477, 1967.
49. **Dawson, R. M. C.,** The formation of phosphatidylglycerol and other phospholipids by the transferase activity of phospholipase D, *Biochem. J.*, 102, 205, 1967.
50. **Clarke, N. G., Irvine, R. F., and Dawson, R. M. C.,** Formation of bis (phosphatidyl) inositol and phosphatidic acid by phospholipase D action on phosphatidylinositol, *Biochem. J.*, 195, 521, 1981.
51. **Stanacev, N. Z. and Stuhne-Seklac, L.,** On the mechanism of enzymatic transphosphatidylation. Biosynthesis of cardiolipin, *Biochim. Biophys. Acta,* 210, 350, 1970.
52. **Walton, P. A. and Goldfine, H.,** Transphosphatidylation activity in *Clostridium butyricum, J. Biol. Chem.*, 262, 10355, 1987.
53. **Darnell, J., Lodish, H., and Baltimore, D.,** *Molecular Cell Biology,* Scientific American Books, New York, 1986, 163.

54. **Alling, C., Gustavsson, L., and Anggard, E.,** An abnormal phospholipid in rat organs after ethanol treatment, *FEBS Lett.,* 152, 85, 1983.
55. **Alling, C., Gustavsson, L., Mansson, J.-E., Benthin, G., and Anggard, E.,** Phosphatidylethanol formation in rat organs after ethanol formation, *Biochim. Biophys. Acta,* 793, 119, 1984.
56. **Milne, R. L. and Kanfer, J. N.,** Phosphatidylethanol formation by phospholipase D by rat brain membranes, *Trans. Am. Soc. Neurochem.,* 16, 434, 1985.
57. **Gustavsson, L. and Alling, C.,** Formation of phosphatidylethanol in rat brain by phospholipase D, *Biochem. Biophys. Res. Commun.,* 142, 958, 1987.
58. **Kobayashi, M. and Kanfer, J. N.,** Phosphatidylethanol formation via transphosphatidylation by rat brain synaptosomal phospholipase D, *J. Neurochem.,* 48, 1597, 1987.
59. **Varma, S. K. and Sharma, B. B.,** Fetal alcohol syndrome, *Prog. Biochem. Pharmacol.,* 18, 122, 1981.
60. **Pai, J. -K., Liebl, E. C., Tettenborn, C. S., Fidelis, I. I., and Mueller, G. C.,** 12-*0*-Tetradecanoylphorbol-13-acetate activates the synthesis of phosphatidylethanol in animal cells exposed to ethanol, *Carcinogen,* 8, 173, 1987.
61. **Tettenborn, C. S. and Mueller, G. C.,** Phorbol esters activate the pathway for phosphatidylethanol synthesis in differentiating HL-60 cells, *Biochim. Biophys. Acta,* 931, 242, 1987.
62. **Pai, J.-K. and Siegel, M. I.,** Activation of phospholipase D by chemotactic peptide in HL-60 granulocytes, *Biochem. Biophys. Res. Commun.,* 150, 355, 1988.
63. **Bocckino, S. B., Wilson, P. B., and Exton, J. H.,** Ca^{2+}-mobilizing hormones elicit phosphatidylethanol accumulation via phospholipase D, *FEBS Lett.,* 225, 201, 1987.
64. **Bocckino, S. B., Blackmore, P. F., Wilson, P. B., and Exton, J. H.,** Phosphatidate accumulation in hormone-treated hepatocytes via a phospholipase D mechanism, *J. Biol. Chem.,* 262, 15309, 1987.
65. **Corradetti, R., Lindmar, R., and Loffelholz, K.,** Mobilization of cellular choline by stimulation of muscarine receptors in isolated chicken heart and rat cortex *in vivo, J. Pharmacol. Exp. Ther.,* 226, 826, 1983.
66. **Lindmar, R., Loeffelholz, K., and Sandman, J.,** Characterization of choline efflux from the perfused heart at rest and after muscarine receptor activation, *Naunyn Schmiedebergs Arch. Pharmacol.,* 332, 224, 1986.
67. **McIntosh, F. C. and Collier, B.,** Neurochemistry of cholinergic terminals, *Handb. Exp. Pharmacol.,* 42, 99, 1976.
68. **Ansell, G. B. and Spanner, S.,** Sources of choline for acetylcholine synthesis in the brain, in *Nutrition and the Brain,* Vol. 5, Barbeau, A., Growdon, J. H., and Wurtman, R. J., Eds., Raven Press, New York, 1979, 35.
69. **Hattori, H. and Kanfer, J. N.,** Synaptosomal phospholipase D: potential role in providing choline for acetylcholine synthesis, *Biochem. Biophys. Res. Commun.,* 124, 945, 1984.
70. **Hattori, H. and Kanfer, J. N.,** Synaptosomal phospholipase D potential role in providing choline for acetylcholine synthesis, *J. Neurochem.,* 45, 1578, 1985.
71. **Weinstein, I. B.,** Current concepts and controversies in chemical carcinogenesis, *J. Supramol. Struct. Cell Biochem.,* 17, 99, 1981.
72. **Guy, G. R. and Murray, A. W.,** Tumor promoter stimulation of phosphatidylcholine turnover in HeLa cells, *Cancer, Res.,* 42, 1980, 1982.
73. **Cockcroft, S. and Allan, D.,** The fatty acid compositon of phosphatidylinositol, phosphatidate and 1,2-diacylglycerol in stimulated human neutrophils, *Biochem. J.,* 222, 557, 1984.
74. **Daniel, L. W., Waite, M., and Wykle, R. L.,** A novel mechanism of diglyceride formation, *J. Biol. Chem.,* 261, 9128, 1986.
75. **Besterman, J. M., Durino, V., and Cuatrecasas, P.,** Rapid formation of diacylglycerol from phosphatidylcholine: a pathway for generation of a second messenger, *Proc. Natl. Acad. Sci. U.S.A.,* 83, 6785, 1986.
76. **Takuwa, N., Takuwa, Y., and Rasmussen, H.,** A tumor promoter, TPA, increases cellular 1,2-diacylglycerol content through a mechanism other than phosphoinositide hydrolysis in Swiss-mouse 3T3 fibroblasts, *Biochem. J.,* 243, 647, 1987.
77. **Ragab-Thomas, J. M.-F., Hullin, F., Chap, H., and Douste-Blazy, L.,** Pathways of arachidonic acid liberation in thrombin and calcium ionophore A23187-stimulated human endothelial cells, *Biochim. Biophys. Acta,* 917, 388, 1987.
78. **Muir, J. S. and Murray, A. W.,** Bombesin and phorbol ester stimulate phosphatidycholine hydrolysis by phospholipase C, *J. Cell. Physiol.,* 130, 382, 1987.
79. **Liscovitch, M., Blusztajn, J. K., Freese, A., and Wurtman, R. J.,** Stimulation of choline release from NG108-15 cells by TPA, *Biochem. J.,* 241, 81, 1987.
80. **Cabot, M. C., Welsh, C. J., Zhang, Z.-C., Cao, H.-T., Chabbott, H., and Lebowitz, M.,** Vasopressin, phorbol diesters and serum elicit glycerophospholipid hydrolysis and diacylglycerol formation in non-transformed cells, *Biochim. Biophys. Acta,* 959, 46, 1988.
81. **Cabot, M., Welsh, C. J., Cao, H.-T., Zhang, Z.-C., and Chabbott, H.,** Agonist-induced diacylglycerol formation from phosphatidylcholine: evidence for a phospholipase D pathway involving protein kinase C, *FASEB J.,* 2, A1364, 1988.

82. **Hubscher, G., Dils, P. R., and Pover, W. F. R.,** *In vitro* incorporation of serine labeled with carbon-14 into mitochondria phospholipids, *Nature (London),* 182, 1806, 1958.
83. **Hubscher, G., Dils, R. R., and Pover, W. F. R.,** Studies on the biosynthesis of phosphatidylserine, *Biochim. Biophys. Acta,* 36, 518, 1959.
84. **Dils, R. R. and Hubscher, G.,** The incorporation *in vitro* of [Me-^{14}C]choline into the phospholipids of rat-liver mitochondria, *Biochim. Biophys. Acta,* 32, 293, 1959.
85. **Dils, R. R. and Hubscher, G.,** Metabolism of phospholipids. III. The effect of Ca ions on the incorporation of labeled choline into rat liver microsomes, *Biochim. Biophys. Acta,* 46, 505, 1961.
86. **Hubscher, G.,** Metabolism of phospholipids. VI. The effect of metal ions on the incorporation of L-serine into phosphatidylserine, *Biochim. Biophys. Acta,* 57, 555, 1962.
87. **Porcellati, G., Arienti, G., Piortta M., and Giorgini, D.,** Base exchange reactions for the synthesis of phospholipids in nervous tissue: the incorporation of serine and ethanolamine into the phospholipids of isolated brain microsomes, *J. Neurochem.,* 18, 1395, 1971.
88. **Kanfer, J. N.,** Base exchange reactions of the phospholipids in rat brain particles, *J. Lipid Res.,* 13, 468, 1972.
89. **Gaiti, A., DeMedio, G. E., Brunetti, N., Amaducci, L., and Porcellati, G.,** Properties and functions of the calcium dependent incorporation of choline, ethanolamine and serine into the phospholipids of isolated rat brain microsomes, *J. Neurochem.,* 23, 1153, 1974.
90. **Bjerve, K. S.,** The Ca^{2+} dependent biosynthesis of lecithin, phosphatidylethanolamine and phosphatidylserine in rat liver subcellular particles, *Biochim. Biophys. Acta,* 296, 549, 1973.
91. **Saito, M., Bourke, E., and Kanfer, J. N.,** Studies on the base exchange reactions of rat brain particles and a "solubilized" system, *Arch. Biochem. Biophys.,* 169, 304, 1975.
92. **Muria, T. and Kanfer, J. N.,** Studies on base exchange reaction of phospholipids in rat brain. Heterogeneity of base-exchange enzymes, *Arch. Biochem. Biophys.,* 175, 654, 1976.
93. **Taki, T. and Kanfer, J. N.,** A phospholipid serine base exchange enzyme, *Biochim. Biophys. Acta,* 528, 309, 1978.
94. **Suzuki, T. and Kanfer, J. N.,** Purification and properties of an ethanolamine-serine base exchange enzyme of rat brain membranes, *J. Biol. Chem.,* 260, 1394, 1985.
95. **Filler, D. A. and Weinhold, P. A.,** Base exchange reactions of the phospholipids in cardiac membranes, *Biochim. Biophys. Acta,* 618, 223, 1980.
96. **Hattori, H. and Kanfer, J. N.,** The base-exchange enzyme activity of sarcolemma and sarcoplasmic reticulum from rat heart, *Biochim. Biophys. Acta,* 835, 542, 1985.
97. **Anderson, R. E. and Keller, P. A.,** Biosynthesis of retinal phospholipids by base exchange reactions, *Exp. Eye Res.,* 32, 729, 1981.
98. **Hattori, H., Bansal, V., Orihel, D., and Kanfer, J. N.,** Presence of phospholipd *N*-methyl transferase and base-exchange enzymes in rat central nervous system axolemma-enriched fractions, *J. Neurochem.,* 43, 1018, 1984.
99. **Bansal, V., Hattori, H., Orihel, D., and Kanfer, J. N.,** Distribution of selected phospholipid modifying enzymes in rat brain microsomal subfractions prepared by density gradient zonal rotor centrifugation, *Neurochem. Res.,* 10, 439, 1985.
100. **Holbrook, P. G. and Wurtman, R. J.,** Presence of base-exchange activity in rat brain nerve endings, *J. Neurochem.,* 50, 156, 1988.
101. **Corazzi, L., Binaglia, L., Roberti, R., Freysz, L., Arienti, G., and Porcellati, G.,** Compartmentation of membrane phosphatidylethanolamine formed by base-exchange reaction in rat brain microsomes, *Biochim. Biophys. Acta,* 730, 14, 1983.
102. **Hutson, J. L. and Higgins, J. A.,** Asymmetric synthesis and trans membrane movement of phosphatidylethanolamine synthesized by base-exchange in rat liver endoplasmic reticulum, *Biochim. Biophys. Acta,* 835, 236, 1985.
103. **Bjerve, K. S.,** Phospholipid substrate specificity of the l-serine base-exchange enzyme in rat liver microsomal fraction, *Biochem. J.,* 219, 781, 1984.
104. **Corazzi, L. and Arienti, G.,** The reaggregation of rat brain microsomal membranes after the treatment with octyl-β-D glucopyranoside. A study on ethanolamine base-exchange, *Biochim. Biophys. Acta,* 875, 362, 1986.
105. **Corazzi, L. and Arienti, G.,** Factors affecting the reaggregation of rat brain microsomes solubilized with octyl glucoside and their relationship with the base-exchange activity of reaggregates, *Biochim. Biophys. Acta,* 903, 277, 1987.
106. **Kanfer, J. N.,** The monomethylethanolamine and dimethylethanolamine base exchange reactions of a rat brain microsomal fraction, *Biochim. Biophys. Acta,* 879, 278, 1986.
107. **Morikawa, S., Taniguchi, S., and Mori, K.,** Effects of calmodulin antagonists and calmodulin on phospholipid base exchange activities in rabbit platelets, *Thrombosis Res.,* 37, 267, 1985.

108. **Fujiwara, M., Morikawa, S., Taniguchi, S., Mori, K., Fujiwara, M., and Takaori, S.,** Effects of calmodulin antagonists on serine phospholipid base exchange reaction in rabbit platelets, *J. Biochem.*, 99, 615, 1986.
109. **Morikawa, S., Taniguchi, S., Mori, H., Fujii, K., Kumada, K., Fujiwara, M., and Fujiwara, M.,** Stimulating effect of calmodulin antagonists on phospholipid base exchange reactions in rabbit platelet membranes, *Biochem. Pharmacol.*, 35, 4473, 1986.
110. **Kiss, Z., Deli, E., and Kuo, J. F.,** Phorbol ester inhibits phosphatidylserine synthesis in human promyelocytic leukemia HL60 cells, *Biochem. J.*, 248, 649, 1987.
111. **Paton, W. D. M. and Wing, D. R.,** Effects of Halothane on the incorporation of [^{14}C]serine into phospholipids of guinea pig ileum, *Br. J. Pharmacol.*, 72, 393, 1981.
112. **Niwa, Y. and Taniguchi, S.,** Phospholipid base exchange in human leukocyte membranes: quantitation and correlation with other phospholipid biosynthetic pathways, *Arch. Biochem. Biophys.*, 250, 345, 1986.
113. **Niwa, Y., Sakane, T., Ozaki, Y., Kanoh, T., and Taniguchi, S.,** Phospholipid base exchange activity in the leukocyte membranes of patients with inflammatory disease, *Am. J. Pathol.*, 127, 317, 1987.
114. **Erkell, L. J., De Medio, G. E., Haglid, K., and Porcellati, G.,** Increased activity of a phospholipid base-exchange system by the differentiation of neoplastic cells from the nervous system, *J. Neurosci, Res.*, 5, 137, 1980.
115. **Yavin, E. and Ziegler, B. P.,** Regulation of phospholipid metabolism in differentiating cells from rat brain cerebral hemispheres in culture. Serine incorporation into serine phosphoglycerides: base exchange and decarboxylation reactions, *J. Biol. Chem.*, 252, 260, 1977.
116. **Bradbury, K.,** Ethanolamine glycerophospholipid formation by decarboxylation of serine glycerophospholipid in myelinating organ cultures of cerebellum, *J. Neurochem.*, 43, 382, 1984.
117. **Bjerve, K. S.,** The biosynthesis of phosphatidylserine and phosphatidylethanolamine from l-[3-^{14}C]serine in isolated rat hepatocytes, *Biochim. Biophys. Acta*, 833, 396, 1985.
118. **Butler, M. and Morell, P.,** The role of phosphatidylserine decarboxylase in brain phospholipid metabolism, *J. Neurochem.*, 41, 1445, 1983.
119. **Kane-Sueoka, T. and King, D. M.,** Phosphatidylethanolamine biosynthesis in rat mammary carcinoma cells that require and do not require ethanolamine for proliferation, *J. Biol. Chem.*, 262, 6074, 1987.
120. **Nishijima, M., Kuge, O., Maeda, M., Nakano, A., and Akamatsu, Y.,** Regulation of phosphatidylcholine metabolism in mammalian cells, *J. Biol. Chem.*, 259, 7101, 1984.
121. **Kuge, O., Nishyima, M., and Akamatsu, Y.,** Phosphatidylserine biosynthesis in cultured CHO cells II, *J. Biol. Chem.*, 261, 5790, 1986.
122. **Voelker, D. R. and Frazier, J. L.,** Isolation and characterization of a CHO cell line requiring either ethanolamine or phosphatidylserine for growth and exhibiting defective phosphatidylserine synthetase activity, *J. Biol. Chem.*, 261, 1002, 1986.
123. **Mersel, M., El-Achkar, P., Hindelang, C., Mandel, P., Van Dorsselan, A., and Freysz, L.,** Ethanolamine base exchange in astrocyte primary cultures: localization and developmental studies, *Neurochem. Res.*, 12, 385, 1987.
124. **Hattori, H. and Kanfer, J. N.,** Effects of base exchange reaction on the Na^+-K^+ ATPase in rat brain microsomes, *Neurochem. Res.*, 8, 1185, 1983.
125. **Heller, M.,** Phospholipase D, *Adv. Lipid Res.*, 16, 268, 1978.

Chapter 6

PHOSPHATIDYLCHOLINE BIOSYNTHESIS FROM LYSOPHOSPHATIDYLCHOLINE

Patrick C. Choy and Gilbert Arthur

TABLE OF CONTENTS

I. INTRODUCTION

Structural studies of phosphatidylcholine (PC) indicate that there is an asymmetric distribution of acyl groups on the molecule. Saturated fatty acids are usually esterified at the C-1 position whereas unsaturated fatty acids are at the C-2 position. The distribution of molecular species are different between species and tissues as well as at cellular and subcellular levels. In addition, the developmental, nutritional, and pathological states of the animal can also influence the molecular composition of PC in tissues. These observations have led to the idea that the distinct molecular compositions of phospholipids are intimately linked to cellular functions. Unfortunately, with a few exceptions, the functions of the individual molecular species are not known. The asymmetric distribution of the acyl groups and membrane-specific molecular composition make it logical to surmise that mechanisms exist in cells to bring about the observed distinctive and nonrandom distribution of acyl groups in PC.

The pioneering work of Lands[1] introduced the concept of remodeling of phospholipid fatty acids by a deacylation-reacylation cycle (Figure 1). This was based on the existence of phospholipases with different positional specificities and the demonstration of enzyme activities that could acylate lysoPC to PC. The discrimination exhibited by these enzymes pointed to their involvement in achieving the observed asymmetry and molecular composition of the phospholipid. Although this mechanism has been known for almost two decades, a number of major questions remain unresolved.

Apart from its role in modifying the molecular composition of PC, the synthesis of PC from lysoPC also serves to reduce the levels of lysoPC which is cytolytic at high concentrations. In certain tissues the acylation process may also serve to control the levels of free fatty acids.[2] The purpose of this chapter is to give a brief overview of the processes involved in the synthesis of PC from lysoPC and to highlight some of the current problems. We will deal mainly with the biosynthesis of PC; however, where appropriate, we will also discuss the synthesis of plasmenylcholine and plasmanylcholine from their respective lysophospholipids. The metabolism of these ether lipids are dealt with in greater detail in Chapter 9.

II. THE FORMATION OF PHOSPHATIDYLCHOLINE FROM LYSOPHOSPHATIDYLCHOLINE *IN VIVO*

In mammalian tissues, the majority of PC is synthesized *de novo* via the CDP-choline pathway.[3] Although the last step of the CDP-choline pathyway, catalyzed by 1,2-diacylglycerol cholinephosphotransferase, has some specificity for certain molecular species of 1,2-diacylglycerol (Chapter 4), the selection of the appropriate acyl groups in the phospholipid is believed to take place mainly after its synthesis, and the deacylation-reacylation mechanism has been generally accepted as the preferred mechanism for the nonrandom distribution of fatty acids in PC and other phospholipids.

Investigations on the deacylation-reacylation mechanism for the remodeling of PC in the last two decades were mostly concerned with *in vitro* studies. Only limited information is available on the *in vivo* studies. Therefore, the exact extent of remodeling of the newly synthesized PC in most mammalian tissues by this mechanism is still a matter of conjecture. Since lysoPC was shown to be present in the plasma and might act as a precursor for PC biosynthesis in mammalian tissues, attempts were made to define the origins of this lysophospholipid. PC is deacylated into lysoPC by the hydrolytic action of phospholipases A_1 and A_2, and these enzymes are found in a variety of mammalian tissues (Chapter 8). Alternatively, cleavage of the 2 acyl ester in the phospholipid may occur by an intermolecular transfer of the acyl group to cholesterol. The earlier work by Stein and Stein[4] showed that when PC labeled with 1-^{14}C linoleic acid at the C-2 position was administered to rats intravenously, a substantial amount of radioactivity was transferred from PC to cholesterol ester in the plasma within 60 min of administration. In the

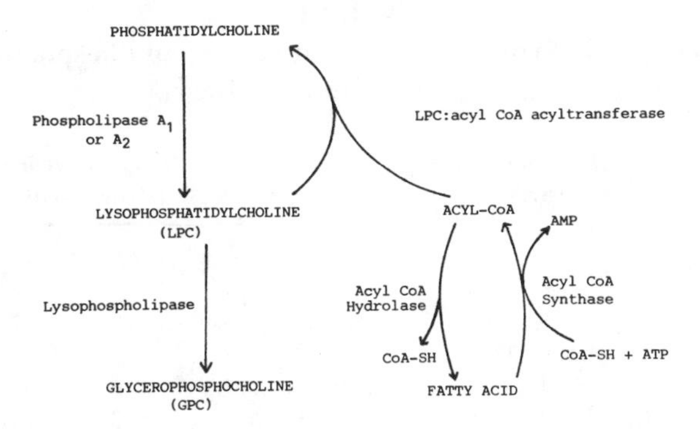

FIGURE 1. The deacylation–reacylation pathway of phosphatidylcholine.

liver, only 16% of the radioactivity associated with PC was transfered to other lipids, and less than 1% of this radioactivity was found in the cholesterol ester. These results provide indirect evidence that lysoPC in the plasma is produced by lecithin:cholesterol acyltransferase, whereas in the liver the lysolipid is produced by phospholipase.

The ability to form PC from lysoPC in intact mammalian tissues was also investigated.[4,5] When labeled lysoPC was injected intravenously to rats, it disappeared from the bloodstream, and labeled PC was recovered in organs such as liver, small intestine, skeletal muscles, lungs, kidneys, and heart. The pathways of formation of PC were studied with lysoPC labeled with ^{32}P and 1-^{14}C-palmitic acid. The labeled PC formed was found to have a $^{32}P/^{14}C$ ratio similar to the lysolipid, suggesting that the acylation reaction, and not the transfer of acyl groups between two lysoPCs, was the predominant pathway. In subsequent studies, the *in vivo* formation of PC from lysoPC via the acylation reaction has been confirmed in other mammalian tissues, including the arterial wall.[6] Using the isolated perfused heart as a model,[7] we estimated that under optimal conditions, the reacylation of exogenous lysoPC might account for a maximum of 14% of the newly formed cardiac PC.

A number of attempts have been made to demonstrate that the newly synthesized PC undergoes remodeling via the deacylation-reacylation mechanism. One of the earlier studies[8] involved the incubation of rat liver slices with labeled glycerol and the analysis of incorporation of label in the molecular species of PC after different incubation periods. Newly synthesized PC species in the intact slices were predominantly monoenes and dienes which were similar to the newly synthesized species of diacylglycerol, but very different from the endogenous PC. From this and other studies,[9,10] it was concluded that 1,2-diacylglycerol:CDP-choline phosphotransferase had very limited ability to select the appropriate molecular species for the biosynthesis of PC, and rearrangement of the acyl groups of PC would occur subsequent to its synthesis. Labeling of the molecular species of PC in the intact liver by a similar approach[11] suggested that there was extensive acylation of monoacylphospholipids, most noticably for the incorporation of arachidonic acid into the phospholipids.

Direct evidence for the occurence of postsynthetic remodeling of PC comes from a pulse-chase study with the isolated hamster heart.[12] Isolated hamster hearts were pulse-labeled with labeled glycerol and then chased for various time periods. The results of this study are depicted in Table 1. The percentage of label in the molecular classes of PC at different periods of the chase were analyzed by silver nitrate impregnated thin-layer chromatographic plates. The molecular classes of diacylglycerol were also determined immediately after pulse-labeling and compared with those obtained from PC at the same time point. With the exception of the dienoic and the tri/tetraenoic groups, the distribution of the molecular groups in the labeled diacylglycerol was not very different from those found in the newly synthesized labeled PC. Since most of the labeled PC was synthesized via the CDP-choline pathway, and the distribution of the molecular species of the labeled PC at this time point is very different from that found in the heart, it appears

TABLE 1
Distribution of Newly Synthesized Diacylglycerol and Phosphatidylcholine in the Perfused Hamster Hearts

Molecular class	Diacylglycerol (% distribution)	Phosphatidylcholine (% distribution)				
Time (min):	0	0	30	60	120	240
Medium chain	11.0 ± 3.9	13.1	8.5	3.9	2.1	1.3
Saturated	5.3 ± 2.4	10.2	10.8	11.5	7.3	3.7
Monoenoic	16.4 ± 1.0	24.9	28.7	27.6	33.9	27.8
Dienoic	49.9 ± 4.8	24.3	26.6	24.2	37.7	48.5
Tri– + tetraenoic	8.0 ± 1.1	22.1	20.2	23.5	16.0	16.4
Polyenoic	9.3 ± 3.0	5.4	5.0	6.7	2.4	2.3

Note: Isolated hamster hearts were perfused for 15 min with 1 m*M* [^{3}H]glycerol (20 μCi/μmol) and subsequently with Krebs–Henseleit buffer up to 240 min. The distribution of radioactivity in the molecular classes of diacylglycerol and phosphatidylcholine was determined at different times during the perfusion. The results are the mean of three experiments for the diacylglycerol and two experiments for the phosphatidylcholine.

Reproduced from Arthur, G. and Choy, P. C., *Biochim. Biophys. Acta,* 795, 221, 1984. With permission.

that 1,2-diacylglycerol:cholinephosphotransferase has only limited ability in the selection of 1,2-diacylglycerol molecular species.

A significant redistribution of label in the molecular species of PC during the chase period was detected.[12] There was a gradual decrease in the medium-chain and saturated species with a concomitant increase in the dienoic species. At 240 min after the chase, the distribution of the labeled molecular species of PC was similar to the endogenous PC. Since over 90% of the radioactivity in diacylglycerol had disappeared within the first 30 min of the chase, the redistribution of label in the molecular species of PC could not be attributed to *de novo* synthesis. The desired molecular species of PC in the hamster heart was not achieved by *de novo* synthesis, but the newly synthesized phospholipid must have undergone extensive remodeling. Since the study was conducted in the isolated heart, it appears that the remodeling process does occur within the tissue and requires no exogenous factors.

III. THE DEACYLATION OF PHOSPHATIDYLCHOLINE — HYDROLYSIS OF ACYL ESTER GROUPS

Phospholipases A_1 and A_2 are responsible for the respective hydrolysis of the acyl ester groups of the phospholipids at the C-1 and the C-2 positions (Chapter 8). Both enzymes are widely distributed in mammalian tissues.[13] It is generally accepted that the enzymes associated with the particulate subcellular fractions have an absolute requirement for calcium, while the cytosolic enzymes may be enhanced by the presence of calcium. Phospholipase A_2 has a high degree of specificity toward the polyunsaturated acyl groups, but the acyl specificity of phospholipase A_1 appears to be somewhat limited. In spite of the recent purifications of phospholipases A_1 and A_2 in mammalian tissues, their control mechanisms remain largely unknown.[13]

IV. *IN VITRO* BIOSYNTHESIS OF PHOSPHATIDYLCHOLINE FROM LYSOPHOSPHATIDYLCHOLINE

In mammalian tissues, lysoPC is reacylated by two types of reactions. The first involves the transfer of an acyl group from acyl-CoA to lysoPC and is catalyzed by acyltransferases. The second type is transacylation, which involves the transfer of an acyl group from another phospholipid directly to the lysoPC, without the release of free fatty acid. These acylation reactions are discussed below.

A. ACYLATION OF LYSOPHOSPHATIDYLCHOLINE BY ACYLTRANSFERASES

The acylation of lysoPC to PC has been reported in homogenates from a large variety of tissues and in the cytosolic, microsomal, mitochondrial, and nuclear fractions.[14-23] The transfer of acyl groups from acyl-CoA to 1-acyl-glycerophosphocholine (1-acyl-GPC) is catalyzed by acyl-CoA:1-acyl-GPC acyltransferase, while the transfer of fatty acids to 2-acyl-GPC is catalyzed by acyl-CoA:2-acyl-GPC acyltransferase. Indirect evidence suggests the two reactions are catalyzed by different enzymes.[24] There are two fundamental questions regarding the role of acyltransferases in the biosynthesis of PC. The first is whether acyltransferases possess the ability to maintain the asymmetric distribution of saturated and unsaturated acyl groups in PC. The second is concerned with the role of acyltransferase in achieving the observed membrane-specific molecular species of PC.

1. The Asymmetrical Distribution of Acyl Groups

If acyltransferases are responsible for the observed selective distribution of saturated and unsaturated acyl groups in PC, then acyl CoA:1-acyl-GPC acyltransferase might be expected to have a preference for unsaturated acyl groups, and acyl CoA:2-acyl-GPC acyltransferase a preference for saturated acyl groups. Using 1-acyl-GPC and 2-acyl-GPC as acyl acceptors, the rate of incorporation of unsaturated acyl groups from acyl-CoAs into 1-acyl-GPC was found to be much greater than the rate of incorporation of saturated acyl-CoA in microsomal preparations.[24,25] Similar results were obtained for the mitochondria[16,22] and nuclear fractions.[26] This selectivity by the enzyme is observed regardless of the use of single acyl-CoAs or a mixture of acyl-CoAs. Indeed preferential acylation with unsaturated acyl-CoAs by acyl-CoA:1-acyl-GPC acyltransferase is more accentuated with mixed populations of saturated and unsaturated acyl-CoA.[27]

Experiments with 2-acyl-GPC have not been as extensive as those with 1-acyl-GPC. When rat liver microsomes were incubated with a mixture of 1-acyl and 2-acyl-GPCs, the saturated acyl groups (stearate and palmitate) were preferentially incorporated at the C-1 position, while the unsaturated acyl groups (linoleate and oleate) were incorporated at the C-2 position.[25] With 2-acyl-GPC as the acyl acceptor, a greater incorporation of saturated fatty acids than unsaturated fatty acids were incorporated into 2-acyl-GPC.[24] The rates of acylation were more pronounced with rat, pig, and beef liver microsomes compared to guinea pig liver microsomes which showed little discrimination.

The best acyl-donor utlized by the guinea pig heart microsomal acyl-CoA:2-acyl-GPC acyltransferase at low concentrations of acyl-CoA (0 to 24 μM) was oleoyl-CoA.[28] This was followed by the saturated species (palmitoyl- and stearoyl-CoA) which were utilized as effectively or better than the linoleoyl- or arachidonoyl-CoA. However, at concentrations above 24 μM, the saturated acyl-CoAs inhibited the reaction, whereas reaction rates with the unsaturated species continued to increase. These observations indicate that the availability as well as the concentration of acyl-CoAs may be important parameters in determining the *in vivo* acylation patterns. The general concensus is that the specificities of acyl-CoA:1-acyl-GPC acyltransferase and acyl-CoA:2-acyl-GPC acyltransferase are indeed sufficient to achieve the observed asymmetric distribution of saturated and unsaturated acyl groups in PC.[29,30]

2. Synthesis of Membrane-Specific Molecular Species of Phosphatidylcholine

Are acyltransferases responsible for attaining and maintaining the observed molecular composition of PC found in specific membranes? If so, one may expect subcellular as well as tissue differences in the properties of the acyltransferases to reflect the differences in PC molecular species in a membrane. Although correlation has been reported in a number of instances,[24,31,32] a prediction of the acyl composition can rarely be obtained from the acyl specificity of the enzyme in the majority of cases. Nevertheless, it is well recognized that acyltransferases play an important role in achieving the acyl composition of the membrane phospholipids, and information on the acyl specificity of the enzyme may contribute to the overall understanding of the mechanism involved in attaining such molecular arrangement.

3. Liver

The acylation of 1-acyl-GPC of defined acyl groups by pig liver microsomal acyl-CoA:1-acyl-GPC acyltransferase was examined with 11 different acyl-CoAs.[33] Although differences in reaction rates were observed among the lysoPC species, the relative rates of utilization of the various acyl-CoAs were not significantly affected. The position to be acylated appeared to be more significant than the acyl species of the lysoPC. Hence, the acyl-CoA specificity of the acyltransferase could not explain the natural pairings of saturated and unsaturated fatty acids on PC isolated from the tissue.[34] A similar study with rat liver microsomes confirmed the above observations, even with mixtures of saturated and unsaturated 1-acyl-GPC or mixtures of acyl-CoAs.[27]

In another attempt to elucidate the role of acyltransferases in the nonrandom pairing of saturated and unsaturated acyl groups in liver PC, the ability of the acyltransferase to acylate preferentially 1-palmitoyl- or 1-stearoyl-GPC (equimolar amounts) with a defined acyl-CoA was examined.[35] Regardless of the acyl-CoA used, the rat liver microsomal acyl-CoA:1-acyl-GPC acyltransferase showed a 3.5-fold preference for 1-palmitoyl-GPC over 1-stearoyl-GPC. Furthermore, decreasing the lysoPC concentration resulted in a decrease in the selectivity for 1-palmitoyl-GPC, while the preference for arachidonoyl-CoA over oleoyl- and linoleoyl-CoA was increased. Although this study was able to demonstrate the ability of the enzyme to discriminate between 1-palmitoyl- and 1-stearoyl-GPC, the results could still not explain the observed pairings of fatty acids in the tissue. In a subsequent study, the fatty acid species of 2-acyl-GPC was varied, and the effect of the changes on the ability of the acyl-CoA:2-acyl-GPC acyltransferase to select preferentially between palmitoyl-CoA and stearoyl-CoA in a mixture of the two acyl donors was investigated.[36] At high concentrations of 2-acyl-GPC (64 μM), and with equimolar mixtures of the two acyl donors, there was no preferential acylation with either acyl-CoA, regardless of the 2-acyl-GPC species examined. At lower concentrations of 2-acyl-GPC (16 μM), stearoyl-CoA was preferred 1.9- to 2.6-fold over palmitoyl-CoA depending on the lysoPC species. It can be concluded that the selectivities of the acyltransferases could not account for the acyl pairings in PC in the liver.

4. Lung

The PC composition of lung surfactant is unusual in having a disaturated molecule (dipalmitoyl-PC) as the predominant species. Since there was evidence that the CDP-choline pathway could not account for the preponderance of this molecular species, the possible involvement of acyltransferases[37-40] or transacylases[41-43] was investigated. The acyl specificities of the lung microsomal acyltransferase indicated the enzyme utilized palmitoyl-CoA at rates similar to unsaturated acyl-CoA such as oleoyl-CoA.[37-39] In the lung 1-palmitoyl-GPC was also shown to be a more effective acyl acceptor (4.2- to 5.7-fold) than 1-stearoyl-GPC regardless of the acyl-CoA species.[40] Although the major pathway for the biosynthesis of the dipalmitoyl-PC is not known with certainty, the enzymes required for its synthesis via the remodeling pathway are present in the tissue.

5. Brain

In the brain, acyl-CoA:1-acyl-GPC acyltransferase activities have been reported in the microsomal and nuclear fractions.[23,26] In general the specific activity of the nuclear enzyme was greater than the microsomal enzyme, although this varied with the composition and concentration of the acyl acceptor as well as the nature of the acyl donor. The acyltransferase activities in the two subcellular fractions differed in both their acyl-CoA specificities and preference for different species of 1-acyl-GPC. Regardless of the acyl composition or concentration of the lysoPC, the nuclear enzyme showed a greater preference for arachidonoyl-CoA than oleoyl-CoA, which was in contrast to the microsomal enzyme.[26] With oleoyl- and arachidonoyl-CoAs as acyl donors, the nuclear enzyme showed a preference for 1-palmitoyl-GPC compared to 1-stearoyl- and 1-oleoyl-GPC. The microsomal enzyme also exhibited a preference for 1-palmitoyl-GPC, although this was not as pronounced as observed for the nuclear enzyme. The molecular composition of PC from the subcellular fractions is not known. Hence, the possible significance of the differences between the nuclear and microsomal enzymes could not be evaluated. Rat and guinea pig brain cerebral cortex mitochondria and plasma membrane were reported to be devoid of acyl-CoA:lysoPC acyltransferase; the activitiy was located in the endoplasmic reticulum.[44] It should be pointed out that these studies were conducted with oleoyl- and palmitoyl-CoA as acyl donors. Therefore, this does not exclude the existence of a linoleoyl-CoA- or arachidonoyl-CoA-specific acyltransferases.

6. Heart

Acyl-CoA:1-acyl-GPC acyltransferases have been reported in cardiac microsomes,[21,45] mitochondria,[22,46] and cytosol.[22,47] In the guinea pig heart microsomes, the enzyme was more active with unsaturated acyl-CoAs compared with the saturated CoAs.[21] However, there was little discrimination among the unsaturated species, and the acyl specificity could not account for the acyl distribution of PC in heart. Experiments with 2-arachidonoyl-GPC and 2-palmitoyl-GPC as acyl acceptors have shown that although the rate of acylation was lower with 2-palmitoyl-GPC, the acyl specificities of the 2-acyl-GPC acyltransferase with the two acyl acceptors were identical.[28] In the reacylation of lysoPC, it appears that the position to be acylated and not the composition of the acyl acceptor is the more significant factor.

An acyl-CoA:1-acyl-GPC acyltransferase activity with the unusual property of utilizing only linoleoyl-CoA as the acyl donor has been reported in guinea pig heart mitochondria.[46] Although linoleate is the major acyl group at the C-2 position of PC in guinea pig mitochondria,[48] significant quantities of other fatty acids are also present. The mitochondrion is incapable of synthesizing PC *de novo* and imports PC from the endoplasmic reticulum by an undefined mechanism.[49] The linoleoyl-CoA:1-acyl-GPC acyltransferase present in this fraction may participate in the remodeling of the fatty acids of the imported PC to give the observed high linoleoyl content. On the other hand, rabbit heart mitochondrial acyltransferase did not exhibit the selectivity for linoleoyl-CoA as in the guinea pig heart. The rabbit heart mitochondrial enzyme was active with a range of acyl CoAs, and the rate of incorporation of the acyl groups was similar to that of the microsomal acyltransferase.[22] In the same study, a unique acyltransferase was identified in the cytosol which was specific for fatty acids that were substrates of lipoxygenase and cyclooxygenase pathways, or were able to reverse symptoms of essential fatty acid deficiency. The ability to acylate lysoPC with these fatty acids was not exclusive to the cytosol. The mitochondria and microsomes were able to utilize these acyl groups in addition to those that were not substrates for the lipoxygenase or cycloxygenase pathways. The enzyme activities in these subcellular fractions were severalfold higher than that found in the cytosol. Taking into consideration the higher activities of the acyltransferases in the membranous fractions where the products of the acylation reactions are likely to be incorporated, the postulated selective role of the cytosolic acyltransferase in the incorporation of arachidonate into the membrane phospholipids is therefore questionable. Indeed, the very low activities and selectivity could be explained by the release of an acyltransferase with the specificities described above, from the particulate fraction into the cytosol during the fractionation procedure. This acyltransferase may be less tightly bound to the membrane than other acyltransferases.

7. Other Tissues

Acyltransferases have been described in other tissues and cells including kidney, testes, pancreas, spleen, epidydimal fat, intestinal mucosa,[15] red blood cells,[18] lymphocytes,[20] and platelets.[50] In view of the role of these enzymes in manipulating the acyl composition of phospholipids, these acyltransferases are in all likelihood present in every tissue and membrane, and reports of their absence probably reflect the use of inappropriate substrates for the assay. This is clearly illustrated in the apparent absence of acyl-CoA:1-acyl-GPC acyltransferase activity in guinea pig or rabbit heart mitochondria when assayed with oleoyl-CoA or palmitoyl-CoA as acyl donors,[21,45] whereas a very active enzyme was detected with linoleoyl-CoA as the acyl donor.[46]

There seems little doubt that acyltransferases are involved in remodeling the acyl chains of PC in membranes. However, the precise mechanism and control of these events have yet to be elucidated. It is worth pointing out that remodeling of acyl groups of PC to achieve a desired molecular composition is not the only function of acyltransferases in all tissues. For example, in the intestinal mucosa, the highest activities of the enzyme were found at the brush border-free microsomes.[15] This enzyme plays an important role in the metabolism of digested fat by converting the lysoPC absorbed from the intestines into PC for chylomicron formation prior to secretion into the lymph.

B. ACYLATION OF 1-ALKENYL-GPC AND 1-ALKYL-GPC

A number of mammalian tissues have significant quantities of plasmenylcholine and/or plasmanylcholine. The 1-alkenyl-GPC and 1-alkyl-GPC formed from the hydrolytic action of phospholipase A_2 can be reacylated by both acyltransferases and transacylases. In this section we will briefly discuss the acyltransferases that acylate 1-alkenyl-GPC and 1-alkyl-GPC to the corresponding diradylGPC. Acyl-CoA:1-alkenyl-GPC acyltransferase activity has been reported in human erythrocytes,[50] rabbit sarcoplasmic reticulum,[51,52] intestinal mucosa,[15] testes,[15] human amnion cells,[53] and the heart.[21] In a comparative study with rat, guinea pig, dog, and rabbit heart microsomes, no correlation between the specific activity of the enzyme and the tissue content of plasmenylcholine was observed.[21] Acyl-CoA:1-alkenyl-GPC acyltransferase has not been detected in liver, an organ with little plasmenylcholine.

The activity of acyl-CoA:1-alkenyl-GPC acyltransferase is generally less than the activity of acyl-CoA:1-acyl-GPC acyltransferases from the same tissue.[21,29,30] The two activities also differ significantly in their acyl-CoA specificities even though they both have a preference for unsaturated fatty acyl-CoAs. These differences led to the suggestion that the reactions are catalyzed by different enzymes.[51] Further evidence was obtained from studies of the guinea pig microsomal enzymes where, apart from differences in acyl specificities, the two enzyme activities also differed with respect to their responses to cations, detergents, and heat.[21] Indeed, kinetic studies revealed that the acylation of 1-acyl-GPC and 1-alkenyl-GPC could not be accommodated by the same catalytic site. Acyl-CoA:1-alkenyl-GPC acyltransferase activity is also present in the guinea pig heart mitochondria.[46]

The recognition of arachidonoyl-containing plasmanylcholine as the precursor for platelet activating factor has provided added stimulation to the research on this subject and has led to the characterization of acyl-CoA:1-alkyl-GPC acyltransferase from several tissues.[52,53] However, the presence of this enzyme activity in human platelets is still the subject of debate. Acyl-CoA:1-alkyl-GPC acyltransferase was not detected in the platelet by one group,[54,55] whereas it was characterized in the same tissue by another group.[56] This platelet enzyme was reported to differ from the acyl-CoA:1-acyl-GPC acyltransferase in its preference for arachidonoyl- and linoleoyl-CoA over oleoyl-CoA.[50,56] In our opinion, acyltransferase activity is present in the tissue, and the conflicting reports by the two groups is probably due to the differences in their assay systems. McKean and Silver[56] incubated their membrane preparation with acyl-CoAs, whereas Kramer et al.[55] incubated the same preparation with factors (CoA, ATP, and Mg) required for fatty acid activation. Since platelet membranes have the ability to synthesize acyl-CoAs from exogenous fatty acids,[57] the inability to detect acyltransferase activity[55] may reflect the absence or very low concentrations of free fatty acids in the platelet membrane preparation.

Acyl-CoA:1-alkyl-GPC acyltransferase activity has also been reported in Ehrlich ascites tumor cells,[58] human erythrocytes,[52] sarcoplasmic reticulum,[52] and intestinal mucosa microsomes.[15] In rat liver microsomes, there was no transfer of fatty acids from oleoyl- or arachidonoyl-CoA to 1-octadecyl-GPC[59] nor was there a transfer from linoleoyl-CoA to 1-alkyl-GPC,[58] suggesting the absence of acyl-CoA:1-alkyl-GPC acyltransferase in the tissue. In most tissues where the alkyl acyltransferase is present, the rate of transfer of fatty acids to 1-alkyl-GPC is lower than the rate of transfer to 1-acyl-GPC.

C. REMODELING OF PHOSPHATIDYLCHOLINE BY TRANSACYLATION

The synthesis of PC from lysoPC can also occur by transacylation. In this process, an acyl group is transferred from a donor phospholipid to the lysoPC. This transfer occurs without the release of fatty acid by phospholipases. Three transacylation reactions — (1) CoA-dependent transacylation, (2) CoA-independent transacylation, and (3) lysoPC-lysoPC transacylation — have all been identified to occur in mammalian membranes.

The transfer of an acyl group to lysoPC by a CoA-dependent transacylation reaction was first described by Irvine and Dawson.[60] This reaction is the reversal of the acyltransferase reaction discussed above. The transacylation process occurs in two stages:

1. Phospholipid + CoA ⟶ Lysophospholipid + Acyl-CoA

2. Lysophosphatidylcholine + Acyl-CoA ⟶ Phosphatidylcholine + CoA

The CoA-independent transacylation occurs by direct transfer of an acyl group from the donor phospholipid to the lysoPC to form PC and may be represented as

Lysophosphatidylcholine + Phospholipid ⟶ Phosphatidylcholine + Lysophospholipid

The third type of transacylation reaction can be regarded as a variation of the CoA-independent transfer. The transfer does not require CoA, and the acyl group is transferred from one lysoPC molecule to another instead of from a diradylglycerophospholipid. This reaction is sometimes referred to as the Marinetti pathway and is represented by the following reaction :

Lysophosphatidylcholine + Lysophosphatidylcholine ⟶ Phosphatidylcholine + Glycerophosphocholine

The CoA-dependent and CoA-independent transacylation reactions have been described in a number of tissues and cells. The overwhelming evidence (see below) indicates that the acyl groups transferred are usually polyunsaturated in nature. This has led to postulations that the high concentrations of polyunsaturated fatty acids found in certain species of phospholipids including the cholinephosphoglycerides are achieved via transacylases rather than acyltransferase.

In rat liver microsomes, a CoA-dependent transfer of fatty acids from phosphatidylinositol to lysoPC to form PC was described which was four times more active with arachidonate than linoleate.[60] There was no transfer of oleate. The transfer of arachidonate, but not oleate from prelabeled rat platelet phosphatidylinositol and phosphatidylethanolamine to lysoPC has been reported.[61] In human platelets, no CoA-dependent or CoA-independent transacylation of acyl groups to lysoPC for the formation of PC was observed,[54,55] but there was a CoA-independent transacylation of arachidonate from endogenous PC to 1-alkyl-GPC to form plasmanylcholine.

Studies with rabbit alveolar macrophages[62] revealed the existence of three acylating systems for lysophospholipids: an acyl-CoA acyltransferase, a CoA-dependent transacylase, and a CoA-independent transacylase. Only the CoA-independent transacylation was specific for arachidonate transfer, and such specificity was maintained with either 1-acyl-GPC or 1-alkyl-GPC as acyl acceptors. The profiles of the molecular composition of choline glycerophospholipids synthesized by each of the three acylating mechanisms were also different.[62] In later studies, the CoA-independent transfer of polyunsaturated fatty acids to 1-acyl-GPC and 1-alkyl-GPC in rabbit alveolar macrophages has been confirmed and extended.[63-65] Such transfer was not limited to arachidonate, and other polyunsaturated acyl groups such as 20:5 (n-3), 22:4(n-6), and 22:6(n-3) were also transferred by the CoA-independent mechanism.[64]

The lysoPC-lysoPC transacylase reaction results in the synthesis of PC from two lysoPC molecules. This reaction occurs in liver homogenates,[66] lung,[41-43] erythrocytes,[67] heart,[68] and polymorphonuclear leukocytes.[69] This transacylation reaction in the lung has been the subject of extensive studies[70-76] since this pathway may play a significant role in the biosynthesis of dipalmitoyl-PC. The purified enzyme possesses both lysophospholipase and transacylase activity.[76]

Cardiac lysoPC-lysoPC transacylase has been purified from rabbit heart cytosol.[68] The purified cardiac enzyme is similar to the lung enzyme in possessing both transacylase and lysophospholipase activities. The transacylase/lysophospholipase ratio of the purified cardiac enzyme was 0.5, which is different from the ratio for the lung enzyme.[68] The role of this cytosolic enzyme in the biosynthesis of cardiac PC is uncertain in view of the fact that the products of the transacylase reaction are likely to be disaturated PC molecules, which are not prominent molecular species in the heart. The enzyme has been postulated to play a role in regulating the lysoPC levels in the heart which may be a biochemical factor for the production of cardiac arrhythmias.[68]

D. PURIFICATION OF ACYLTRANSFERASES

The purification of acyltransferases and transacylases is essential to understand how the activities and specificities of these enzymes are regulated. The majority of acyltransferases appear to be intrinsic membrane proteins and therefore require solubilization prior to purification. Unfortunately, enzyme activities are easily inhibited by low concentrations of detergents. For example, although only about 20% of the membrane proteins were solubilized by 1% cholate, 0.2% glycholate, 0.25% deoxycholate, 0.05% Triton X-100, 0.25% Nonidet P-40, and 0.25% Berol-043, 90% of the acyltransferase activity was inhibited.[77] Therefore, it is not suprising that the purification of these enzymes has been a difficult task. Nevertheless, considerable progress has been made in the last decade.

Acyl-CoA:LPC acyltransferase was selectively solubilized from thymocyte plasma membranes with synthetic ether analogues of lysoPC without loss of activity,[77] but no attempt was made to purify the enzyme further. The same enzyme in rat liver microsomes was solubilized with Triton-X100 and purified fourfold by gel filtration chromatography, followed by density gradient centrifugation.[78] The partially purified preparation differed from the microsomal enzyme in its high specificity for arachidonoyl-CoA to the virtual exclusion of other acyl donors. In a separate study, a 30-fold purification of the rat liver microsomal acyl-CoA:1-acyl-GPC acyltransferase was obtained after solubilization of the washed microsomes with lysoPC, followed by gel filtration chromatography.[79] The molecular weight of the enzyme was estimated to be about 225,000. This enzyme preparation had the ability to acylate not only 1-acyl-GPC, but also 1-acyl-glycerophosphoethanolamine and 1-acyl-glycerophosphoinositol. However, it was not active toward 2-acyl-GPC. The acyl specificity of the partially purified enzyme was found to be 20:4 >> 18:2 = 18:1 > 16:0. It is intriguing that substantially different results were obtained from two studies in which the partially purified enzyme preparations were derived from the same source. It is possible that the apparently contradictory observations can be reconciled if one assumes the existence of different acyltransferases in the tissue, each specific for a given acyl acceptor. Thus, the treatment by Yamashita et al.[78] might have inhibited all except the arachidonoyl-CoA-specific acyltransferase activity, or alternatively, there was a selective solubilization of the arachidonoyl CoA-specific acyltransferase.

Gavino and Deamer[80,81] reported an ingenious procedure for the purification of the acyltransferase. The native membrane was solubliized with high concentrations of oleoyl-CoA and lysoPC which resulted in the formation of mixed micromicelles that would ideally contain a single protein per micelle. On prolonged incubation, the lysoPC in any acyltransferase-containing micromicelle would be gradually converted to PC. Since PC micelles are larger and less dense than the lysoPC/oleoyl-CoA micromicelles, the two populations may be separated by density gradient flotation. Application of these ideas led to a 150-fold purification of the rat liver microsomal enzyme. This procedure appears to have the potential for the separation of acyl-CoA-specific acyltransferases, if they do, indeed, exist, by using different acyl-CoAs for solubilization.

Rat brain microsomal acyl-CoA:1-acyl-GPC acyltransferase has been purified 3,000-fold to near homogenity.[82] The molecular weight of the denatured enzyme was 43,000, but it is not clear if this is the molecular weight of the subunit or the native enzyme. The enzyme obtained after DEAE-chromatography utilized 1-acyl-GPC, 1-acyl-glycerophosphoinositol, 1-acyl-glycerophosphoric acid, and 1-acyl-glycerophosphoserine as acyl acceptors but the preparation obtained after the Matrex green column was specific for 1-acyl-GPC as acyl acceptor. Furthermore, the broad acyl-CoA specificity of the partially purified enzyme was also lost after chromatography on the Matrex green column. Whether these differential properties reflect changes in the catalytic properties of a single enzyme or the separation of multiple acyltransferases in the final purification step is yet to be established. The purified enzyme, unlike that found in the microsomal fraction, had a very high degree of specificity for arachidonoyl-CoA and had no ability to utilize palmitoyl- or stearoyl-CoA. Decanoyl- and myristoyl-CoA were utilized at about half the rate for arachidonoyl-CoA, while the values for linoleoyl- and oleoyl-CoA were about a tenth of the maximum activity. When the purified soluble enzyme was incorporated into liposomes of similar composition to brain microsomal membrane, changes in some characteristics of the enzyme were observed. Upon reconstitution, the specificity of the enzyme for lysoPC was maintained, but the ability of the enzyme to utilize decanoyl and myristoyl-CoA was lost. The liposomal-bound acyltransferase was also more active at much lower concentrations of lysoPC than the soluble enzyme. These studies clearly demonstrate that the properties of the enzyme can be modulated by its environment.

V. CONCLUDING REMARKS

Although there have been numerous studies on acyltransferases since they were initially reported, there have been few major advances in our understanding of the role of these enzymes in the biosynthesis of PC. One reason has been the difficulty in purifying the enzymes. Purified enzymes are required to support or refute the suggestion that cells have acyltransferases that are specific for each acyl group,[29,30] rather than the current view of a single enzyme with different affinities for different acyl groups. Preliminary evidence for the existence of acyl-CoA-specific acyltransferases comes from a study where the aging of microsomal acyl-CoA:1-acyl-GPC acyltransferase at 4°C resulted in a differential loss of acyl specificity toward different acyl donors.[83] The idea of the existence of a number of acyl-specific acyltransferases is also supported by the observation that at various stages of purification, acyltransferases display different specificities from those found in the subcellular fraction. In the mitochondria, we have reported the presence of a linoleoyl-CoA-specific acyl-CoA:1-acyl-GPC acyltransferase.[46] If there are, indeed, multiple acyl-CoA-specific acyltransferases, each of which is responsible for a given acyl acceptor, a radical change in our perception and conception of how acyltransferases operate will be warranted. Thus, the acyl specificities previously determined in subcellular fractions would simply reflect the quantitative distribution of the different acyltransferases in the fractions. There would also be a need to reevaluate the reported characteristics of acyltransferases since they may be applicable to only one of the numerous enzymes present.

It can be postulated that the acyltransferases are only partially responsible for attaining and maintaining the observed molecular composition of PC in the tissues. The rationale is that PC is synthesized by multiple pathways which may be responsible for the synthesis of particular molecular species.[3] In addition, the acyltransferase activities are affected by the reaction substrates lysoPC and acyl-CoA, the concentrations of which are controlled by a number of enzymes including phospholipases A_1 and A_2, acyl-CoA synthase, and acyl-CoA hydrolase (see Figure 1). These enzymes exhibit acyl specificities of their own and may be regulated independently of the acyltransferases. Apart from the intrinsic acyl-CoA specificities of the enzyme, the intracellular flux and composition of lysoPC and acyl-CoA will undoubtedly contribute significantly to determining the final composition of the PC. Until a better understanding of these dynamic processes are achieved, correlating the acyl specificity of acyltransferases to the acyl composition represents an overly simplistic approach to understanding the role of these enzymes in attaining the molecular composition of PC.

ACKNOWLEDGMENT

This work was supported by the Medical Research Council of Canada (PCC) and the Manitoba Heart Foundation (GA). PCC is an MRC Scientist and GA is an MRC Scholar.

REFERENCES

1. **Lands, W. E. M.,** Metabolism of glycerolipids. II. The enzymatic acylation of lysolecithin, *J. Biol. Chem.*, 253, 2233, 1960.
2. **Irvine, R. F.,** How is the level of free arachidonic acid controlled in mammalian cells?, *Biochem. J.*, 204, 3, 1982.
3. **Pelech, S. L. and Vance, D. E.,** Regulation of phosphatidylcholine biosynthesis, *Biochim. Biophys. Acta,* 779, 217, 1984.
4. **Stein, Y. and Stein, O.,** Metabolism of labeled lysolecithin, lysophosphatidylethanolamine and lecithin in the rat, *Biochim. Biophys. Acta,* 116, 95, 1968.
5. **Stein, Y., Widnell, C., and Stein, O.,** Acylation of lysophosphatides by plasma membrane fractions of rat liver, *J. Cell Biol.*, 39, 185, 1968.
6. **Portman, O. W. and Illingworth, D. R.,** Metabolism of lysolecithin *in vivo* and *in vitro* with particular emphasis on the arterial wall, *Biochim. Biophys. Acta,* 348, 136, 1974.
7. **Savard, J. D. and Choy, P. C.,** Phosphatidylcholine formation from exogenous lysophosphatidylcholine in isolated hamster heart, *Biochim. Biophys. Acta,* 711, 40, 1982.
8. **Hill, E. E., Husbands, D. R., and Lands, W. E. M.,** The selective incorporation of ^{14}C-glycerol into different species of phosphatidic acid, phosphatidylethanolamine and phosphatidylcholine, *J. Biol. Chem.*, 243, 4440, 1968.
9. **Hill, E. E., Lands, W. E. M., and Slakey, P. M.,** The incorporation of ^{14}C-glycerol into different species of diglycerides and triglycerides in rat liver slices, *Lipids,* 3, 411, 1968.
10. **Kanoh, H.,** Biosynthesis of molecular species of phosphatidylcholine and phosphatidylethanolamine from radioactive precursors in rat liver slices, *Biochim. Biophys. Acta,* 176, 756, 1969.
11. **Akesson, B., Elovson, J., and Arvidson, G.,** Initial incorporation into rat liver glycerolipids of intraportally injected [^3H]glycerol, *Biochim. Biophys. Acta,* 210, 15, 1970.

12. **Arthur, G. and Choy, P. C.,** Acyl specificity of hamster heart CDP-choline 1,2-diacylglycerol phosphocholinetransferase in phosphatidyl choline biosynthesis, *Biochim. Biophys. Acta,* 795, 221, 1984.
13. **Van den Bosch, H.,** Phospholipases, in *Phospholipids,* Hawthorne, J.N. and Ansell, G.B., Eds., Elsevier Biomedical Press, Amsterdam, 1982, 313.
14. **Eibl, H., Hill, E. E., and Lands, W. E. M.,** The subcellular distribution of acyltransferases which catalyse the synthesis of phosphoglycerides, *Eur. J. Biochem.,* 9, 250, 1969.
15. **Subbiah, P. V., Sastry, P. S., and Ganguly, J.,** Acylation of lysolecithin in the intestinal mucosa of rats, *Biochem. J.,* 118, 241, 1970.
16. **Sarzala, M. G., Van Golde, L. M. G., de Kruyff, B., and Van Deenen, L. L. M.,** The intramitochondrial distributution of some enzymes involved in the biosynthesis of rat liver phospholipids, *Biochim. Biophys. Acta,* 202, 106, 1970.
17. **Eisenberg, S., Stein, Y., and Stein, O.,** The role of placenta in lyso-lecithin metabolism in rats and mice, *Biochim. Biophys. Acta,* 137, 115, 1967.
18. **Robertson, A. F. and Lands, W. E. M.,** Metabolism of phospholipids in normal and spherocytic human erythrocytes, *J. Lipid Res.,* 5, 88, 1964.
19. **Waite, M., Sisson, P., and Blacknell, E.,** Comparison of mitochondrial with microsomal acylation of monoacyl phosphoglycerides, *Biochemistry,* 9, 746, 1970.
20. **Resch, K., Ferber, E., Odenthal, J., and Fischert, H.,** Early changes in the phospholipid metabolism of lymphocytes following stimulation with phytohemagglutinin and with lysolecithin, *Eur. J. Immunol.,* 1, 162, 1971.
21. **Arthur, G. and Choy, P. C.,** Acylation of 1-alkenyl-glycerophosphocholine in guinea-pig heart, *Biochem. J.,* 236, 481, 1986.
22. **Needleman, P., Wyche, A., Spreche, H., Elliot, W. J., and Evers, A.,** A unique cardiac cytosolic acyltransferase with preferential selectivity for fatty acids that form cyclooxygenase/lipoxygenase metabolites and reverse essential fatty acid deficiency, *Biochim. Biophys. Acta,* 836, 267, 1985.
23. **Baker, R. R. and Chang, H.-Y.,** The acylation of 1-acyl-*sn*-glycero-3-phosphorylcholine by glial and neuronal nuclei and derived neuronal nuclei envelope: a comparison of nuclear and microsomal membranes, *Can. J. Biochem.,* 59, 848, 1981.
24. **Lands, W. E. M. and Hart, P.,** Metabolism of glycerophospholipids VI. Specificities of acyl coenzyme A:phospholipid acyltransferases, *J. Biol. Chem.,* 240, 1905, 1965.
25. **Lands, W. E. M. and Merkl, I.,** Metabolism of glycerophospholpids III. Reactivity of various acyl esters of coenzyme A with a′-acylglycerophosphorylcholine and positional specificities of lecithin synthesis, *J. Biol. Chem.,* 238, 898, 1963.
26. **Baker, R. R. and Chang, H.-Y.,** A comparison of lysophosphatidylcholine acyltransferase activities in neuronal nuclei and microsomes isolated from immature rabbit cerebral cortex, *Biochim. Biophys. Acta,* 666, 223, 1981.
27. **Van den Bosch, H., Van Golde, L. M. G., Eibl, H., and Van Deenen, L. L. M.,** The acylation of 1-acyl-glycero-3-phosphorylcholines by rat liver microsomes, *Biochim. Biophys. Acta,* 144, 613, 1967.
28. **Arthur, G.,** Acylation of 2-acyl-glycerophosphocholine in guinea pig heart microsomes and mitochondria, *Biochem. J.,* in press.
29. **Hill, E. E. and Lands, W. E. M.,** Phospholipid metabolism, in *Lipid Metabolism,* Wakil, S. J., Ed., Academic Press, New York, 1970, 185.
30. **Lands, W. E. M. and Crawford, C. G.,** Enzymes of membrane phospholipid metabolism in animals, in *The Enzymes of Biological Membranes,* Vol 2., Martonosi, A., Ed., Plenum Press, New York, 1976, 3.
31. **Lands, W. E. M., Blank, M. L., Nutter, L. J., and Privett, O. S.,** A comparison of acyltransferase activities *in vitro* with the distribution of fatty acids in lecithins and triglycerides *in vivo, Lipids,* 1, 224, 1966.
32. **Waku, K. and Lands, W. E. M.,** Control of lecithin biosynthesis in erythrocyte membranes, *J. Lipid Res.,* 9, 12, 1968.
33. **Brandt, A. E. and Lands, W. E. M.,** The effect of acyl-group composition on the rate of acyltransferase catalysed synthesis of lecithin, *Biochim. Biophys. Acta,* 144, 605, 1967.
34. **Holub, B. J. and Kuksis, A.,** Metabolism of molecular species of diacyl glycerophospholipids, *Adv. Lipid Res.,* 16, 1, 1978.
35. **Holub, B. J., Macnaughton, J. A., and Piekarski, J.,** Synthesis of 1-palmitoyl and stearoyl phosphatidylcholines from mixtures of acyl acceptors via acyl-CoA:1-acyl-*sn*-glycero-3-phosphorylcholine acyltransferase in liver microsomes, *Biochim. Biophys. Acta,* 572, 413, 1979.
36. **Holub, B. J.,** The suitability of different acyl acceptors as substrates for the acyl-CoA:2-acyl-*sn*-glycero-3-phosphorylcholine acyltransferase in rat liver microsomes, *Biochim. Biophys. Acta,* 664, 221, 1981.
37. **Vereyken, J. M., Montfoort, A., and Van Golde, L. M. G.,** Some studies on the biosynthesis of the molecular species of phosphatidylcholine from rat lung and phosphatidylcholine and phosphatidylethanolamine from rat liver, *Biochim. Biophys. Acta,* 260, 70, 1972.

38. **Frosolono, M. F., Slivka, S., and Charms, B. C.,** Acyltransferase activities in dog lung microsomes, *J. Lipid Res.,* 12, 96, 1971.
39. **Tansey, F. A. and Frosolono, M. F.,** Role of 1-acyl-2-lyso-phosphatidyl choline acyltransferase in the biosynthesis of pulmonary phosphatidyl choline, *Biochem. Biophys. Res. Commun.,* 67, 1560, 1975.
40. **Holub, B. J., Piekarski, J., and Possmayer, F.,** Relative suitability of 1-palmitoyl and 1-stearoyl homologues of 1-acyl-*sn*-glycerylphosphoryl choline and different acyl donors for phosphatidylcholine synthesis via acyl-CoA:1-acyl-*sn*-glycero-3-phosphorylcholine acyltransferase in rat lung microsomes, *Can. J. Biochem.,* 58, 434, 1980.
41. **Abe, M., Ohno, K., and Sato, R.,** Possible identity of lysolecithin acylhydrolase with lysolecithin-lysolecithin acyltransferase in rat lung soluble fraction, *Biochim. Biophys. Acta,* 369, 361, 1974.
42. **Van den Bosch, H. and Brumley, G. W.,** Properties of a lysophospholipase-transacylase from rat lung: its possible involvement in the synthesis of lung surfactant, *Adv. Exp. Med. Biol.,* 101, 343, 1978.
43. **Oldenborg, V. and Van Golde, L. M. G.,** Activity of cholinephosphotransferase, lysolecithin:lysolecithin acyltransferase and lysolecithin acyltransferase in the developing lung, *Biochim. Biophys. Acta,* 441, 433, 1976.
44. **Fisher, S. K. and Rowe, C. E.,** The acylation of lysophosphatidylcholine by subcellular fractions of guinea pig cerebral cortex, *Biochim. Biophys. Acta,* 618, 231, 1980.
45. **Gross, R. W. and Sobel, B. E.,** Lysophosphatidylcholine metabolism in the rabbit heart:characterisation of metabolic pathways and the partial purification of myocardial lysophospholipase-transacylase, *J. Biol. Chem.,* 257, 6702, 1982.
46. **Arthur, G., Page, L., Zaborniak, C., and Choy, P. C.,** The acylation of lysophosphoradylglycerocholines in guinea pig heart mitochondria, *Biochem. J.,* 242, 171, 1987.
47. **Severson, D. L. and Fletcher, T.,** Regulation of lysophosphatidylcholine-metabolising enzymes in isolated myocardial cells from rat heart, *Can. J. Physiol. Pharmacol.,* 63, 944, 1985.
48. **Arthur, G., Mock, T., Zaborniak, C., and Choy, P. C.,** The distribution and acyl compositon of plasmalogens in guinea pig heart, *Lipids,* 20, 693, 1985.
49. **McMurray, W. C.,** Origins of phospholipids in animal mitochondria, *Biochem. Cell Biol.,* 64, 1115, 1986.
50. **McKean, M. L., Smith, J. B., and Silver, M. J.,** Phospholipid biosynthesis in human platelets. Formation of phosphatidylcholine from 1-acyl lysophosphatidylcholine by acyl-CoA:1-acyl-*sn*-glycero-3-phosphocholine acyltransferase, *J. Biol. Chem.,* 257, 11278, 1982.
51. **Waku, K. and Lands, W. E. M.,** Acylcoenzyme A:1-alkenyl-glycero-3-phosphorylcholine acyltransferase action in plasmalogen biosynthesis, *J. Biol. Chem.,* 243, 2654, 1968.
52. **Waku, K. and Nakazawa, Y.,** Acyltransferase activity to 1-*O*-alkyl-glycero-3-phosphocholine in sarcoplasmic reticulum, *J. Biochem.,* 68, 459, 1970.
53. **Matsumoto, M. and Suzuki, Y.,** Acylation of lysophospholipids including plasmalogen by cultured human amnion cells (FL cells), *J. Biochem.,* 73, 793, 1973.
54. **Kramer, R. M. and Deykin, D.,** Arachidonoyl transacylase in human platelets: coenzyme A-independent transfer of arachidonate from phosphatidylcholine to lysoplasmenylethanolamine, *J. Biol. Chem.,* 258, 13806, 1983.
55. **Kramer, R. M., Pritzker, C. R., and Deykin, D.,** Coenzyme A-mediated arachidonic acid transacylation in human platelets, *J. Biol. Chem.,* 2403, 1984.
56. **McKean, M. C. and Silver, M. J.,** Phospholipid biosynthesis in human platelets: the acylation of lyso-platelet-activating factor, *Biochem. J.,* 225, 723, 1985.
57. **Wilson, D. B., Prescott, S. M., and Majerus, P. W.,** Discovery of an arachidonoyl coenzyme A synthetase in human platelets, *J. Biol. Chem.,* 257, 3510, 1982.
58. **Waku, K. and Nakazawa, Y.,** Regulation of the fatty acid composition of alkyl ether phospholipid in Ehrlich Ascites tumour cells, *J. Biochem.,* 82, 1779, 1977.
59. **Neumuller, W., Fleer, E. A. M., Unger, C., and Eibl, H.,** Enzymatic acylation of ether and ester lysophospholipids in rat liver microsomes, *Lipids,* 22, 808, 1987.
60. **Irvine, R. F. and Dawson, R. M. C.,** Transfer of arachidonic acid between phospholipids in rat liver microsomes, *Biochem. Biophys. Res. Commun.,* 91, 1399, 1979.
61. **Colard, O., Breton, M., and Bereziat, G.,** Induction by lysophospholipids of CoA-dependent arachidonyl transfer between phospholipids in rat platelet homogenates, *Biochim. Biophys. Acta,* 793, 42, 1984.
62. **Robinson, M., Blank, M. L., and Snyder, F.,** Acylation of lysophospholipids by rabbit alveolar macrophages, *J. Biol. Chem.,* 260, 7889, 1985.
63. **Sugiura, T., Katayama, O., Fukui, J., Nakagawa, Y., and Waku, K.,** Mobilization of arachidonic acid between diacyl and ether phospholipids in rabbit alveolar macrophages, *FEBS Lett.,* 165, 273, 1984.
64. **Sugiura, T., Masuzawa, Y., and Waku, K.,** Transacylation of 1-*O*-alkyl-*sn*-glycero-3-phosphocholine (lysoplatelet activating factor) and 1-*O*-alkenyl-*sn*-glycero-3-phosphoethanolamine with docosahexaenoic acid (22:6 w3), *Biochem. Biophys. Res. Commun.,* 133, 574, 1985.

65. **Sugiura, T., Masuzawa, Y., Nakagawa, Y., and Waku, K.,** Transacylation of lysoplatelet activating factor and other lysophospholipids by macrophage microsomes:distinct donor and acceptor selectivities, *J. Biol. Chem.,* 262, 1199, 1987.
66. **Erbland, J. and Marinetti, G. V.,** The enzymatic acylation and hydrolysis of lysolecithin, *Biochim. Biophys. Acta,* 106, 128, 1965.
67. **Mulder, E., Van den Berg, J. W. O., and Van Deenen, L. L. M.,** Metabolism of red cell lipids. II. Conversion of lysophosphoglyceride, *Biochim. Biophys. Acta,* 106, 118, 1965.
68. **Gross, R. W., Drisdel, R. C., and Sobel, B. E.,** Rabbit myocardial lysophospholipase-transacylase. Purification and characterisation and inhibition by endogenous cardiac amphiphiles, *J. Biol. Chem.,* 258, 15165, 1983.
69. **Elsbach, P., Van den Berg, J. W. O., Van den Bosch, H., and Van Deenen, L. L. M.,** Metabolism of phospholipids by polymorphonuclear leukocytes, *Biochim. Biophys. Acta,* 106, 338, 1965.
70. **Brumley, G. and Van den Bosch, H.,** Lysophospholipase-transacylase from rat lung:isolation and partial purification, *J. Lipid Res.,* 18, 523, 1977.
71. **Vianen, G. M. and Van den Bosch, H.,** Lysophospholipase and lysophosphatidylcholine:lysophosphatidylcholine transacylase from rat lung, evidence for a single enzyme and some aspects of its specificity, *Arch. Biochem. Biophys.,* 190, 373, 1978.
72. **Batenburg, J. J., Langmore, W. J., Klazinga, W., and Van Golde, L. M.,** Lysolecithin acyltransferase and lysolecithin:lysolecithin acyltransferase in adult rat lung alveolar type II epithelial cells, *Biochim. Biophys. Acta,* 573, 136, 1979.
73. **De Vries, A. C., Batenburg, J. J., and Van Golde, L. M.,** Lysophosphatidylcholine acyltransferase and lysophosphatidylcholine:lysophophatidylcholine acyltransferase in alveolar type II cells from fetal rat lung, *Biochim. Biophys. Acta,* 833, 93, 1985.
74. **Van Heudsen, G. P. H., Noteborn, H. P., and Van den Bosch, H.,** Selective utilisation of palmitoyl lysophosphatidylcholine in the synthesis of disaturated phosphatidylcholine in rat lung: a combined *in vitro* and *in vivo* approach, *Biochim. Biophys. Acta,* 664, 49, 1981.
75. **Van Heusden, G. P. H., Reutelingsperger, C. P. M., and Van den Bosch, H.,** Substrate specificity of lysophospholipase transacylase from rat lung and its action on various physical forms of lysophosphatidylcholine, *Biochim. Biophys. Acta,* 633, 22, 1981.
76. **Van den Bosch, H., Vianen, G. M., and Van Heudsen, G. P. H.,** Lysophospholipase-transacylase from rat lung, *Methods Enzymol.,* 71, 513, 1981.
77. **Weltzien, H. U., Richter, G., and Ferber, E.,** Detergent properties of water-soluble choline phosphatides: selective solubilization of acyl-CoA: lysolecithin acyltransferase from thymocyte plasma membrane, *J. Biol. Chem.,* 254, 3652, 1979.
78. **Yamashita, S., Hosaka, K., and Numa, S.,** Acyl donor specificities of partially purified 1-acylglycerophosphate acyltransferase, 2-acyl glycerophosphate acyltransferase and 1-acyl glycerophosphocholine acyltransferase from rat liver microsomes, *Eur. J. Biochem,* 38, 25, 1973.
79. **Hasegawa-Saski, H. and Ohno, K.,** Extraction and partial purification of acyl-CoA:1-acyl-*sn*-glycero-3-phosphocholine acyltransferase from rat liver microsomes, *Biochim. Biophys. Acta,* 617, 205, 1980.
80. **Gavino, V. C. and Deamer, D. W.,** Purification of acyl-CoA:1-acyl-*sn*-glycero-3-phosphorylcholine acyltransferase, *J. Bioeng. Biomembr.,* 14, 513, 1982.
81. **Deamer, D. W. and Gavino, V. C.,** Lysophosphatidylcholine acyltransferase: purification and application in membrane studies, *Ann. N.Y. Acad. Sci.,* 414, 90, 1983.
82. **Deka, N., Sun, G. Y., and MacQuarrie, R.,** Purification and properties of acyl-CoA:1-acyl-*sn*-glycero-3-phosphocholine-*O*-acyltransferase from bovine brain microsomes, *Arch. Biochem. Biophys.,* 246, 554, 1986.
83. **Reitz, R. C., El-Sheikh, M., Lands, W. E. M., Ismail, I. A., and Gunstone, F.D.,** Effects of ethylenic bond position upon acyltransferase activity with isomeric *cis*-octadecenoyl coenzyme A thiol esters, *Biochim. Biophys. Acta,* 176, 480, 1969.

Chapter 7

PHOSPHATIDYLETHANOLAMINE *N*-METHYLTRANSFERASE

Neale D. Ridgway

TABLE OF CONTENTS

I. INTRODUCTION

Phosphatidylcholine (PC), the major phospholipid of mammalian cells, is synthesized primarily by the CDP-choline pathway.[1] There is a second route for PC synthesis which involves *N*-methylation of phosphatidylethanolamine (PE). Successive transfer of methyl groups from *S*-adenosyl-L-methionine (AdoMet) generates the intermediates phosphatidyl-*N*-monomethylethanolamine (PMME) and phosphatidyl-*N,N*-dimethylethanolamine (PDME), and finally PC. PE methylation is quantitatively important only in liver where it accounts for 20 to 40% of PC synthesis.[2] PE methylation activity in other organs is very low.[3] The reason for this disparity in distribution and the function of PE methylation in other organs is unknown.

PE *N*-methyltransferase (EC. 2.1.1.17), like other lipid biosynthetic enzymes, catalyzes a reaction involving an amphiphilic substrate at a membrane interface and is itself tightly associated with phospholipid bilayers. These two features make the methyltransferase extremely refractory to biochemical analysis. However, purification of the enzyme[4] and the advent of models for studying enzyme catalysis in phospholipid/detergent mixed micelles[5,6] should make study of this enzyme more amenable.

II. CHARACTERIZATION OF PE *N*-METHYLTRANSFERASE

A. ELUCIDATION OF THE PE METHYLATION PATHWAY

As early as 1941 it was recognized that the methyl group of methionine was utilized for the synthesis of choline.[7] Stetten later demonstrated that [^{15}N]ethanolamine feeding to rats resulted in production of labeled "choline phosphatide".[8] The result was interpreted to indicate that ethanolamine was methylated to choline prior to incorporation into PC. In 1959, Bremer and Greenberg showed that the primary acceptor of methyl groups was phosphatidylethanolamine and not the water-soluble precursors of the CDP-ethanolamine pathway.[9] In addition, AdoMet was demonstrated to be the immediate donor of methyl groups to PE.[10,11] The majority of PE methylation activity was associated with the $100,000 \times g$ fraction of rat liver homogenate,[10,11] and >95% of the radiolabeled product was PC, with the remainder in PMME and PDME.[9,12] On the basis of product distribution, the addition of the first methyl group to PE was deemed to be rate limiting.[10] Rehbinder and Greenberg found that methylation could be stimulated two- to threefold by the addition of PMME and PDME to deoxycholate solubilized microsomes.[13] Oddly, PE did not stimulate methylation.[10,13,14] In retrospect, this was due to improper delivery of substrate to the enzyme, since it has been reported that the Triton X-100-solubilized enzyme methylates exogenous PE.[4,15]

Enzyme activity is dependent on a free sulfhydryl(s) for activity.[13,14] Another interesting feature of PE *N*-methyltransferase is its alkaline pH optimum (10 to 10.5).[12,13] This optimum could be a reflection of the ionization state of the ethanolamine headgroup of PE (pKa approximately 9.5) or of catalytic residues in the active site of the enzyme. Whether this alkaline pH optimum has physiological significance is unknown, but methylation at pH 7.5 occurs at about one tenth the rate of that at pH 10.[13,16,17] The microsomal enzyme is insensitive to EDTA[14] and does not require Mg^{2+} for activity.[18]

B. TISSUE AND SUBCELLULAR LOCALIZATION

Bremer and Greenberg were the first to show that liver had the highest methyltransferase specific activity.[14] Kidney, heart, lung and testis microsomal fractions all contained measurable but low activity (2 to 6% of liver). Brain, spleen, and intestine were devoid of methylation activity.[14] Since this original observation, numerous studies on extrahepatic PE *N*-methyltransferase have been published. In all instances the specific activity is low compared to liver. The localization, specific activity, and cofactor requirements of these extrahepatic activities have been summarized elsewhere.[19] The low activity expressed in extrahepatic tissues may be the result of differential gene expression. However, no reports exist on enzyme mass in these tissues, so tissue-specific effectors or inhibitors cannot be eliminated. It is clear that PE methylation in extrahepatic tissues contributes insignificantly to PC synthesis.

Subcellular analysis of PE *N*-methyltransferase activity in rat liver has often produced conflicting results. Similar to earlier findings,[13,14] Van Golde et al. reported that the majority of PDME methylation activity was confined to endoplasmic reticulum.[20] Unfortunately, no information was provided on endogenous PE methylation. Contrary to this report, Higgins and Fieldsend found twice the PE *N*-methyltransferase activity in "*cis*-enriched" Golgi compared to endoplasmic reticulum or "*trans*-enriched" Golgi.[21] A systematic evaluation of PE *N*-methyltransferase activity in Golgi revealed that results varied depending on the fractionation techniques used.[22] Thus, all fractionation procedures enriched for *trans*-Golgi, but methyltransferase specific activities ranged from 0 to 0.8 nmol methyl groups transferred per minute per milligram protein. It was estimated that Golgi contained 2.3% of the total cellular methyltransferase activity.[22] In this same report, mitochondria and plasma membranes had PE *N*-methyltransferase activity above that due to contamination from endoplasmic reticulum. Localization of PE *N*-methyltransferase in subcellular organelles by immunocytochemistry would be the method of choice for future studies.

C. TOPOLOGY IN MEMBRANES

The enzymes of glycerol lipid synthesis are asymmetrically distributed to the cytosolic surface of the microsomal membrane.[23,24] PE-, PMME-, and PDME-dependent PE *N*-methyltransferase activities were all degraded by trypsin treatment of intact microsomes.[25] Contrary to this, it was reported that PE was converted to PMME on the microsomal lumen and subsequently converted to PC by a second methyltransferase on the cytosolic surface.[26] This conclusion relied on the use of phospholipase C as a probe for localizing PC, PMME, and PDME, but the authors gave no evidence that PMME or PDME were hydrolyzed by the phospholipase C.[26] The use of phospholipases as probes for membrane asymmetry has drawbacks which include perturbing effects of products on membrane structure and differential hydrolysis due to the heterogenous nature of membranes.[27] The fact that PMME and PDME were "sequestered" from phospholipase C,[26] even when the microsome was disrupted, suggests that these intermediates remain bound to PE *N*-methyltransferase until completely methylated. Because of the technical problems associated with the use of phospholipases, conclusions on asymmetric distribution of two methyltransferases in liver are unfounded.

D. PURIFICATION

PE *N*-methyltransferase is tightly associated with membranes, no doubt due to one or more hydrophobic membrane-spanning domains. Because of its integral nature, purification requires the use of amphiphiles or detergents to solubilize PE *N*-methyltransferase from membranes in a form free from lipids and other proteins. Any purification of a bioactive protein requires that a balance be struck between optimal solubilization with detergent and maintenance of biological activity. Selection of the proper detergent to achieve this end is largely empirical.

A variety of detergents have been used to solubilize PE *N*-methyltransferase.[18] Most purifications have utilized liver microsomes as starting material for the logical reason that liver contains the highest methyltransferase specific activity. Schneider and Vance reported that sonication of rat liver microsomes in the presence of 0.2% Triton X-100 released 44% of methyltransferase activity.[18] Attempts at purifying the soluble enzyme were partially successful (28-fold purification from homogenate). A similar solubilization of PE *N*-methyltransferase from mouse liver microsomes (using 0.2% Triton X-100) was reported at the same time.[15] Percy et al. also showed that the calf brain enzyme could be solubilized with Triton X-100.[28] Partial purification of PE *N*-methyltransferase from mouse thymocytes using Triton X-100 has also been reported.[29] The activity in thymocytes is so low that the 1500-fold purified (PE-dependent activity) enzyme had a specific activity similar to that in rat liver microsomes.

The first substantial purification (200-fold) of methyltransferase from 3-([3-cholamidopropyl]dimethylammonio)-1-propane-sulfonate (CHAPS)-solubilized rat liver microsomes was achieved by Pajares et al.[30] The alleged methyltransferase had a molecular mass of 25 kDa and was shown to be photoaffinity labeled with [*methyl*-^{3}H]8-azido-AdoMet. Oddly, the authors showed that the enzyme also methylated fatty acids.[30] Shortly after this, a paper was published by the same group now claiming a 50-kDa protein was PE *N*-methyltransferase.[31] By a series of experiments showing association of denatured, sodium dodecyl sulfate(SDS)-solubilized proteins they concluded that the 50-kDa protein was actually a dimer of the 25-kDa protein described in the earlier report.[30] In both of these publications little direct evidence (such as cochromatography of protein and activity) was provided that either a 25- or 50-kDa protein is PE *N*-methyltransferase.[30,31] Because CHAPS was removed after the first DEAE chromatography step in these purifications, the result was isolation of a large lipid-protein aggregate or "membrane fragment".

More recently, PE *N*-methyltransferase has been purified to homogeneity from Triton X-100-solubilized rat liver microsomes.[4] The enzyme had a molecular mass of 18.3 kDa as determined by SDS-polyacrylamide gel electrophoresis, and gel filtration analysis indicated a stoichiometry of one methyltransferase per Triton X-100 micelle.[4] This single protein catalyzed the methylation of PMME and PDME as well as the complete conversion of PE to PC. The pH optimum for methylation of these three lipid substrates was 10. Also, it was found that the enzyme had an extremely basic pI (>9.5).

In relation to previous purification attempts,[30,31] it was noted that the major protein in the partially pure preparations had a molecular mass of 50 kDa.[4] However, this 50-kDa protein was completely absent from the purified enzyme. In addition, a polyclonal antibody raised against the 18.3-kDa methyltransferase did not cross-react with the 50-kDa contaminant when rat liver microsomes were analyzed by immunoblotting.[32] Thus, it is doubtful that the 50-kDa protein bears any relation to the actual PE *N*-methyltransferase.

E. MOLECULAR STRUCTURE AND KINETICS

The fact that PE methylation occurs via a three-step process immediately raises the question as to the number of enzymes involved. Those researchers that originally characterized liver PE methylation[13,14] and enzyme purification results inferred that a single enzyme was responsible for all methylation events.[4,18,30,31] Prior to purification of the enzyme,[4] Axelrod and co-workers provided evidence for a two-enzyme model in bovine adrenal medulla,[33] erythrocyte membranes,[34] rat liver microsomes,[35] rat basophilic leukemia cells,[36,37] rat brain synaptosomes,[38] and reticulocyte ghosts.[39] The model proposed that a unique methyltransferase catalyzed the synthesis of PMME from PE (methyltransferase I) and that a second enzyme added the last two methyl groups to form PC (methyltransferase II). The majority of the evidence was derived from kinetic data, pH optima, and divalent cation requirements. Methyltransferase I had a reported pH optimum of 6.5 to 8, a K_m for AdoMet of 0.6 to 4 μM and a requirement for Mg^{2+}. Methyltransferase II showed a pH optimum of 10 to 10.5, a K_m for AdoMet of 67 to 110 μM, and no cation requirement.[34-35,38,40-42] Lineweaver-Burke plots of initial velocity data at increasing AdoMet concentrations (at pH 8) had a high- and low-K_m component.[35,38,40,41] The Mg^{2+} requirement was often variable and, depending on the tissue, absolute requirements or only 15% stimulation in activity was demonstrated.[33,38] It was reported by Prasad and Edwards that sonication of rat pituitary extracts partially solubilized methyltransferase I.[40] Contrary to this, Crews et al.[38] reported that sonication of rat brain synaptosomes released methyltransferase II into a 100,0000 × *g* supernatant fraction.

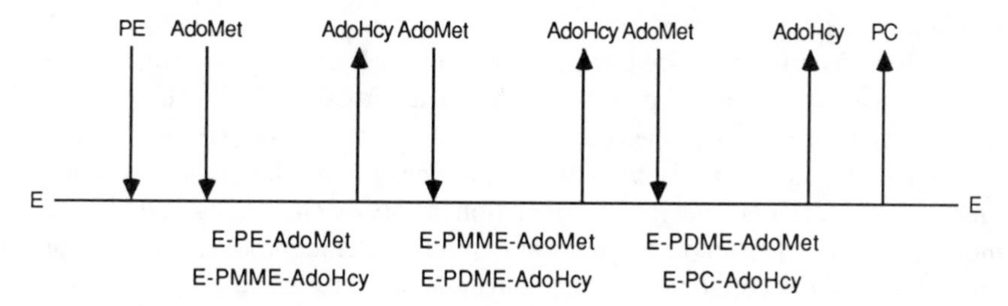

FIGURE 1. A concerted mechanism foe conversion of PE to PC.

The majority of studies indicating two methyltransferases based conclusions on the accumulation of radioactivity in PMME and PDME, both of which are intermediates in PE to PC conversion. Audubert and Vance pointed out that PE methylation is a steady-state process and intermediates in the pathway (PMME and PDME) rapidly reach constant levels in 1 min, after which time only the amount of the end product (PC) increases.[17] Thus, measurements of radioactivity in PMME and PDME are only an assessment of their steady-state levels. Based on this concept, formulas were derived that estimate the actual amounts of PMME, PDME, and PC formed during methylation of microsomal PE. K_m values for AdoMet for the methylation of PE, PMME, and PDME were reported to be 50 to 100 μM, and the pH optimum for all three methylations was 10.5.[17] Although PE *N*-methyltransferase has not been purified from extrahepatic tissues, the failure to recognize the steady-state nature of the PE methylation pathway, and the resultant inaccurate measurements of PMME and PDME, means that the two-enzyme model is incorrect.

Kinetic analysis of a multistep pathway, whether catalyzed by one or more enzymes, is very complex. While the formulas to determine the flux of label through intermedates will determine the K_m for AdoMet, elucidation of the kinetic mechanism will require a different approach. The best approach would involve assaying the individual steps with PMME and PDME, since in both cases only one methyl addition occurs to yield PDME and PC, respectively.[17,18,28] Assaying the methylation of PE to PMME poses a problem, for no discrete assay is yet available, due to the rate-limiting nature of this first step. The kinetic mechanism for purified rat liver PE *N*-methyltransferase in phospholipid/Triton X-100 mixed micelles has recently been solved.[43] As shown in Figure 1 the pathway follows a concerted mechanism with a common substrate binding site for all three lipid substrates. The individual methylation of PMME and PDME follows an ordered Bi-Bi mechanism, with lipid substrate the first ligand to bind and lipid product the last ligand to dissociate.[43] PE and the methylated intermediates remain bound to the enzyme throughout the entire reaction, and all that is necessay is repeated binding of AdoMet and dissociation of AdoHcy. Similar to the liver enzyme, a Bi-Bi mechanism for PMME and PDME methylation has been proposed for red blood cell membrane PE *N*-methyltransferase, but it was observed that AdoMet was the first substrate bound.[44]

The kinetic properties of purified PE *N*-methyltransferase are quite similar to those found for other enzymes that act on mixed micellar substrates:[5,6] methylation is subject to surface dilution inhibition,[4] double reciprocal plots at variable lipid substrate were nonlinear,[4,43] and the enzyme was activated by its substrates and by PC.[43] The activation by PC, and perhaps by other lipids, seems to be related to a requirement for a full boundary layer of phospholipid.[43]

F. SUBSTRATE SPECIFICITY

Specificity can imply both selectivity for different molecular species of the primary substrates PE, PMME, and PDME (i.e., varying in fatty acid composition) or for compounds with widely divergent structural properties, but with a primary-, secondary-, or tertiary-amino group. Molecular species specificity will be discussed first.

1. Molecular Species Specificity

All reports to date have utilized argentation thin-layer chromatography as a method for separating molecular species of phospholipids. Thus "fractions" with the same number of double bonds, and not individual species, are resolved. Most reports on specificity have involved *in vivo* or *in vitro* labeling with [*methyl*-^{3}H]-methionine or -AdoMet, respectively, and analysis of the labeled PC molecular species by argentation thin-layer chromatography. *In vivo*, di- and tetraenoic or tetra- and hexaenoic fractions of PC were the chief products of PE methylation in rat liver.[45-48] Arvidson reported that [1,2-^{14}C]ethanolamine labeled hexaenoate-rich PE was the primary substrate for methylation.[49] This conclusion has since been shown to be incorrect, and the general consensus is that methylation of any particular PE species is dependent on its concentration and not fatty acid composition.[45,48,50] The hexaenoic fraction of PC synthesized by methylation had the highest initial specific activity and appeared to turn over rapidly.[45] Fractionation of the molecular species products of purified PE *N*-methyltransferase by reverse-phase high performance liquid chromatography revealed that, like the microsomal activity, the enzyme displays no specificity.[50] However, in intact hepatocytes the labeling pattern of newly formed PC via PE methylation varied slightly from total hepatocyte PE and hepatocyte microsomal PE.[50] This PE-derived PC, which is enriched in 1-palmitoyl-2-docosahexaenoyl-PC, underwent rapid remodeling of its acyl chains such that 18 h following synthesis it resembled cell PC in molecular species composition.[50]

Analysis of PE *N*-methyltransferase molecular specificity in rat brain revealed polyunsaturated-rich species of PC to be the major methylation products. *In vitro* labeling of synaptosomal PC and *in vivo* labeling of whole rat brain PC showed tetra-, penta-, and hexaenoic species of PC to be the primary products of methylation.[51,52] Similar to liver studies, there was a pronounced turnover of hexaenoic PC species and a concomitant increase in percent label in more saturated PC fractions.[52]

2. Head-Group Specificity

There is a paucity of information on potential substrates other then PE, PMME, and PDME that are methylated by PE *N*-methyltransferase. The problem has been the unavailability of pure enzyme; thus, activities in crude membranes cannot be ascribed to PE *N*-methyltransferase with complete certainty. It has been reported that rat liver microsomes possess an activity that methylates ceramide-phosphoethanolamine.[53] A similar activity exists in rat brain microsomes.[53] Contrary to this report, LeKim et al.[54] reported no hepatic methylation in rats of intravenously injected ceramide-phospho-*N*,*N*-dimethylethanolamine. Lyso-PMME and lyso-PDME, but not lysoPE, are methylated by liver microsomes.[57] These two lyso lipids compete with their diacyl analogues, suggesting that methylation is catalyzed by the same enzyme. Clearly, more work is required (preferably with purified enzyme) to assign these activities to PE *N*-methyltransferase.

Modification of the base moiety of PE has given some insight into PE *N*-methyltransferase specificity. Incubation of hepatocytes with monoethylethanolamine, 2-aminopropanol, and 2-aminobutanol resulted in their incorporation into phosphatides and the resultant addition of a single methyl group.[56] Phosphatidyl-diethanolamine, -diethylethanolamine, and -dimethylaminopropan-2-ol were not methylated by microsomes.[56] It was reported that *N*-isopropylethanolamine is converted to phosphatidyl-*N*-isopropylethanolamine *in vivo* and subsequently undergoes a single methylation.[57]

The structural features of AdoMet essential for methyl transfer are not well defined. An early study indicated that the ethyl sulfonium analogue of AdoMet, *S*-adenosyl-L-ethionine (AdoEt), was a competitive inhibitor of PE methylation.[14] Incorporation of ethyl groups into PE was not assessed. When [*ethyl*-^{14}C]ethionine was injected into rats, label was recovered in liver PC, but only 1.5% compared to that recovered in PC following [*methyl*-^{14}C]methionine injection.[58]

Apparently, AdoEt can occupy the AdoMet active site on PE *N*-methyltransferase, but ethyl transfer occurs only slowly.

III. FUNCTIONS OF PE *N*-METHYLTRANSFERASE IN LIVER

A. CONTRIBUTION TO PC SYNTHESIS

It is commonly accepted that PE methylation accounts for 20 to 40% of PC synthesis in the hepatocyte.[2] An earlier report that monitored PE methylation from [^{3}H]ethanolamine-labeled PE estimated that 3 to 10% of PC was derived by methylation.[59] Because of difficulties in estimating rates of PC formation by various pathways, these figures must be considered soft estimates.

B. AS A SOURCE OF PC IN LIPOPROTEINS

Since methylation of PE contributes significantly to the total cellular pool, it is feasible that this PC has some preordained function. Both very low density lipoproteins (VLDL) and high density lipoproteins (HDL) are actively secreted by hepatocytes.[60] The surface monolayer of these lipoproteins is composed primarily of the choline-containing lipids, PC and sphingomyelin, which function to solubilize the apolar core (triglyceride and cholesterol ester) components of these particles.[61] The PC required for assembly and secretion of VLDL and HDL from hepatocytes in monolayer culture seems to be synthesized by both the CDP-choline and methylation pathways.[62] PC made via methylation of ethanolamine-derived PE is secreted in lipoproteins. However, cellular specific activity of this PC was found to be twofold greater than that secreted into hepatocyte culture medium.[62] The relative lack of input into lipoprotein PC by PE methylation was further demonstrated by treating hepatocytes with deazaadenosine (DZA). This adenosine analogue raises cellular AdoHcy pools and inhibits methylation of PE, but does not affect lipoprotein secretion.[63] Oddly, DZA did not inhibit the conversion of [3-^{3}H]serine-labeled PE into secreted PC,[64] and this serine-derived PC was preferentially and rapidly (30 min) secreted in lipoproteins.[62] These results indicate that methylation of phosphatidylserine-derived PE is performed by a AdoHcy-insensitive enzyme, or the methylation activity is somehow sequestered from this inhibitor. This rapid secretion of phosphatidylserine-derived PE and PC is even more peculiar since methylation and decarboxylation occur in separate organelles, and transit half-time from endoplasmic reticulum to mitochondria is estimated at 6 h.[65] This is not the only interpretation since the labeled serine used in these studies will label the AdoMet pool via tetrahydrofolate.[62,63] Therefore, a large proportion of the label in PC could be in the methyl group and not the ethanolamine moiety.[66]

It appears that PE methylation is not required for lipoprotein secretion. However, it is worth noting that impaired VLDL secretion from choline- and methionine-deficient hepatocytes is reversed by supplementation of methionine alone.[67]

C. SYNTHESIS OF SPECIFIC POOLS OF PC

Saturates, monoenes, dienes, and trienes are the major molecular species of PC formed by the CDP-choline pathway.[68,69] However, as discussed in a previous section, PC formed via methylation of PE is rich in 1-palmitoyl-2-docosahexaenoyl-PC.[50] It is tempting to speculate that PE *N*-methyltransferase supplies the cell with PC enriched in unsaturated species by virtue of the highly unsaturated nature of PE fatty acids.[69] There is still little direct proof for this hypothesis apart from specificity data on PE *N*-methyltransferase and cholinephosphotransferase (Chapter 4). The deacylation-reacylation and CDP-choline pathways in microsomes can synthesize most of the cellular arachidonate- and docosahexaenoate-containing PC molecular species, respectively.[70] It seems unlikely that PE methylation, with its relatively minor contribution to PC synthesis, is responsible for maintenance of any one pool of PC molecular species. However, a specific, localized function for PE-derived PC cannot be discounted.

IV. REGULATION OF PE *N*-METHYLTRANSFERASE

A. REGULATION BY SUBSTRATE LEVELS

Incubation of hepatocytes with 0.5 m*M* ethanolamine produced a 75% increase in methylation of PE and an increase in cellular PE from 20 to 30% of total phospholipid.[71] The effect of ethanolamine supplementation on PE *N*-methyltransferase activity assayed *in vitro* was less striking, but some stimulation in activity was observed.[71] Experiments showed that reduction of the endogenous PE concentration (by treatment with the amino group-blocking reagent methylacetimidate) in microsomes caused a simultaneous reduction in methylation rates.[56] A similar regulation of PE *N*-methyltransferase activity by PE is found in the choline-deficient rat liver and in choline-deficient hepatocytes.[32] Alterations in PE *N*-methyltransferase activity in microsomes from choline-deficient rat livers measured *in vitro* correlated with increased PE concentrations, and not to alterations in enzyme protein mass (as determined by immunoblotting with an anti-PE *N*-methyltransferase antibody).[32] Similar experiments in choline-deficient hepatocytes showed that methionine supplementation decreased PE-dependent methyltransferase activity by twofold.[32] The mass of PE *N*-methyltransferase protein and activity assayed in the presence of saturating concentrations of methyl acceptor (PMME) were not altered by methionine supplementation. The PE/PC ratio is probably the important parameter for determining the actual mole percent or surface concentration of PE in membranes. The effect of PC on PE methylation in intact membranes has not been demonstrated. However, PC was shown to stimulate the methylation activities of the pure enzyme in Triton X-100 micelles.[43]

Supplementation of cultured cells with monomethylethanolamine and dimethylethanolamine resulted in the synthesis of the PE *N*-methyltransferase substrates PMME and PDME.[2,72-74] In LM cells cultured with either of these two bases up to 60% of cellular lipid was PMME or PDME.[72] Because LM cells are deficient in PE methylation, little of the PMME and PDME is converted to PC. In hepatocytes, monomethylethanolamine and dimethylethanolamine supplementation raised the phosphatides of these bases to 10 to 20% of the total phospholipid and stimulated methylation by 40%.[2] Since PMME and PDME are trace components in hepatic phospholipids,[75] these studies have little physiological relevance.

In addition to PE, AdoMet and AdoHcy (or more importantly the ratio of the two) modulate PE methylation. Incubation of hepatocytes with 0.1 mM methionine stimulated PE to PC conversion by twofold.[2] Higher concentrations had no further effect. Normal perfused rat liver has an AdoMet concentration of 32 nmol/g of tissue.[76] This level was elevated to 120 and 300 nmol/g tissue when 0.05 and 2.25 m*M* methionine, respectively, were included in the perfusate.[76] This increase in AdoMet levels was not accompanied by increased AdoHcy. However, perfusion of livers with 3.4 m*M* homocysteine and 4.0 m*M* adenosine elevates AdoHcy levels from 8 to 4000 nmol/g tissue and AdoMet to 1250 nmol/g tissue.[76] This change in AdoMet/AdoHcy ratios from 5.6 to 0.3 resulted in a complete abolition of PE methylation.

Further insights on the influence of AdoMet/AdoHcy on PE methylation have been made using a variety of AdoHcy analogues or compounds that raise intracellular AdoHcy levels. The most widely used of these compounds is DZA, a compound that potently inhibits AdoHcy hydrolase (I_{50} = 0.008 m*M*) and elevates cellular AdoHcy.[77,78] Intraperitoneal injection of DZA into rats decreased the AdoMet/AdoHcy ratio from 4.5 to 1.6 in 4 h.[78] 3-DZA-AdoHcy levels were found to be similar to AdoHcy (151 compared to 125 nmol/g liver, respectively). Administration of DZA to rats caused a 90% decrease in [*methyl*-^{3}H]methionine labeling of PC.[78] Thus, inhibition of PE methylation *in vivo* is due to appearence of two competetive inhibitors (AdoHcy and DZA-AdoHcy) of PE *N*-methyltransferase.[78,79] Activity of PE *N*-methyltransferase in isolated microsomes from DZA-treated rats was not affected.[79]

The AdoHcy analogues *S*-7-deazaadenosyl-homocyteine and 5′-deoxy-5′(1,4-diaminopentanoic acid)adenosine (Sinefungin) also inhibit PE methylation (the latter only *in vitro*).[80] Other adenosine analogues, 9-β-D-arabinofuranosyladenine, 5′-deoxy-5′-isobutylthioadenosine (SIBA), and A([-]-9-*(trans*-2,*trans*-3-dihydroxy-4-[hydroxymethyl]cyclopent-4-enyl)adenine (Neplanocin A), are potent inhibitors of PE methylation *in vivo* and appear to raise cellular AdoHcy levels in a manner analogous to DZA.[63,80]

B. HORMONAL EFFECTS ON PE *N*-METHYLTRANSFERASE IN LIVER

The reported effects of various hormones on PE methylation in hepatocytes vary greatly. The majority of the studies have centered on the effects of glucagon, insulin, vasopressin, and angiotensin on microsomal PE *N*-methyltransferase activity and *in vivo* conversion of PE to PC measured by [*methyl*-^{3}H]methionine labeling.

Geelen et al.[81] were the first to demonstrate that pretreatment of hepatocytes with glucagon for 3 h caused a small (20%) increase in PE to PC conversion. Since glucagon also enhanced PE synthesis, the effect on methylation may have been due to an expanded substrate pool.[81] Experiments performed in a similar manner showed a 33% reduction in PE to PC conversion after 2.5 h in the presence of 100 n*M* glucagon.[82] Incubation of hepatocytes with the cAMP analogue chlorophenylthio-cAMP reduced by 50% the incorporation of [1-^{3}H]ethanolamine-labeled PE into PC.[83] cAMP analogues and glucagon also inhibit PC production from choline.[82,84] In the two reports that showed *in vivo* inhibition of PE methylation by glucagon and cAMP analogues,[82,83] PE *N*-methyltransferase activity in a microsomal fraction was unchanged or stimulated twofold, respectively. Castano et al.[85] reported a twofold stimulation of PE *N*-methyltransferase activity in homogenates from hepatocytes treated with glucagon. In accord with a role for cAMP-dependent phosphorylation, treatment of rat liver microsomes with cAMP and ATP caused a twofold stimulation in enzyme activity.[86] In opposition to this result, Pelech et al.[87] reported no effect of cAMP-dependent protein kinase, protein phosphatase 1 or 2A, casein kinase II, nor calmodulin-dependent protein kinase on PE *N*-methyltransferase activity in microsomes. However, ATP or GTP and cytosol caused a 50% increase in activity.[87]

Results from various laboratories, though quite contradictory, suggested that cAMP-dependent phosphorylation was activating PE N-methyltransferase. A 50-kDa protein was phosphorylated *in vitro* on a serine residue by cAMP-dependent kinase with a resultant fourfold increase in PE methylation activity.[88] AdoMet seemed to activate phosphorylation or inhibit dephosphorylation of methyltransferase in these preparations.[89] Also, the 50-kDa protein was immunoprecipitated from glucagon-treated hepatocytes and shown to have incorporated ^{32}P.[90] Protein kinase C also activated PE *N*-methyltransferase and phosphorylated the 50-kDa protein.[91] Since we now know that this 50-kDa protein is a contaminant and not PE *N*-methyltransferase, it would be wise to reexamine the role of reversible phosphorylation in regulation of PE methylation. It is still possible that the enzyme is phosphorylated, but due to the impurity of enzyme prepartions, no label could be detected in the 18.3-kDa PE *N*-methyltransferase. Indeed, we have recently found that cAMP-dependent protein kinase will phosphorylate the 18.3-kDa methyltransferase *in vitro*.[92]

Vasopressin, angiotensin, and the Ca^{2+} ionophore A23187 inhibited choline labeling of PC,[93] but activated by twofold PE *N*-methyltransferase in hepatocyte homogenates.[94] Ca^{2+} and ATP also stimulated microsomal methyltransferase activity twofold, presumably through the mediation of calmodulin.[95] Apparently, the β-adrenergic receptor agonist isoprenaline also stimulated PE *N*-methyltransferase activity in hepatocytes from adrenalectomized rats to a greater extent than controls.[96,97] Again, this activition appeared to be mediated by cAMP.

Insulin had no effect on PC synthesis from choline[82] or via methylation of PE.[98] However, insulin was reported to inhibit the glucagon-dependent stimulation of methyltransferase activity in hepatocytes.[98]

C. HORMONAL REGULATION OF PE *N*-METHYLTRANSFERASE IN THE ADIPOCYTE

The PE *N*-methyltransferase of adipocytes seems to be activated by cAMP-dependent mechanisms in a manner similar to that of hepatocytes. Accordingly, adrenocorticotropin,[99] epinephrine,[100] isoproterenol,[100] and forskolin[100,101] were all shown to stimulate methylation activity in intact adipocytes. Administration of oxytocin, a hormone that does not act via cAMP,[102] to adipocytes resulted in a time- and dose-dependent 1.5-fold activation of PE methylation.[100] Phorbol 12-myristate-13-acetate was reported to stimulate methylation activity twofold in rat adipocytes, presumably via protein kinase C.[101]

Insulin was reported to abolish the effects of various hormones acting through cAMP-dependent mechanisms,[99,100] and by itself inhibited PE methylation in intact adipoctyes by 40%.[99] Strangely, the same group reported that insulin stimulated PE *N*-methyltransferase activity in isolated adipocyte plasma membranes.[103] Isoproterenol, as well as stimulating activity, was recently shown to enhance phosphorylation of a 50-kDa protein in a manner analogous to hepatocytes.[104] Insulin and a phospho-oligosaccharide (prepared from rat liver membranes by PI-specific phospholipase C digestion) both inhibited isoproterenol-dependent phosphorylation.[104] The phospho-oligosaccharide, presumed to be cleaved from its lipid moiety by an insulin-responsive phospholipase C, is proposed to serve as a second messenger of insulin action.[105] Considering that the 50-kDa protein is not PE *N*-methyltransferase, one should be wary of these reports on enzyme phosphorylation and effects on activity.

D. OTHER EFFECTORS

There appear to be cytosolic factors that influence PE *N*-methyltransferase activity, but they remain unidentified.[87] A heat-stable, low molecular weight inhibitor has been identified in rat liver cytosol.[87,106] This inhibitor may be similar to a 25-amino acid peptide isolated from rabbit liver cytosol that inhibits various methyltransferases.[107]

Unsaturated fatty acids are potent inhibitors of PE methylation in intact hepatocytes and microsomes.[108] Long-chain fatty acyl-CoA esters were also effective inhibitors, and inhibition was reversed upon addition of bovine serum albumin.

V. DEVELOPMENTAL REGULATION OF PE METHYLATION

Two reports on PE *N*-methyltransferase in pre- and postnatal rat liver demonstrated a steady rise in activity from –5 to +15 or +20 d and a slow decline to adult values thereafter.[109,110] Activities in prenatal rabbit livers were about 33% lower than values at birth and reached a maximum at +14 d.[111] Similarly, rat brain PE *N*-methyltransferase activity was demonstrated to be the highest between day 5 and 30.[112] Since enzyme mass was not correlated with activity, it is feasible that activity changes noted after birth could be due to altered PE or AdoMet/AdoHcy levels.

VI. COORDINATE REGULATION WITH THE CDP-CHOLINE PATHWAY

Several examples of coordinate regulation of the two major pathways for PC synthesis have been demonstrated. Unsaturated fatty acids, while potently inhibiting PE methylation,[108] were found to stimulate PC production from choline in cultured rat hepatocytes.[113] This enhanced synthesis via the CDP-choline pathway was correlated with translocation of phosphocholine cytidylyltransferase activity from cytosol to microsomes.

The elevated cellular levels of AdoHcy caused by DZA treatment inhibited PE methylation,[78] but caused a two- to threefold increase in PC synthesis via the CDP-choline pathway and a threefold increase in microsomal cytidylyltransferase activity.[79]

Maintenance of rats on a choline-deficient diet caused various metabolic perturbations including fatty liver, decreased circulating lipoprotein levels, increased hepatic PE, and decreased hepatic PC.[114,115] Accompanying these changes was a twofold elevation in *in vitro* PE *N*-methyltransferase activity and a twofold reduction in cytosolic cytidylyltransferase activity.[116,117] However, Yao et al. showed that microsomal cytidylyltransferase activity was elevated in choline- and methionine-deficient hepatocytes.[118] As mentioned previously, the elevated *in vitro* activity of PE *N*-methyltransferase is the direct result of increased PE concentrations in choline deficient liver membranes.[32] However, the *in vivo* activity of PE *N*-methyltransferase is depressed in choline deficiency since most of the diets are methionine poor.[119] The direct result is low AdoMet levels and an elevated AdoHcy/AdoMet ratio.[120]

VII. PE METHYLATION IN OTHER ORGANISMS

A. *NEUROSPORA CRASSA* AND *DICTYOSTELIUM DISCOIDEUM*

Two mutant strains of the mold *Neurospora crassa* were shown to have phospholipid compositions consistent with a requirement for two enzymes in PE to PC conversion.[121,122] Microsomes from one of these mutant strains appeared to be defective in PE methylation to PMME, while the second strain accumulated PMME, and thus may have a dèfect in the conversion of PMME to PC.[123] Partial purification of the PMME- and PDME-dependent activities was also consistent with the notion of two enzymes for PE to PC conversion in *N. crassa*.[124] It is possible that, like *Saccharomyces cerevisiae* (refer to Chapter 10 in this book), *N. crassa* has a PE methyltransferase that converts PE to PC and a second enzyme forms PC from PMME.

The slime mold *Dictyostelium discoideum* has been found to possess enzyme systems for the methylation of both neutral lipids and PE.[125] PE *N*-methyltransferase of *D. discoideum* appears to be composed of two enzymes, a conclusion based on the biphasic nature of reciprocal plots of AdoMet initial velocity data.[126] The major product of PE *N*-methyltransferase was reported to be PMME.[125,126] It has been suggested that cell aggregation (a stage in mold development), mediated by cellular cGMP levels that are elevated in response to occupation of cAMP receptors on the cell surface, is accompanied by a parallel stimulation of PE methylation.[127] PE *N*-methyltransferase activity in cell homogenates was enhanced about twofold upon treatment with cGMP and calmodulin.[126,127] However, PE methylation activity did not correlate with developmental changes mediated by cAMP, and instead seemed to be involved in membrane synthesis.[125]

B. *TETRAHYMENA PYRIFORMIS*

The ciliate protozoan *Tetrahymena pyriformis* is peculiar owing to the presence of 20% 2-aminoethylphosphonolipid (an analogue of PE with 2-aminoethylphosphonate as the base) in its phospholipid.[128] *Tetrahymena* contains about 29% PC, of which 60% is synthesized by PE methylation.[128] Interestingly, 2-aminoethylphosphonolipid (AEP-lipid) and *N,N*-dimethylaminoethylphosphonolipid are not substrates for methylation,[129,130] and *N,N,N*-trimethylaminoethylphosphonolipid (TMAEP-lipid) is not a natural component of *Tetrahymena* lipids.[128] *Tetrahymena* grown on medium containing TMAEP incorporates this base into TMAEP-lipid such that it represents 20% of the total phospholipid.[131] At this proportion of the total lipid, cell growth is seriously inhibited.[131] The toxicity of TMAEP-lipid makes it imperative that AEP-lipid is not methylated. Whether the mammalian PE *N*-methyltransferase requires a phosphoester bond between the head group and phosphate is yet unknown. AEP injected into rats was found in AEP-diglyceride, but the trimethyl derivative was not observed indicating the O-P bond is necessary for methylation.[132]

C. PE METHYLATION IN BACTERIA

As a general rule, bacteria do not contain PC or the partially methylated intermediates of the

PE methylation pathway.[133] There are some notable exceptions to this rule. Twelve strains of hydrogen-oxidizing bacteria were found to contain 23 to 47% PC,[134] 6 strains of the phototrophic purple bacteria (family Chromatiaceae) contained 10 to 17% of their total phospholipid as PC,[134] and several strains of methane- and methanol-utilizing bacteria also contain PC as well as PMME and PDME.[136,137] Phospholipids of thiobacilli were shown to be composed of PC, PMME, and PDME.[138] Eight strains of *Clostridium beijerinckii,* a butyric acid-producing bacteria, contained PMME but no other methylated phospholipids.[139] The source of the methyl groups in bacterial PC is methionine and AdoMet appears to be the direct methyl donor.[138,140] It has been proposed that the occurrence of PC among these bacteria is correlated with the presence of intracytoplasmic membrane systems or photosynthetic pigments.[134] Such structure-function relationships are tenuous at best, since within a particular bacterial genus some species contain PC while others do not. Goldfine noted that PC and PDME were found in bacteria rich in unsaturated fatty acids, a combination that would enhance membrane bilayer stability.[141]

Characterization of the PE methylating system of *Agrobacterium tumefaciens* revealed the presence of a soluble enzyme catalyzing the synthesis of only PMME and a particulate enzyme that produced primarily PC from endogenous PE.[142] The PE *N*-methyltransferase of *Rhodopseudomonas sphaeroides* was found to be exclusively (>90%) associated with a soluble cell fraction.[143] A single enzyme catalyzing all three methylations to PC was purified 1300-fold (PE-dependent activity) from Triton X-100-solubilized membranes of *Zymomonas mobilis.*[144] The putative enzyme is composed of a single subunit of 42 kDa.[144]

As yet, no choline nucleotide pathway has been identified for the synthesis of bacterial PC. This would indicate that of the two pathways for PC synthesis, PE methylation was the first to evolve. The prevalence of PMME and PDME (often in the absence of PC) in some bacterial strains suggests that the pathway evolved in a stepwise manner.[134,141]

VIII. PE METHYLATION AND TRANSMEMBRANE SIGNALING

During the course of studies on the role of protein methylation in the chemotactic response in various eukaryotic cells, the methyl label from methionine or AdoMet was observed in phospholipids, presumably by PE methylation, and the association between PE methylation and signal transmission across membranes was formulated.[33,145] Since the volume of reports linking PE methylation with signal transduction are too large to summarize individually, it will suffice to say that these reports based the second-messenger function of PE methylation on the observation that synthesis of methyl-labeled lipids preceded the appearance of cAMP, arachidonate, Ca^{2+}, or prostaglandins.[145] Also, DZA, 3-deaza-SIBA, and homocysteine-thiolactone, all inhibitors of PE methylation, appeared to attenuate the response of cells to various agonists.[145] Results from other laboratories tend to refute most of the conclusions reached by these workers.[146] The evidence against PE methylation in signal transduction was based on either no observed stimulation of methyl-labeled lipid synthesis or a lack of effect of methyltransferase inhibitors on response to various stimuli.

An interesting report by Moore et al.[147] demonstrated that stimulation of rat basophilic leukemia cells, mast cells, and mouse thymocytes, and subsequent release of histamine by a Ca^{2+}-dependent signal, was not coincident with PE methylation. Instead, the initial response to agonists was phosphatidylinositol (PI) hydrolysis to inositol polyphosphates and diglyceride, which are known to mediate intracellular release of Ca^{2+} [148,149] or activate protein kinase C,[149] respectively. Very recently, AdoHcy was shown to be a competitive inhibitor of PI kinase.[150] This inhibition translated into a 37% decrease in inositol bisphosphate levels and abolition of the chemotactic response in human polymorphonuclear leukocytes.[150] Thus, the inhibition of response to various effectors by elevation of AdoHcy levels is probably not related to an inhibition of PE methylation, but may be the result of blockage of the PI cascade in these cells.

IX. SUMMARY

More than 20 years of research on PE *N*-methyltransferase has not solved the question as to the function for this route of PC synthesis. PC is, of course, an essential component of a eukaryotic cell. However, does PE-derived PC have a more defined role in cellular function? A mundane, yet obvious role for PE methylation is to supply the cell with choline necessary for synthesis of PC (via the CDP-choline pathway) and the neurotransmitter acetylcholine. The conversion of PE to PC is the only well-documented pathway discovered for the formation of choline.

The recent purification of the methyltransferase and elucidation of its molecular properties should initiate studies on enzyme regulation. It would be interesting if modes of short-term regulation exist beyond that due to fluctations in substrate and product levels. Cloning of the methyltransferase gene CDNA and through the use of polyclonal antibodies or oligonucleotide probes to the *N*-terminal region of the enzyme. Should answer why expression of the enzyme is tissue specific and give insights into the structure of an apparently constitutively expressed liver gene.

REFERENCES

1. **Kennedy, E. P.,** The metabolism and function of complex lipids, *Harvey Lect.*, 57, 143, 1962.
2. **Sundler, R. and Åkesson, B.,** Regulation of phospholipid biosynthesis in isolated rat hepatocytes, *J. Biol. Chem.*, 250, 3359, 1975.
3. **Vance, D. E., Audubert, F., and Pritchard, P. H.,** The conversion of phosphatidylethanolamine to phosphatidylcholine in animal cells, in *Biochemistry of S-Adenosylmethionine and Related Compounds*, Usdin, E., Borchardt, R. T., and Creveling, C. P., Eds., MacMillan Press, London, 1982, 119.
4. **Ridgway, N. D. and Vance, D. E.,** Purification of phosphatidylethanolamine *N*-methyltransferase from rat liver, *J. Biol. Chem.*, 262, 17231, 1987.
5. **Dennis, E. A.,** Phospholipase A_2 activity towards phosphatidylcholine in mixed micelles:surface dilution kinetics and the effect of thermotropic phase transitions, *Arch. Biochem. Biophys.*, 158, 485, 1973.
6. **Hendrickson, H. S. and Dennis, E. A.,** Kinetic analysis of the dual phospholipid model for phospholipase A_2 action, *J. Biol Chem.*, 259, 5734, 1984.
7. **Du Vigneaud, V., Cohn, M., Chandler, J. P., Schenck, J. R., and Simmonds, S.,** The utilization of the methyl group of methionine in the biological synthesis of choline and creatine, *J. Biol. Chem.*, 140, 625, 1941.
8. **Stetten, D.,** Biological relationships of choline, ethanolamine and related compounds, *J. Biol. Chem.*, 138, 437, 1941.
9. **Bremer, J. and Greenberg, D. M.,** Mono- and dimethylethanolamine isolated form rat-liver phospholipids, *Biochim. Biophys. Acta*, 35, 287, 1959.
10. **Bremer, J. and Greenberg, D. M.,** Biosynthesis of choline *in vitro*, *Biochim. Biophys. Acta*, 37, 173, 1960.
11. **Bremer, J., Figard, P. H., and Greenberg, D. M.,** The biosynthesis of choline and its relation to phospholipid methylation, *Biochim. Biophys. Acta*, 43, 477, 1960.
12. **Gibson, K. D., Wilson, J. D., and Udenfriend, S.,** The enzymatic conversion of phospholipid ethanolamine to phospholipid choline in rat liver, *J. Biol. Chem.*, 236, 673, 1961.
13. **Rehbinder, D. and Greenberg, D. M.,** Studies on the methylation of ethanolamine phosphatides by liver preparations, *Arch. Biochem. Biophys.*, 108, 110, 1965.
14. **Bremer, J. and Greenberg, D. M.,** Methyl transferring enzyme system of microsomes in the biosynthesis of lecithin (phosphatidylcholine), *Biochim. Biophys. Acta*, 46, 205, 1961.
15. **Tanaka, Y., Doi, O., and Akamatsu, Y.,** Solubilization and properties of a phosphatidylethanolamine-dependent methyltransferase system for phosphatidylcholine synthesis from mouse liver microsomes, *Biochem. Biophys. Res. Commun.*, 87, 1109, 1979.
16. **Hoffman, D. R. and Cornatzer, W. E.,** Microsomal phosphatidylethanolamine methyltransferase: some physical and kinetic properties, *Lipids*, 16, 533, 1981.
17. **Audubert, F. and Vance, D. E.,** Pitfalls and problems in studies on the methylation of phosphatidylethanolamine, *J. Biol. Chem.*, 258, 10695, 1983.
18. **Schneider, W. J. and Vance, D. E.,** Conversion of phosphatidylethanolamine to phosphatidylcholine in rat liver, *J. Biol. Chem.*, 254, 3886, 1979.

19. **Vance, D. E. and Ridgway, N. R.,** The methylation of phosphatidylethanolamine, *Prog. Lipid Res.,* 27, 61, 1988.
20. **Van Golde, L. M. G., Raben, J., Batenburg, J. J., Fleischer, B., Zambrano, F., and Fleisher, S.,** Biosynthesis of lipids in golgi complex and other subcellular fractions from rat liver, *Biochim. Biophys. Acta,* 360, 179, 1974.
21. **Higgins, J. A. and Fieldsend, J. K.,** Phosphatidylcholine synthesis for incorporation into membranes or for secretion as plasma lipoproteins by Golgi membranes of rat liver, *J. Lipid Res.,* 28, 268, 1987.
22. **Vance, J. E. and Vance, D. E.,** Does rat liver Golgi have the capacity to synthesize phospholipids for lipoprotein secretion?, *J. Biol. Chem.,* 263, 5898, 1988.
23. **Vance, D. E., Choy, P. C., Farren, S. F., Lim, P. H., and Schneider, W. J.,** Asymmetry of phospholipid biosynthesis, *Nature,* 270, 268, 1977.
24. **Bell, R. M., Ballas, L. M., and Coleman, R. A.,** Lipid topogenesis, *J. Lipid Res.,* 22, 391, 1981.
25. **Audubert, F. and Vance, D. E.,** Evidence that the enzymes involved in the methylation of phosphatidylethanolamine are on the external side of the microsomal vesicles, *Biochim. Biophys. Acta,* 792, 359, 1984.
26. **Higgins, J. A.,** Biogenesis of endoplasmic reticulum phosphatidylcholine: translocation of intermediates across the membrane bilayer during methylation of phosphatidylethanolamine, *Biochim. Biophys. Acta,* 640, 1, 1981.
27. **Op den Kamp, J. A. F.,** Lipid asymmetry in membranes, *Annu. Rev. Biochem.,* 48, 47, 1979.
28. **Percy, A. K., Moore, J. F., and Waechter, C. J.,** Properties of particulate and detergent-solubilized phospholipid *N*-methyltransferase activity from calf brain, *J. Neurochem.,* 38, 1404, 1982.
29. **Makishima, F., Toyoshima, S., and Osawa, T.,** Partial purification and characterization of phospholipid *N*-methyltransferase from murine thymocyte microsomes, *Arch. Biochem. Biophys.,* 238, 315, 1985.
30. **Pajares, M. A., Alemany, S., Varela, I., Marin Cao, D., and Mato, J.,** Purification and photoaffinity labelling of lipid methyltransferase from rat liver, *Biochem. J.,* 223, 61, 1984.
31. **Pajares, M., Villalba, M., and Mato, J.,** Purification of phospholipid methyltransferase from rat liver microsomal fraction, *Biochem. J.,* 237, 699, 1986.
32. **Ridgway, N. D., Yao, Z., and Vance, D. E.,** Phosphatidylethanolamine levels and regulation of phosphatidylethanolamine *N*-methyltransferase, *J. Biol. Chem.,* 264, 1203, 1989.
33. **Hirata, F., Viveros, O. H., Diliberto, E. J., and Axelrod, J.,** Identification and properties of two methyltransferases in conversion of phosphatidylethanolamine to phosphatidylcholine, *Proc. Natl. Acad. Sci. U.S.A.,* 75, 1718, 1978.
34. **Hirata, F. and Axelrod, J.,** Enzymatic synthesis and rapid translocation of phosphatidylcholine by two methyltransferases in erythrocyte membranes, *Proc. Natl. Acad. Sci. U.S.A.,* 75, 2348, 1978.
35. **Sastry, B. V. R., Statham, C. N., Axelrod, J., and Hirata, F.,** Evidence for two methyltransferases involved in the conversion of phosphatidylethanolamine to phosphatidylcholine in the rat liver, *Arch. Biochem. Biophys.,* 211, 762, 1981.
36. **McGivney, A., Crews, F. T., Hirata, F., Axelrod, J., and Siraganian, R. P.,** Rat basophilic leukemia cell lines defective in phospholipid methyltransferase enzymes, Ca^{2+} influx and histamine release; reconstitution by hybridization, *Proc. Natl. Acad. Sci. U.S.A.,* 78, 6178, 1981.
37. **Crews, F. T., Morita, Y., McGivney, A., Hirata, F., Siraganian, R. P., and Axelrod, J.,** IgE-mediated histamine release in rat basophilic leukemia cells: receptor activation, phospholipid methylation, Ca^{2+} flux and release of arachidonic acid, *Arch. Biochem. Biophys.,* 212, 561, 1981.
38. **Crews, F. T., Hirata, F., and Axelrod, J.,** Identification and properties of two methyltransferases that synthesize phosphatidylcholine in rat brain synaptosomes, *J. Neurochem.,* 34, 1491, 1980.
39. **Hirata, F., Strittmatter, W. J., and Axelrod, J.,** β-Adrenergic receptor agonists increase phospholipid methylation, membrane fluidity and β-adrenergic receptor-adenylate cyclase coupling, *Proc. Natl. Acad. Sci. U.S.A.,* 76, 368, 1979.
40. **Prasad, C. and Edwards, R. M.,** Synthesis of phosphatidylcholine from phosphatidylethanolamine by at least two methyltransferases in rat pituitary extracts, *J. Biol. Chem.,* 256, 13000, 1981.
41. **Dudeja, P. K., Foster, E. S., and Brasitus, T. A.,** Synthesis of phosphatidylcholine by two distinct methyltransferases in rat colonic brush-border membranes: evidence for extrinsic and intrinsic membrane activities, *Biochim. Biophys. Acta,* 875, 493, 1986.
42. **Blusztajn, J. K., Zeisel, S. H., and Wurtman, R. J.,** Developmental changes in the activity of phosphatidylethanolamine *N*-methyltransferase in rat brain, *Biochem. J.,* 232, 505, 1985.
43. **Ridgway, N. D. and Vance, D. E.,** Kinetic mechanism of phosphatidylethanolamine *N*-methyltransferase, *J. Biol. Chem.,* 263, 16864, 1988.
44. **Reitz, R. C.,** personal communication.
45. **Lyman, R. L., Hopkins, S. M., Sheehan, G., and Tinoco, J.,** Incorporation and distribution of [*Me*-^{14}C]methionine methyl into liver phosphatidylcholine fractions from control and essential fatty acid deficient rats, *Biochim. Biophys. Acta,* 176, 86, 1969.

46. **Salerno, D. M. and Beeler, D. A.,** The biosynthesis of phospholipids and their precursors in rat liver involving *de novo* methylation and base exchange, *Biochim. Biophys. Acta,* 326, 325, 1973.
47. **Fex, G.,** Metabolism of phosphatidylcholine, phosphatidylethanolamine and sphingomyelin in regenerating liver rat liver, *Biochim. Biophys. Acta,* 231, 161, 1971.
48. **Glenn, J. L. and Austin, W.,** The conversion of phosphatidylethanolamines to lecithins in normal and choline deficient rats, *Biochim. Biophys. Acta,* 231, 153, 1971.
49. **Arvidson, G. A. E.,** Structural and metabolic heterogeneity of rat liver glycerophospholipids, *Eur. J. Biochem.,* 4, 478, 1968.
50. **Ridgway, N. D. and Vance, D. E.,** Specificity of rat hepatic phosphatidylethanolamine *N*-methyltransferase for molecular species of diacyl phosphatidylethanolamine, *J. Biol. Chem.,* 263, 16856, 1988.
51. **Tacconi, M. and Wurtman, R. J.,** Phosphatidylcholine produced in rat synaptosomes by *N*-methylation is enriched in polyunsaturated fatty acids, *Proc. Natl. Acad. Sci. U.S.A.,* 82, 4828, 1985.
52. **Lakher, M. B. and Wurtman, R. J.,** Molecular composition of the phosphatidylcholines produced by the phospholipid methylation pathway in rat brain *in vivo, Biochem. J.,* 244, 325, 1987.
53. **Malgat, M., Maurice, A., and Baraud, J.,** Sphingomyelin and ceramide-phosphoethanolamine synthesis by microsomes and plasma membranes from rat liver and brain, *J. Lipid Res.,* 27, 251, 1986.
54. **LeKim, D., Betzing, H., and Stoffel, W.,** Studies *in vivo* and *in vitro* on the methylation of phosphatidyl-*N,N*-dimethylethanolamine to phosphatidylcholine in rat liver, *Hoppe-Seyler's Z. Physiol. Chem.,* 354, 437, 1973.
55. **Audubert, F. and Bereziat, G.,** The specificity of rat liver phospholipid methyltransferase for lyso derivatives and diacyl derivatives of phosphatidylethanolamine, *Biochim. Biophys. Acta,* 920, 26, 1987.
56. **Åkesson, B.,** Structural requirements of the phospholipid substrate for phospholipid *N*-methylation in rat liver, *Biochim, Biophys. Acta,* 752, 460, 1983.
57. **Moore, C., Blank, M. L., Lee, T-C., Benjamin, B., Piantadosi, C., and Snyder, F.,** Membrane lipid modifications: biosynthesis and identification of phosphatidyl-*N*-methyl-*N*-isopropylethanolamine in rat liver microsomes, *Chem. Phys. Lipids,* 21, 175, 1978.
58. **Natori, Y.,** Studies on ethionine, *J. Biol. Chem.,* 238, 2075, 1963.
59. **Bjornstad, P. and Bremer, J.,** *In vivo* pathways for the biosynthesis of lecithin in the rat, *J. Lipid Res.,* 7, 38, 1966.
60. **Gotto, A. M., Jr., Pownall, H. J., and Havel, R. J.,** Introduction to the plasma lipoproteins, *Methods Enzymol.,* 128, 1, 1986.
61. **Chapman, M. J.,** Comparative analysis of plasma lipoproteins, *Methods Enzymol.,* 128, 70, 1986.
62. **Vance, J. E. and Vance, D. E.,** Specific pools of phospholipids are used for lipoprotein secretion by cultured rat hepatocytes, *J. Biol. Chem.,* 261, 4486, 1986.
63. **Vance, J. E., Nguyen, T. M., and Vance, D. E.,** The biosynthesis of phosphatidylcholine by methylation of phosphatidylethanolamine is not required for lipoprotein secretion by cultured rat hepatocytes, *Biochim. Biophys. Acta,* 875, 501, 1986.
64. **Vance, J. E. and Vance, D. E.,** A deazaadenosine-insensitive methylation of phosphatidylethanolamine is involved in lipoprotein secretion, *FEBS Lett.,* 204, 243, 1986.
65. **Voelker, D. R. and Frasier, J. L.,** Isolation and characterization of a Chinese hamster ovary cell line requiring ethanolamine or phosphatidylserine for growth and exhibiting defective phosphatidylserine synthase activity, *J. Biol. Chem.,* 261, 1002, 1986.
66. **Yeung, S. K. F. and Kuksis, A.,** Utilization of L-serine in the *in vivo* biosynthesis of glycerophospholipids by rat liver, *Lipids,* 11, 498, 1976.
67. **Yao, Z. and Vance, D. E.,** The active synthesis of phosphatidylcholine is required for very low density lipoprotein secretion from rat hepatocytes, *J. Biol. Chem.,* 263, 2998, 1988.
68. **Sundler, R. and Åkesson, B.,** Biosynthesis of phosphatidylethanolamines and phosphatidylcholines from ethanolamine and choline in rat liver, *Biochem. J.,* 146, 309, 1975.
69. **Holub, B. J. and Kukis, A.,** Structural and metabolic interrrelationships among glycerophosphatides of rat liver *in vivo, Can. J. Biochem.,* 49, 1347, 1971.
70. **Ridgway, N. D. and Vance, D. E.,** unpublished observations.
71. **Åkesson, B.,** Autoregulation of phospholipid *N*-methylation by the membrane phosphatidylethanolamine content, *FEBS Lett.,* 92, 177, 1978.
72. **Schroeder, F., Holland, J. F., and Vagelos, P. R.,** Physical properties of membranes isolated from tissue culture cells with altered phospholipid composition, *J. Biol. Chem.,* 251, 6747, 1976.
73. **Dainous, F. and Kanfer, J. N.,** Effect of modification of membrane phospholipid composition on phospholipid methylation in aggregating cell culture, *J. Neurochem.,* 46, 1859, 1986.
74. **McKenzie, P. C., Gillespie, C. S., and Brophy, P. J.,** The effect of polar head-group substitution on phospholipid methylation and the β–adrenergic response in C6 glial cells, *Biochem. J.,* 231, 769, 1985.
75. **Katyal, S. L. and Lombardi, B.,** Quantitation of phosphatidyl *N*-methyl- and *N,N*-dimethylaminoethanol in rat liver, *Lipids,* 11, 513, 1976.

76. **Hoffman, D. R., Marion, D. W., Cornatzer, W. E., and Duerre, J. A.,** *S*-Adenosylmethionine and *S*-adenosylhomocysteine metabolism in isolated rat liver, *J. Biol. Chem.*, 255, 10822, 1980.
77. **Chiang, P. K., Richards, H. H., and Cantoni, G. L.,** *S*-Adenosyl-L-homocysteine hydrolase: analogues of *S*-adenosyl-L-homocysteine as potential inhibitors, *Mol. Pharmacol.*, 13, 939, 1977.
78. **Chiang, P. K., and Cantoni, G. L.,** Perturbations of biochemical transmethylations by 3-deazaadenosine *in vivo*, *Biochem. Pharmacol.*, 28, 1897,1979.
79. **Prichard, P. H., Chiang, P. K., Cantoni, G. L., and Vance, D. E.,** Inhibition of phosphatidylethanolamine *N*-methylation by 3-deazaadenosine stimulates the synthesis of phosphatidylcholine via the CDP-choline pathway, *J. Biol. Chem.*, 257, 6362, 1982.
80. **Schanche, J.-S., Schanche, T., and Ueland, P. M.,** Inhibition of phospholipid methylation in isolated rat hepatocytes by analogues of adenosine and *S*-adenosylhomocysteine, *Biochim. Biophys. Acta* , 721, 399, 1982.
81. **Geelen, M. J. H., Groener, J. E. M., De Hass, C. G. M., and Van Golde, L. M. G.,** Influence of glucagon on the synthesis of phosphatidylcholines and phosphatidylethanolamines in monolayer cultures of rat hepatocytes, *FEBS Lett.*, 105, 27, 1979.
82. **Pelech, S. L., Prichard, H. P., Sommerman, E. F., Percival-Smith, A., and Vance, D. E.,** Glucagon inhibits phosphatidylcholine biosynthesis and transmethylation pathways in cultured rat hepatocytes, *Can. J. Biochem. Cell Biol.*, 62, 196, 1984.
83. **Pritchard, P. H., Pelech, S. L., and Vance, D. E.,** Analogues of cyclic AMP inhibit phosphatidylethanolamine *N*-methylation by cultured rat hepatocytes, *Biochim. Biophys. Acta*, 666, 301, 1981.
84. **Pelech, S. L., Pritchard, P. H., and Vance, D. E.,** cAMP analogues inhibit phosphatidylcholine biosynthesis in cultured rat hepatocytes, *J. Biol. Chem.*, 256, 8283, 1981.
85. **Castano, J. G., Alemany, S., Nieto, A., and Mato, J. M.,** Activation of phospholipid methyltransferase by glucagon in rat hepatocytes, *J. Biol. Chem.*, 255, 9041, 1980.
86. **Mato, J. M., Alemany, S., Gil, M. G., Cao, D. M., Varela, I., and Castano, J. G.,** Receptor mediated regulation of phospholipid methyltransferase activity, in *Biochemistry of S-Adenosylmethionine and Related Compounds*, Usdin, F., Borchardt, R. T., and Creveling, C. R., Eds., Humana Press, Clifton, NJ, 1982, 187.
87. **Pelech, S. L., Ozen, N., Audubert, F., and Vance, D. E.,** Regulation of rat liver phosphatidylethanolamine *N*-methyltransferase by cytosolic factors. Examination of a role for reversible protein phosphorylation, *Biochem. Cell Biol.*, 64, 565, 1986.
88. **Varela, I., Merida, I., Pajares, M., Villalba, M., and Mato, J. M.,** Activation of partially purified rat liver lipid methyltransferase by phosphorylation, *Biochem. Biophys Res. Commun.*, 122, 1065, 1984.
89. **Villalba, M., Varela, I., Merida, I., Pajares, M. A., Martinez del Pozo, A., and Mato, J. M.,** Modulation by the ratio *S*-adenosylmethionine/*S*-adenosylhomocysteine of cyclic AMP-dependent phosphorylation of the 50 kDa protein of rat liver phospholipid methyltransferase, *Biochim. Biophys. Acta*, 847, 273, 1985.
90. **Varela, I., Merida, I., Villalba, M., Vivanco, F., and Mato, J. M.,** Phospholipid methyltransferase phosphorylation by intact hepatocytes: effect of glucagon, *Biochem. Biophys. Res. Commun.*, 131, 477, 1985.
91. **Villalba, M., Pajares, M. A., Renhart, M. F., and Mato, J. M.,** Protein kinase C catalyzes the phosphorylation and activation of rat liver phospholipid methyltransferase, *Biochem. J.*, 241, 911, 1987.
92. **Ridgway, N. D. and Vance D. E.,** unpublished observations.
93. **Alemany, S., Varella, I., and Mato, J. M.,** Inhibition of phosphatidylcholine synthesis by vasopressin and angiotensin in rat hepatocytes, *Biochem. J.*, 208, 453, 1982.
94. **Alemany, S., Varela, I., and Mato, J. M.,** Stimulation by vasopressin and angiotensin of phospholipid methyltransferase in isolated rat hepatocytes, *FEBS Lett.*, 135, 111, 1981.
95. **Alemany, S., Varela, I., Harper, J. F., and Mato, J. M.,** Calmodulin regulation of phospholipid and fatty acid methylation by rat liver microsomes, *J. Biol. Chem.*, 257, 9249, 1982.
96. **Marin-Cao, D., Alveriz Chiva, V., and Mato, J. M.,** β-Adrenergic control of phosphatidylcholine synthesis by transmethylation in hepatocytes from juvenile, adult and adrenalectomized rats, *Biochem. J.*, 216, 675, 1983.
97. **Mato, J. M. and Alemany, S.,** What is the function of phospholipid *N*-methylation?, *Biochem. J.*, 213, 1, 1983.
98. **Merida, I., and Mato, J. M.,** Inhibition by insulin of glucagon-dependent phospholipid methyltransferase phosphorylation in rat hepatocytes, *Biochim. Biophys. Acta*, 928, 92, 1987.
99. **Kelly, K. L., Wong, E. H-A., and Jarett, L.,** Adrenocorticotropic stimulation and insulin inhibition of adipocyte phospholipid methylation *J. Biol. Chem.*, 260, 3640, 1985.
100. **Kelly, K. L. and Wong, E. H-A.,** Hormonal regulation of phospholipid methyltransferase by 3′,5′-cyclic adenosine monophosphate-dependent and independent mechanisms, *Endocrinology*, 120, 2421, 1987.
101. **Kelly, K. L.,** Stimulation of adipocyte phospholipid methyltransferase activity by phorbol 12-myristate 13-acetate, *Biochem. J.*, 241, 917, 1987.

102. **Bonne, D. and Cohen, P.,** Characterization of oxytocin receptors on isolated rat fat cells, *Eur. J. Biochem.,* 56, 295, 1975.
103. **Kelly, K. L., Kiechle, F. L., and Jarett, L.,** Insulin stimulation of phospholipid methylation in isolated rat adipocyte plasma membranes, *Proc. Natl. Acad. Sci. U.S.A.,* 81, 1089,1984.
104. **Kelly, K. L., Merida, I., Wong, E. H.-A., DiCenzo, D., and Mato, J. M.,** A phospho-oligosaccharide mimics the effects of insulin to inhibit isoproterenol-dependent phosphorylation of phospholipid methyltransferase in isolated adipocytes, *J. Biol. Chem.,* 262, 15285, 1987.
105. **Saltiel, A. R. and Cuatrecasas, P.,** Insulin stimulates the generation from hepatic plasma membranes of modulators derived from an inositol glycolipid, *Proc. Natl. Acad. Sci. U.S.A.,* 83, 5793, 1986.
106. **Alvarez Chiva, V. and Mato, J. M.,** Inhibition of phospholipid methylation by a cytosolic factor, *Biochem. J.,* 218, 637, 1984.
107. **Lyon, E. S., McPhie, P., and Jakoby, W. B.,** Methinin: a peptide inhibitor of methylation, *Biochem. Biophys. Res. Commun.,* 108, 846, 1982.
108. **Audubert, F., Pelech, S. L., and Vance, D. E.,** Fatty acids inhibit *N*-methylation of phosphatidylethanolamine in rat hepatocytes and liver microsomes, *Biochim. Biophys. Acta,* 792, 348, 1984.
109. **Hoffman, D. R., Cornatzer, W. E., and Duerre, J. A.,** Relationship between tissue levels of *S*-adenosylmethionine, *S*-adenosylhomocysteine and transmethylation reactions, *Can. J. Biochem. Cell Biol.,* 57, 56, 1979.
110. **Pelech, S. L., Power, E., and Vance, D. E.,** Activities of the phosphatidylcholine biosynthetic enzymes of rat liver during development, *Can. J. Biochem. Cell Biol.,* 61, 1147, 1983.
111. **Cornatzer, W. E., Hoffamn, D. R., and Haning, J. A.,** The effect of embryological development on phosphatidylethanolamine methyltransferase, phosphatidyldimethylethanolamine methyltransferase and cholinephosphotransferase of rabbit liver microsomes, *Lipids,* 19, 1, 1984.
112. **Blusztajn, J. K., Zeisel, S. H., and Wurtman, R. J.,** Developmental changes in the activity of phosphatidylethanolamine *N*-methyltransferase in rat brain, *Biochem. J.,* 232, 505, 1985.
113. **Pelech, S. L., Prichard, P. H., Brindley, D. N., and Vance, D. E.,** Fatty acids promote translocation of CTP:phosphocholine cytidylyltransferase to the endoplasmic reticulum and stimulate rat hepatic phosphatidylcholine synthesis, *J. Biol. Chem.,* 258, 6782, 1983.
114. **Mookerjea, S., Park, C. E., and Kuksis, A.,** Lipid profiles of plasma lipoproteins of fasted and fed normal and choline deficient rats, *Lipids,* 10, 374, 1975.
115. **Haines, D. S. M. and Rose, C. I.,** Impaired labeling of liver phosphatidylethanolamine from ethanolamine-^{14}C in choline deficiency, *Can. J. Biochem. Cell Biol.,* 48, 885, 1970.
116. **Schneider, W. J. and Vance, D. E.,** Effects of choline deficiency on the enzymes that synthesize phosphatidylcholine and phosphatidylethanolamine in rat liver, *Eur. J. Biochem.,* 85, 181, 1978.
117. **Hoffman, D. R., Uthus, E. O., and Cornatzer, W. E.,** Effect of diet on choline phosphotransferase, phosphatidylethanolamine methyltransferase and phosphatidyldimethylethanolamine methyltransferase in liver microsomes, *Lipids,* 15, 439, 1980.
118. **Yao, Z., Jamil, H., and Vance, D. E.,** personal communication.
119. **Blumenstein, J.,** Studies in phospholipid metabolism. II. Incorporation of metabolic precursors of phospholipids in choline deficiency, *Can. J. Physiol. Pharmacol.,* 46, 487, 1968.
120. **Hoffman, D. R., Haning, J. A., and Cornatzer, W. E.,** Effects of a methyl-deficient diet on rat liver phosphatidylcholine biosynthesis, *Can. J. Biochem.,* 59, 543, 1981.
121. **Hall, M. O. and Nyc, J. F.,** Lipids containing mono- and dimethylethanolamine in a mutant strain of *Neurospora crassa, J. Am. Chem. Soc.,* 81, 2275, 1959.
122. **Crocken, B. J. and Nyc, J. F.,** Phospholipid variations in mutant strains of *Neurospora crassa, J. Biol. Chem.,* 239, 1727, 1964.
123. **Scarborough, G. A. and Nyc, J. F.,** Methylation of ethanolamine phosphatides by microsomes from normal and mutant strains of *Neurospora crassa, J. Biol. Chem.,* 242, 238, 1967
124. **Scarborough, G. A. and Nyc, J. F.,** Properties of a phosphatidylmonomethylethanolamine *N*-methyltransferase from *Neurospora crassa, Biochim. Biophys. Acta,* 146, 111, 1967.
125. **van Waarde, A. and van Hoof, P. J. M.,** Nonpolar lipid methylation during development of *Dictyostelium discoideum, Biochim. Biophys. Acta,* 836, 27, 1985.
126. **Garcia Gil, M., Alemany, S., Marin Cao, D., Castano, J. G., and Mato, J. M.,** Calmodulin modulates phospholipid methyltransferase of *Dictyostelium discoideum, Biochem. Biophys. Res. Commun.,* 94, 1325, 1980.
127. **Alemany, S., Garcia Gil, M., and Mato, J. M.,** Regulation by guanosine 3′:5′-cyclic monophosphate of phospholipid methylation during chemotaxis in *Dictyostelium discoideum, Proc. Natl. Acad. Sci. U.S.A.,* 77, 6996, 1980.
128. **Smith, J. D.,** Phosphatidylcholine homeostasis in phosphatidylethanolamine-depleted *Tetrahymena, Arch. Biochem. Biophys.,* 246, 347,1986.

129. **Smith, J. D. and Law, J. H.,** Phosphatidylcholine biosynthesis in *Tetrahymena pyriformis*, *Biochim. Biophys. Acta,* 202, 141, 1970.
130. **Smith, J. D.,** Effect of dimethylaminoethylphosphonate on phospholipid methabolism in *Tetrahymena*, *Biochim. Biophys. Acta,* 878, 450, 1986.
131. **Smith, J. D. and Giegel, D. A.,** Effect of a phosphonic acid analogue of choline phosphate on phospholipid metabolism in *Tetrahymena*, *Arch. Biochem. Biophys.,* 213, 595, 1982.
132. **Smith, J. D.,** Metabolism of phosphonates, in *The Role of Phosphonates in Living Systems,* Hilderbrand, R. L., Ed., CRC Press, Boca Raton, FL, 1983, 31.
133. **Ikawa, M.,** Bacterial phosphatides and natural relationships, *Bacteriol. Rev.,* 31, 54, 1967.
134. **Thiele, O. W. and Oulevey, J.,** Occurrence of phosphatidylcholine in hydrogen-oxidizing bacteria, *Eur. J. Biochem.,* 118, 183, 1981.
135. **Asselineau, J. and Truper, H. G.,** Lipid composition of six species of the phototrophic bacterial genus ectothiorhodospira, *Biochim. Biophys. Acta,* 712, 111, 1982.
136. **Makula, R. A.,** Phospholipid composition of methane-utilizing bacteria, *J. Bacteriol.,* 134, 771, 1978.
137. **Goldberg, I. and Jensen, A. P.,** Phospholipid and fatty acid composition of methanol-utilizing bacteria, *J. Bacteriol.,* 130, 535, 1977.
138. **Barridge, J. K. and Shively, J. M.,** Phospholipids of thiobacilli, *J. Bacteriol.,* 95, 2182, 1968.
139. **Johnston, N. C. and Goldfine, H.,** Lipid composition in the classification of the butyric acid-producing Clostridia, *J. Gen. Microbiol.,* 129, 1075, 1983.
140. **Goldfine, H. and Hagen, P.-O.,** *N*-methyl groups in Bacterial lipids. III. Phospholipids of Hyphomicrobia, *J. Bacteriol.* 95, 367, 1968.
141. **Goldfine, H.,** Bacterial membranes and lipid packing theory, *J. Lipid Res.,* 25, 1501, 1984.
142. **Kaneshiro, T. and Law, J. H.,** Phosphatidylcholine synthesis in *Agrobacterium tumefaciens*, *J. Biol. Chem.,* 239, 1705, 1964.
143. **Cain, B. D., Donohue, T. J., Sheperd, W. D., and Kaplan, S.,** Localization of phospholipid biosynthetic enzyme activities in cell-free fractions derived from *Rhodopseudomonas sphaeroides*, *J. Biol. Chem.,* 259, 942, 1984.
144. **Tahara, Y., Ogawa, Y., Sakakibara, T., and Yamada, Y.,** Phosphatidylethanolamine *N*-methyltransferase from *Zymomonas mobilis*: purification and characterization, *Agric. Biol. Chem.,* 50, 257, 1986.
145. **Hirata, F. and Axelrod, J.,** Phospholipid methylation and biological signal transmission, *Science,* 209, 1082, 1980.
146. **Vance, D. E. and Ridgway, N. D.,** in *Biological Methylation and Drug Design,* Borchardt, R. T., Creveling, C. R., and Ueland, P. M., Eds., Humana Press, Clifton, NJ, 1986, 75.
147. **Moore, J. P., Johannsson, A., Hesketh, T. R., Smith, G. A., and Metcalfe, J. C.,** Calcium signals and phospholipid methylation in eukaryotic cells, *Biochem J.,* 221, 675, 1984.
148. **Berridge, M. J.,** Inositol trisphosphate and diacylglycerol as second messengers, *Biochem. J.,* 220, 345, 1984.
149. **Kishimoto, A., Takai, Y., Mori, T., Kikkawa, U., and Nishizuka, Y.,** Activation of calcium and phospholipid-dependent protein kinase by diacylglycerol, its possible relation to phosphatidylinositol turnover, *J. Biol. Chem.,* 255, 2273, 1980.
150. **Pike, M. C. and DeMeester, C. A.,** Inhibition of phosphoinositide metabolism in human polymorphomnuclear leukocytes by *S*-adenosylhomocysteine, *J. Biol. Chem.,* 263, 3592, 1988.

Chapter 8

THE ROLE OF PHOSPHOLIPASES IN PHOSPHATIDYLCHOLINE CATABOLISM

Mary Fedarko Roberts and Edward A. Dennis

TABLE OF CONTENTS

I. INTRODUCTION

The catabolism of phosphatidylcholine (PC) and other phospholipids is carried out by phospholipases.[1,2] These phospholipases are classified according to which phospholipid bond they hydrolyze, as shown in Figure 1. The catabolism of PC as shown in Figure 2 can be initiated by any one of four phospholipases: phospholipase A_1 (PLA_1) hydrolyzes the *sn*-1 fatty acyl bond, phospholipase A_2(PLA_2) the *sn*-2 fatty acyl bond, phospholipase C (PLC) the glycerophosphate ester bond, and phospholipase D (PLD) the choline phosphate ester bond. The products of these hydrolyses are free fatty acid, lysophospholipid, diglyceride, phosphatidic acid, phosphorylcholine, and choline. These products can either be used for synthesis or be degraded further. The lysoPC is degraded by lysophospholipase (lysoPLA) which converts it to fatty acid and glycerol phosphorylcholine (GPC); the latter may be subsequently hydrolyzed to choline and glycerol phosphate, or glycerol and phosphorylcholine by a phosphodiesterase (PDE). The fatty acid is catabolized further by the well-established pathways of β-oxidation. Diglyceride is converted to fatty acid and monoglyceride and perhaps on to glycerol and fatty acid by various lipases. The glycerol, choline, and inorganic phosphate are considered to be the final catabolic products, and may be used as building blocks in various anabolic pathways.

Since most of the enzymes of PC catabolism must act on aggregated substrate, they exhibit several characteristics that are not displayed by soluble enzymes that act on soluble substrates. The presence of aggregated substrate creates a set of unique conditions that are common to all of these enzymes and that have been found to affect dramatically their activities and to complicate kinetic analysis. Therefore, this review will not focus on the detailed pathways of PC catabolism, but rather on the general issues that affect the activity of these enzymes and that distinguish them from other soluble esterases. The emphasis will be on understanding how the substrate aggregation state modulates activity and the role, if any, of product inhibition (or activation) on these activities. Since other recent reviews[1-3] deal with positional specificity, stereochemistry of bond cleavage, and fatty acyl chain preferences, these will not be specifically addressed here. We will briefly discuss the PLA_1, the lysoPLA, PLD, and the PDE enzymes, but the main thrust of our discussion will involve the PLA_2 and the PLC enzymes because they have been extensively studied and because they dramatically demonstrate the principles and pitfalls of phospholipase enzymology as it relates to the study of PC catabolism.

II. BIOLOGICAL FUNCTIONS OF PHOSPHOLIPASES

Phospholipases are abundant, ubiquitous enzymes present in all cells that have been examined, and while their catabolic roles are unambiguous, the mechanisms by which these catabolic events are controlled is less clear cut. There is considerable ambiguity as to whether many of them are soluble or membrane-bound, and, in fact, control of activity could involve modulating phospholipase partitioning to the membrane through either specific lipid binding or phosphorylation of the enzyme.

In addition to their roles as simple catabolic enzymes, the phospholipases may also play an important role in the remodeling of existing phospholipids. Several of the degradation products can be reincorporated into new phospholipids. For example, lysophospholipids, the product of either PLA_1 or PLA_2, may be reacylated as discussed in another chapter[4] or be degraded further. Also, a comparison of the mammalian PLD and the base exchange enzyme is considered in a separate chapter.[5] Many membrane functions depend on the exact composition of the phospholipids in the membrane, and, thus, remodeling may play an important role in many cellular functions.

It has now become clear that in addition to their roles in PC catabolism, these enzymes also play very important functions in other cellular processes. They have been implicated as key controlling enzymes in many important metabolic processes including the production of

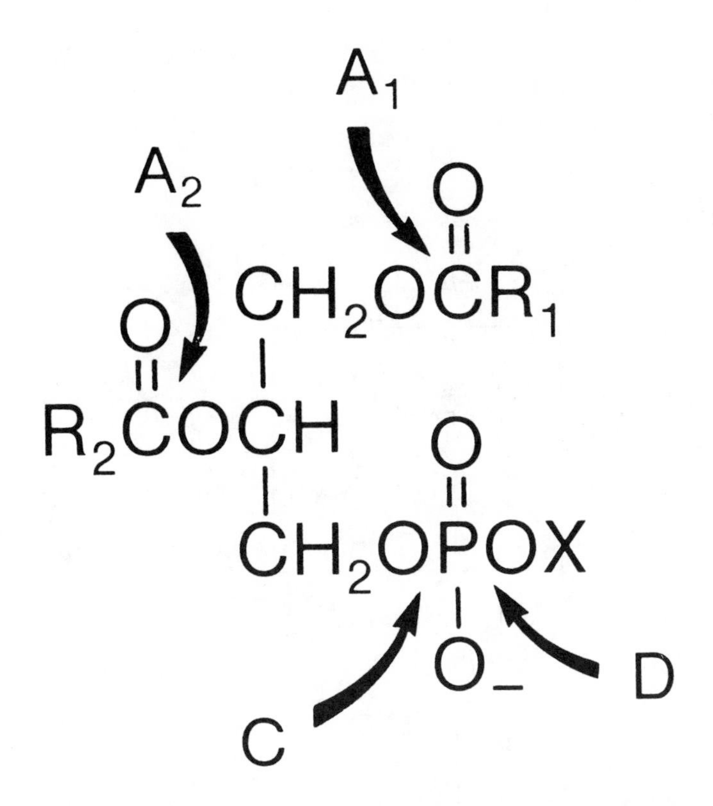

FIGURE 1. Site of action of the phospholipases. R_1COO and R_2COO refer to the fatty acids on the *sn*-1 and *sn*-2 positions, respectively, of the stereospecifically numbered phospholipid, and X refers to the polar groups.

arachidonic acid for eicosanoid biosynthesis, the production of inositol phosphates for intracellular Ca^{2+} mobilization, and the generation of diglyceride for activation of protein kinase C, all of which play regulatory roles in cellular activation. A more detailed review of this subject has been presented elsewhere.[3,6] The remainder of this section is a more specific synopsis of the various roles that each of these phospholipases may play in the cell.

Phospholipase A_2 is the major enzyme in the catabolism of phospholipids in digestion, and pancreatic PLA_2 from various sources has been studied in considerable detail.[7] Similar enzymes, present in snake and other venoms, are equally well studied and, presumably, also have a predigestive function.[1,6] The intracellular PLA_2 is considered by many to be the most likely enzyme to control arachidonic acid release and, thus, prostaglandin synthesis.[3]

Recent work of Dratz and co-workers[8] has suggested a role for PLA_2 in the detoxification of peroxidized phospholipids. Oxygen can spontaneously oxidize unsaturated fatty acids in membranes to yield toxic lipid hydroperoxides which cause cell damage. Unsaturated fatty acids are usually esterified to the *sn*-2 position in phospholipids. Glutathione peroxidase reduces the hydroperoxide fatty acids to hydroxy fatty acids and, reportedly, does not work on intact phospholipids. Excision of the peroxidized fatty acid by PLA_2 (or via PLC followed by a diglyceride lipase) is, then, a prerequisite for reduction of these damaging lipid species.

While numerous PLA_1 and PLA_2 enzymes have been identified in various tissues and cells, it has only recently become apparent that for many cells, the activity of lysoPLA in lysed cell preparations is much greater than either PLA_1 or PLA_2.[9-11] Lysophospholipids are potent biological detergents which can alter cell permeability and, hence, are lytic to cells.[12] For example, myocardial ischemia results in the depletion of phospholipids[13] with the accumulation of lysophospholipids.[14] Lysophospholipids alter the physical characteristics of myocardial

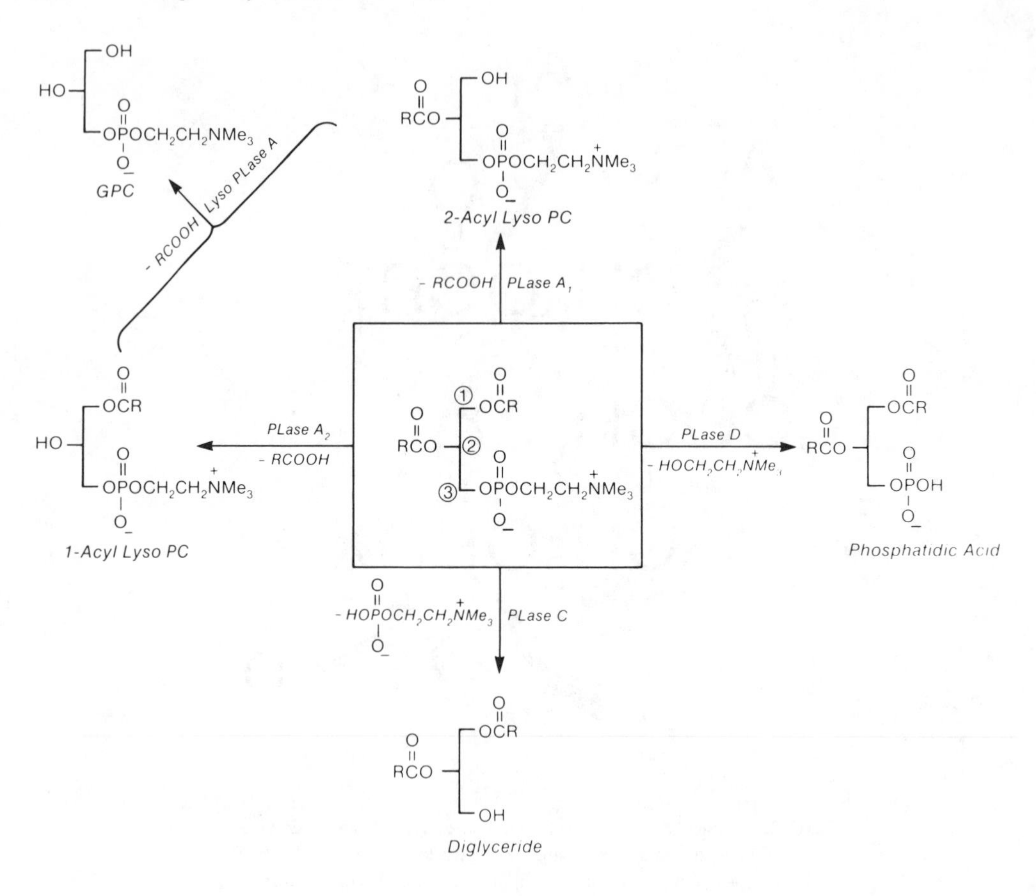

FIGURE 2. The catabolic pathways for phosphatidylcholine are shown. RCOO refers to the fatty acid esterified to the *sn*-1 and *sn*-2 positions of the stereospecifically numbered phosphatidylcholine. There are four possible first steps, each catalyzed by a phospholipase. Subsequent steps are shown as well.

sarcolemmal membranes[15] and have been implicated as biochemical mediators of electrophysiological alterations. LysoPLA activities may function in the rapid removal of the toxic lysophospholipid products of PLA_1 and PLA_2 action. The GPC and related products of the lysoPLA reaction are further metabolized to water-soluble phosphates. Whether specific or nonspecific PDEs are involved has not been examined in detail.

PLC is secreted in the growth media of many bacterial cells[1] where it is assumed to serve a lytic function toward other cells.[16] However, the major PLC enzymes identified and studied in mammalian cells have shown specificity for phosphatidylinositol (PI). PI-specific PLC activities have been found and purified (at least partially) in bovine brain,[17-20] platelets,[21] macrophages,[22] lymphocytes,[23] rat liver plasma membranes,[24] skin fibroblasts,[25] bovine adrenal chromaffin cells,[26] myocardium,[27] sheep seminal vesicles,[28] fetal membranes, and uterine decidua.[29] These PI-specific enzymes usually require Ca^{2+} for activity, and much of the total cellular PLC activity is associated with cellular membranes. The relation between membrane-bound and soluble forms is unclear. Protein phosphorylation and/or GTP binding may be involved in partitioning PI-PLC between membrane and cytosol. A PLC activity specific against 1-*O*-alkyl PC species (including PAF) has been reported.[30] This may have important physiological implications since diacylglyercides are activators of protein kinase C, while the alkylacylglycerides appear to be inhibitors.[31] Thus, the products of PLC with differing specificities could be involved in a complex regulation of protein kinase C. It is also worth noting that in addition to the mammalian PLD,[5] a glycan-PI-specific PLD[32] similar to the PI-specific PLC,[33] may be important in metabolism.

Mammalian PLC enzymes that act on PC and have a broad specificity for other phospholipids have recently been identified in canine myocardium,[34] a human monocyte cell line,[35] gallbladder,[36] and lysosomes.[37-39] While most of the published work has dealt with the characterization of these proteins, substrate specificity, pH optimum, etc., an interesting study with peroxidized phospholipids suggests a critical role for one particular class of nonspecific PLCs. Lysosomal PLC enzymes have been shown to hyrolyze preferentially peroxidized phospholipids. While bacterial PLC is, in fact, inhibited by peroxidized lipids, mammalian acidic PLCs with general specificity (sphingomyelin, PC, phosphatidylethanolamine, PI, PS are all substrates) show a pronounced activation.[40] Similar studies with venom and pancreatic PLA_2 show no such clear activating effect, and, in fact, factors such as the extent and/or mode of peroxidation of the phospholipids appear to influence the susceptibility of peroxidized lipid to these phospholipases. It has been proposed that lipid peroxidation leading to, or associated with, activation of phospholipases may result in irreversible myocardial cell injury.[41] During ischemia, the reduction of molecular oxygen to form oxygen free radicals is accelerated by a developing acidosis. These conditions promote lipid peroxidation and destabilize membranes, resulting in the activation of lysosomal acid hydrolases. The lysosomal PLC, by hydrolyzing peroxidized phospholipids, may then cause membrane dysfunction and/or irreversible damage.

The action of the PDE on GPC, the product of PLA_2 and lysoPLA action on PC, and its regulation in the general scheme of phospholipid catabolism has been less well studied. One technique that can be used to follow this enzyme *in vivo* is ^{31}P NMR spectroscopy. Observation of this nucleus provides a window where key water-soluble intermediates of phospholipid turnover can be detected in metabolizing cells.[42,43] Phosphomonoesters are products of choline and ethanolamine kinases, the first steps in phospholipid synthesis, as well as products of PLC action; phosphodiesters are substrates of GPC phosphodiesterase, the last step in phospholipid catabolism. Both of these phosphate compounds are elevated in tumor and other rapidly proliferating tissues.[44] A recent study by Daly et al.[45] showed that both phosphomono- and phosphodiester resonances could be modulated in cancer cells using a perfusion system to add specific compounds. This allows definition of the effect of different molecules on the final stages of lipid degradation in a wide range of cells and tissues. Similar ^{31}P NMR studies should provide important information on abnormalities of phospholipid turnover by using GPC as a marker of phospholipid catabolism.

In summary, numerous intracellular and extracellular PLA_1, PLA_2, PLC, PLD, and lysoPLA enzymes capable of hydrolyzing PC have been identified and in some cases purified. The complete delineation of their roles and their regulation in PC catabolism and general cellular metabolism must still be elucidated. A major obstacle to accomplishing this is the nature of the PC substrate. Regardless of whether or not the enzymes are membrane bound, their substrate always is. This fact complicates even the simplest kinetic analysis.

Kinetic studies of enzymes acting on lipid/water interfaces require special approaches that studies of soluble enzymes acting on soluble substrates do not require.[1] Such approaches have been developed in studies on pure, extracellular phospholipases and are only now being extended and developed for the relevant intracellular phospholipases involved in phospholipid turnover and metabolism. In the following section, we will discuss these issues, but will necessarily focus on snake venom PLA_2 and bacterial PLC as models for how such enzymes should be studied. Then, we will review what is known about the role of the products of phospholipase action on the regulation and inhibition of PC catabolism. Besides traditional considerations of the potential role of products, they, too, can affect the aggregation state of the substrate and have profound effects on activity, as will be considered.

III. SUBSTRATE REQUIREMENTS

A. PHOSPHOLIPID BINDING REQUIREMENTS FOR PHOSPHOLIPASE A_2 AND C

1. Phospholipase A_2

Phospholipase A_2 from *Naja naja naja* is the best studied of the venom phospholipases.[6] High resolution X-ray crystal structures have been determined for both pancreatic[46] and snake venom (*Crotalus atrox*[47] and *Agkistrodon piscivorus piscivorus*[48]) PLA_2 enzymes. There is a striking similarity, although the venom enzyme exists as a dimer, while the pancreatic version is monomeric in aqueous solution. In the future, detailed mechanistic questions about the role of specific amino acids in the pancreatic PLA_2 can be addressed since the cDNA for porcine $proPLA_2$ has been cloned and expressed in yeast.[49] Ca^{2+} binds to the enzyme and causes a conformational change which is necessary for catalysis.[50] This change is not necessary for phospholipid binding to the enzyme if the substrate is a monomeric short-chain PC, micellar short-chain PC, or long-chain PC in vesicles.[51] (Pancreatic PLA_2 also binds PC molecules without Ca^{2+}, although a Ca^{2+}-dependent activation of the initially formed complex is necesssary for enzyme activity).[52] If, on the other hand, the substrate PC is solubilized in Triton® X-100 mixed micelles, it appears that Ca^{2+} must first bind to the enzyme in order for PC to bind.[50,51]

Evidence has been accumulating that this enzyme binds two or more phospholipids for optimal catalysis. The two lipids bind to functionally distinct sites: the catalytic site, which is nonspecific as to phospholipid head group, but which requires two fatty acyl chains; and the activator site, which has a specific requirement for a phosphorylcholine moiety.[53-56] The existence of two lipid sites has led to a model called the "Dual Phospholipid Model" where both sites must be occupied for effective catalysis.[57,58] This is illustrated and explained in Figure 3. Much of our understanding is derived from kinetics with PE solubilized in Triton® X-100 micelles. The PE alone is a poor substrate for PLA_2, but becomes an excellent substrate if lysoPC or related molecules are added.[54] Support for multiple lipid binding by PLA_2 has also been provided by studies with branched short-chain PC micelles which are analogues of diheptanoyl-PC, an excellent substrate.[59] Arguments for this model have been reviewed elsewhere.[6,56]

2. Phospholipase C

Soluble PLC activities have been isolated and purified from several bacterial sources, most notably *Bacillus cereus*.Three different PLC activities are produced in this organism: a PLC (containing two tightly bound Zn^{2+} ions) with broad phospholipid specificity,[60] a PI-specific PLC (without bound Zn^{2+}),[61] and a sphingomyelinase. Biochemical studies of these soluble PLC species are not as extensive as for pancreatic or venom PLA_2, although many of the kinetic trends are the same for both types of phospholipases. The Zn^{2+}-containing PLC has been the most studied. Both metal ion sites must be occupied for PLC to be active:[62] one site is specific for Zn^{2+} while the second can be replaced by a wide range of ions, including lanthanide ions.[63] This second metal ion site must play a structural role since it is greater than 10 Å from atoms of diglyceride and detergents, which are competitive inhibitors bind to the active site of PLC.[64] Detailed studies on the broad specificity PLC are more significant since antibodies against this enzyme cross-react with PC-preferring PLC in mammalian cells.[65] Although it has not been examined, a similar cross-reactivity may exist between PI-PLC of *B. cereus* and some mammalian enzymes.

As with any interfacially active enzyme, care must be taken to separate the structural requirements for substrate binding and catalysis from kinetic effects related to differences in interface structure. For PLC, this has been done using short-chain PCs. In order for a

Two Site Single Subunit Model

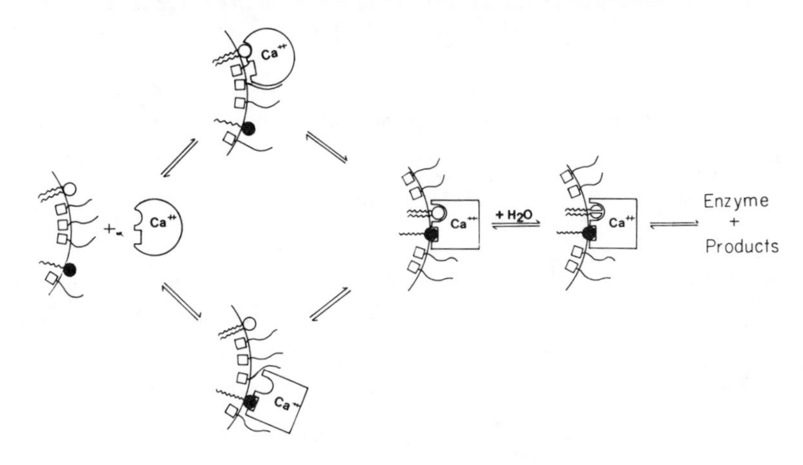

Two Site Dimer Model

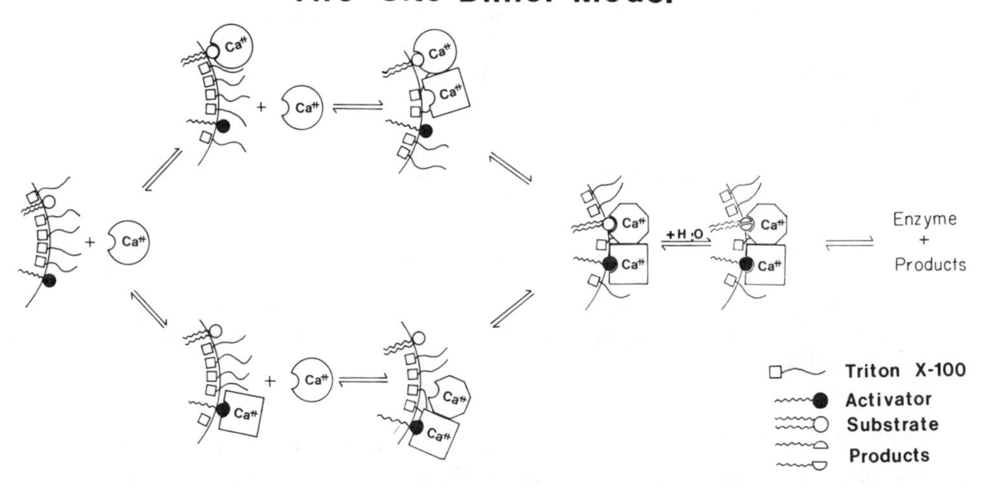

FIGURE 3. The "Dual Phospholipid Model" for phospholipase A_2 action is illustrated schematically. There are two possible ways in which the requirement for two kinds of phospholipid can be visualized. In the "Two Site Single Subunit Model", the monomeric enzyme has two functional sites, an activator site (binds activator phospholipid) and a catalytic site (binds substrate phospholipid). For purposes of illustration, the activator and substrate phospholipids are localized in a mixed micelle interface containing the nonionic detergent Triton X-100, but they could also be shown in a bilayer membrane vesicle. When an enzyme binds an activator phospholipid in its activator site, this causes a conformational change in the enzyme indicated by a circle to a square. When a second phospholipid in the micelle interface is bound to the catalytic site, the Michaelis complex is formed, and it goes on to products. A random mechanism of binding of activator and substrate is assumed here. In the "Two Site Dimer Model", each enzyme monomer has only one functional site. When the monomeric enzyme binds an activator phospholipid, it causes a conformational change, again indicated by a circle to a square. However, this conformational change induces aggregation of the otherwise monomeric enzyme on the micelle interface. When a substrate phospholipid is bound to the second enzyme subunit (indicated by an octagon), then the Michaelis complex is formed and proceeds to products. Again, a random order for binding of activator and substrate is assumed. The aggregated dimeric enzyme thus provides an activator site on one subunit and a catalytic site on another subunit. The critical feature is that the Michaelis complex is by definition an asymmetric dimer or higher order aggregate in which one subunit constitutes the activator subunit and the second subunit constitutes the catalytic subunit at any given time. Experiments have shown that the enzyme is aggregated when bound to phospholipid at the lipid-water interface. The critical question is whether the functional subunit is a monomer or dimer in the aggregated enzyme. In either case, the Dual Phospholipid Model can account for the observed kinetics of substrate hydrolysis. (Reproduced from Dennis, E. A. and Pluckthun, A., in *Enzymes of Lipid Metabolism II,* Freysz, L., Dreyfus, H., Massarelli, R., and Gatt, S., Eds., Plenum Press, New York, 1986, 121. With permission.)

TABLE 1
Kinetic Parameters for Phospholipase C *(Bacillus cereus)* in Different Assay Systems

Substrate	V_{max} (mmol min^{-1} mg^{-1})	K_m (m*M*)	Ref.
Dibutyroyl–PC			
(Monomer)	100	37	116
Dihexanoyl–PC			
(Monomer)	1000	0.36	66
(Micelle)	2900	0.012	66
Diheptanoyl–PC			
(Monomer)	1340	0.20	66
(Micelle)	2650	0.02	66
Dimyristoyl–PC/Triton® X–100 micelles	2000	2.0	92
	2100	4.0 + 0.2[S][a]	95
Dimyristoyl–PC SUVs[b]	15	6.7	66, 95
Dipalmitoyl–PC SUVs	25	6	66
Diheptanoyl–PC/dipalmitoyl–PC SLUVs[c]	2400	0.2	101

[a] In the Langmuir kinetic model, K_m is not a constant but proportional to substrate concentration,[S].
[b] SUVs = small unilamellar vesicles, see Figure 5.
[c] Ratio of diheptanoyl–PC to dipalmitoyl–PC is 1:4; SLUV = short chain PC/long–chain phospholipid unilamellar vesicles.

phospholipid to bind to PLC and have the glycerol phosphoester bond efficiently hydrolyzed, an ester linkage and greater than four carbons in the fatty acyl chain are necessary.[66] This can be seen in Table 1 by comparing K_m and V_{max} for dibutyroyl-PC to monomeric dihexanoyl-PC. If bulky groups, either a branch methyl or a phenyl group, are adjacent to the carbonyl moiety, K_m increases and V_{max} drops.[66] Figure 4 shows the region of dioctanoyl-PC required for high PLC activity. Therefore, even though PLC is responsible for cleavage of a phosphoester bond in the lipid head group, more hydrophobic segments and a fatty acyl linkage are important for substrate binding. In fact, the lipid segments necessary for binding to PLC are the same as those required for PC binding to the PLA_2 site.[1] Perhaps some of the common kinetic characteristics of the two enzymes are related to the similarity in PC regions necessary for substrate binding.

B. SUBSTRATE PHYSICAL FORMS

Although the substrate for these enzymes is present *in vivo* as membranes or as micellar emulsions with bile salts, choosing the correct physical form of the substrate for *in vitro* assays and kinetic analysis is not straightforward. *In vitro,* phospholipids can assume a variety of physical forms depending on the chemical structure, temperature, and presence of other additives in the system (Figure 5). Each of these forms has its own characteristics and offers different advantages and disadvantages for the kineticist. The main problem is that the form of the substrate dramatically affects the activity of the phospholipases. A balance must be drawn between the membrane, which is the natural form of the substrate but which is particularly

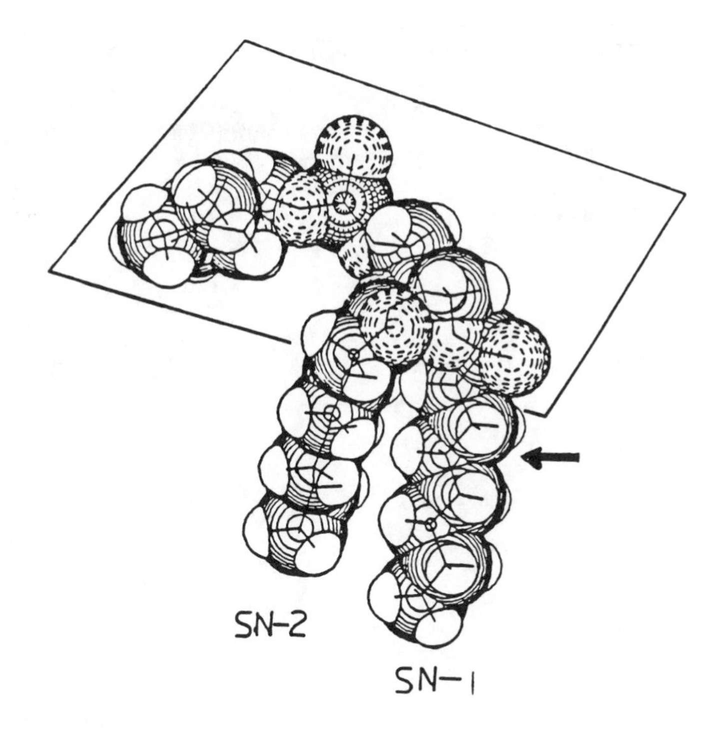

FIGURE 4. Space filling representation of dioctanoyl-PC based on an intramolecular energy minimized conformation for long-chain phospholipids. The boxed region surrounds portions of the phospholipid molecule important for binding to phospholipase C *(B. cereus)*. The arrow indicates where a four carbon fatty acyl chain at the *sn*-1 position would end. (Reproduced from El-Sayed, M. Y., Debose, C. D., Coury, L. A., and Roberts, M. F., *Biochim, Biophys. Acta,* 837, 325, 1985. With permission.)

intractable for kinetic analysis, and the micellar or vesicular forms that are easier to deal with kinetically, but are not the native form of PC. The following are the various forms of phospholipids that are currently used to assay these enzymes.

Naturally occurring long-chain phospholipids hydrated in aqueous solution form multilamellar vesicles (MLV). Upon the application of ultrasonic energy or mechanical manipulation, these MLVs can be converted to unilamellar vesicles (LUVs, large unilamellar vesicles; SUVs, small unilamellar vesicles). Unilamellar vesicles are not thermodynamically stable and over some time scale (that depends on the identity of the PC and incubation temperature) will convert back to multilamellar structures.

Long-chain PCs by themselves do not form micelles. If synthetic PC molecules with chain lengths of 6 to 8 carbons are dispersed in water, micelles form. The critical micelle concentration (the point at which molecules dispersed as hydrated monomers begin to aggregate) and micelle size depend on chain length.[67,68] A short-chain PC molecule in a micelle has conformational, dynamic, and packing features analogous to those of naturally occurring PCs in bilayers.[69-71] In order to form micelles from long-chain PCs, detergents such as Triton X-100, deoxycholate, zwittergent, etc., need to be added. Physical characterization of detergent-mixed micelles show that PC in these micelles has conformational properties similar to bilayer PC, but phospholipid-phospholipid interactions are weaker, if not nonexistent, at high detergent concentrations.[72-75] The new detergent-phospholipid hydrophobic interactions may be similar to what is observed in pure phospholipid systems (in fact, phase transitions from a gel state chain orientation to a liquid crystalline-like chain can be detected in micelles, although the enthalpy of the transition

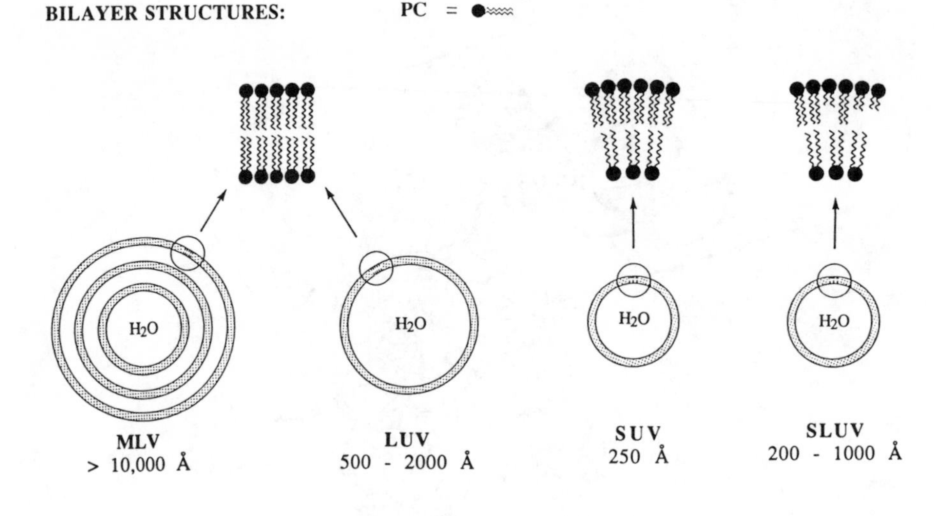

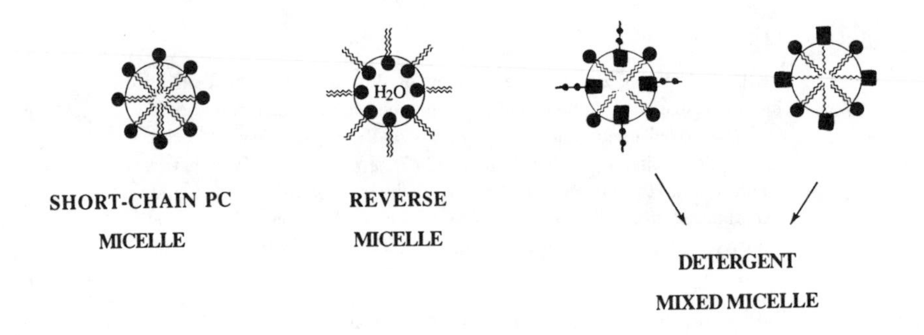

FIGURE 5. Structures formed by phosphatidylcholine (PC) that have been used as substrates for the phospholipases are illustrated schematically. For bilayer membrane vesicles their diameters are also indicated. Multilamellar vesicles (MLV) and large unilamellar vesicles (LUV) made from a single long-chain PC or from short-chain PC combined with liquid crystalline long-chain PC molecules (SLUV) have such large diameters that the outer surface, compared to the cross-sectional area of a phospholipid molecule, is relatively flat. For small unilamellar vesicles (SUV and SLUVs made from gel state long-chain PC), however, the outer surface is highly curved as illustrated. Synthetic phospholipids with short fatty acid chains can form monomers at low concentrations below their critical micelle concentration (cmc), and above it they can form micelles in aqueous solution without the addition of detergents. In certain mixed-solvent systems, long-chain phospholipids form reverse micelles, or microemulsions with a small amount of water in the central core. In the presence of detergent, the phospholipids are solubilized in mixed micelles, shown here schematically for the nonionic detergent Triton® X-100 and a charged or zwitterionic detergent with a well-defined head group.

is invariably smaller), but polar head-group interactions will definitely be different. For example, specific charge interactions of the PC head group cannot form in a nonionic detergent matrix such as Triton X-100. If the components are dispersed in an organic solvent with small amounts of water, inverted micelles form, but these are harder to characterize and form a poor substrate assay system for detailed kinetics.

A special type of unilamellar vesicle forms spontaneously by adding 20 mol% short-chain PC to long-chain phospholipid multibilayers.[76] These SLUVs (short-chain PC/long-chain phospholipid unilamellar vesicles) have unique properties and are useful in understanding the interfacial behavior of water-soluble phospholipases. Residual multibilayers, monomers, or micelles are not detected in these SLUV mixtures.[77] Vesicle sizes range from 150 to 1000 Å in diameter with most (>75%) of the short-chain PC in the SLUV outer monolayer. These vesicles are stable and impermeable to ions and fluorescent dyes.[76,78,79] The PC in each type of aggregate will have slightly different packing properties and interactions with other components. Careful kinetic studies can determine if these parameters are important for phospholipase activity.

C. PHOSPHOLIPASE ACTIVITY AS A FUNCTION OF AGGREGATION STATE

1. Monomer vs. Micelle

The activity of phospholipases toward these substrates depends dramatically on the physical form of the PC. A striking difference exists between monomer and micelle. Figure 6 shows the *B. cereus* PLC-catalyzed hydrolysis of monomeric and micellar dihexanoyl-PC and the *Naja naja naja* PLA_2 activity toward diheptanoyl-PC: for both enzymes, monomers are poorer substrates (with higher apparent K_m and lower V_{max} values as seen in Table 1 for PLC) compared to micelles.[66,80] Similar kinetic behavior is seen for other PLA_2 enzymes.[1,2] This increase in V_{max} has been termed "interfacial activation", and a variety of explanations have been proposed.

Several of these focus on aggregation-induced changes in the substrate. NMR studies have indicated a small conformational change does occur upon lipid aggregation,[69,70,81] This involves constraining the fatty acyl chains to a more parallel orientation. The resultant PC has a head group on average parallel to the interface and a pronounced kink at the beginning of the *sn*-2 chain. Recent work by Thuren[82] using dipyrene-labeled lipids is consistent with substrate conformational changes as an explanation for interfacial activation, at least in that specific system. Most of the evidence linking substrate conformational changes with enzyme activation is circumstantial. A more direct test as to whether or not these lipid changes are catalytically relevant is to examine a phospholipid molecule constrained in the eclipsed conformation. For example, with a cyclopentanoid PC[83] there is no change in PLA_2 activity on going from monomer to micelle — evidence that the observed conformational change may, indeed, be critical to interfacial activation.[84]

Other workers have attributed interfacial activation to phospholipid aggregation inducing conformational changes in the enzyme, dimerization of the enzyme, or a change in polypeptide orientation such that lipid binding is altered, leading to more effective hydrolysis.[85-87] Yet another possibility is that the lipid matrix may affect product release.[88,89] Almost all water-soluble phospholipases exhibit this interfacial activation, although the increase exhibited by each type of phospholipase may be caused by a different mechanism.

2. Surface Dilution Kinetics

Both PLA_2 and PLC exhibit an interesting type of kinetic behavior toward PC solubilized in detergent micelles.[90-92] As more and more detergent is added to a fixed concentration of PC, enzyme activity decreases. Two types of explanations have been proposed, one for PLA_2 and another for PLC. In the case of Triton X-100 as the detergent matrix, the detergent has only very weak affinity for PLA_2.[58] PLA_2 requires two phospholipid molecules for optimal catalysis. As more and more detergent is added to PC/Triton X-100 mixed micelles, the surface concentration of the PC decreases. After one PC molecule is bound to the enzyme, it must search the surface for another. Thus, the enzyme follows a two-step model where initial contact of enzyme and PLA_2 depends on bulk PC and enzyme concentrations, but once bound to the micelle surface, the enzyme must bind a second substrate molecule of which the two-dimensional concentration is important.[93,94] Thus, the PLA_2 activity does indeed depend on the surface concentration; therefore, it shows "surface dilution kinetics".

For PLC, the dependence of activity on detergent concentration has been explained solely by

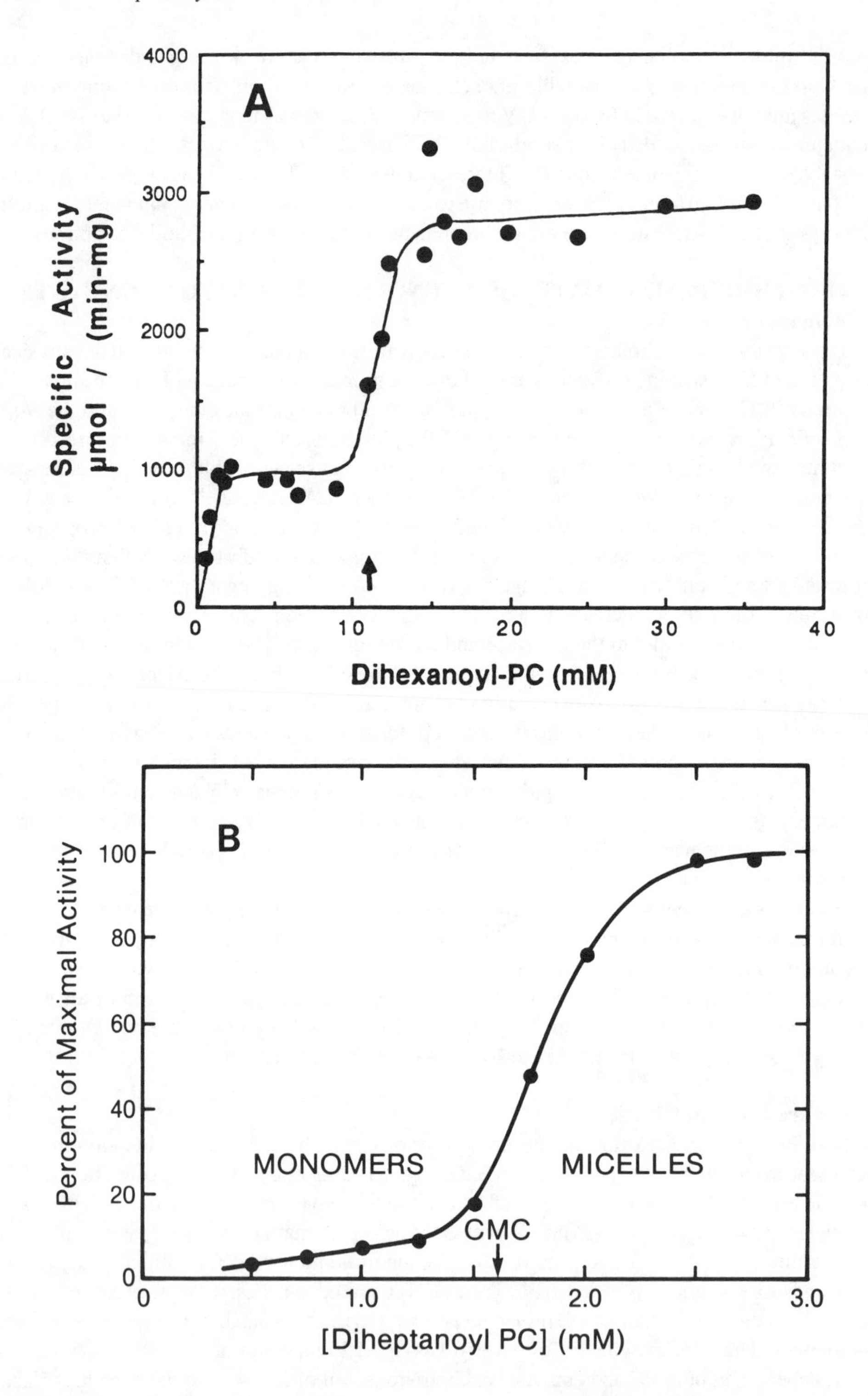

FIGURE 6. (A) Specific activity of phospholipase C *(B. cereus)* toward dihexanoyl-PC as a function of PC concentration. The solid line drawn through the data is derived from the Michaelis-Menten treatment with phase separation for micellization and the following parameters: V_{max} (monomer) = 1000 µmol min^{-1} mg^{-1}, K_m (monomer) = 0.36 m*M*, cmc = 11.1 m*M*, V_{max} (micelle) = 2900 µmol min^{-1} mg^{-1}, K_m (micelle) = 0.04 m*M*. (B) Percent of maximal activity of phospholipase A_2 *(Naja naja naja)* toward diheptanoyl-PC as a function of PC concentration. In both (A) and (B) the cmc of the short-chain phospholipid is indicated by an arrow. (Adapted from Hazlett, T. L., Ph.D. dissertation, University of California at San Diego, 1986, 65. With permission.)

assuming competitive inhibition by the detergents and treating kinetic expressions in a manner following a Langmuir adsorption isotherm.[95] For this enzyme, there is no evidence for multiple phospholipid binding required for optimal activity. The magnitude of the effect of different detergents can be directly related to the K_D of the detergent.

3. Bilayer vs. Micelle

Yet another kind of interfacial phenomenon is seen in comparing enzyme activity toward bilayers and micelles. For all the water-soluble phospholipases examined, bilayer structures are much poorer substrates than micellar species. Studies with pancreatic PLA_2 enzymes often detect lag phases for hydrolysis of the bilayer aggregate. It has been reported that this enzyme shows a pronounced activation around the T_m (transition temperature) of PCs.[96] This may be related to reducing the lag phase to the point where appreciable hydrolysis is detected and/or accumulation of products, which appear to accelerate PC hydrolysis by introducing defects into the interface that facilitate PLA_2 binding.[97,98] Cobra venom PLA_2 does not show lags in activity toward PC vesicles. Its activity is higher toward vesicles of gel state PC (although at best it is still five- to tenfold lower than micelle V_{max} values) than liquid crystalline PC,[99] and a discontinuity occurs around the T_m of the PC. When cobra PLA_2 interacts with small unilamellar vesicles composed of the substrate analogue dihexadecyl-PC, it causes leakage of small ions, but retention of larger species (such as carboxyfluorescein).[51] The K_D for PC binding is temperature invariant over the range 15 to 50°C , even though the PC changes from a gel to the liquid crystalline state.[100] In view of the fact that multiple phospholipids are needed for optimal activity, the nature of the ether-linked substrate analogue binding is unclear. It is possible that the observed leakiness and insensitivity to temperature reflect only activator or catalytic site occupation.

The activity difference for PLC toward bilayers and micelles (either Triton® X-100 mixed micelles or short-chain PC micelles) is roughly 50-fold.[66,88] Interestingly, the enzyme V_{max} values for different micellar substrate systems are similar, while K_m values are much higher for the long-chain PC solubilized in micelles than for pure short-chain PC micelles (Table 1). V_{max} values for pure PC vesicles are the lowest of any assay system, while K_m values are similar to those for PC in Triton® X-100 micelles. These differences could be related to the action/regulation of membrane-bound phospholipases. Such phospholipases are anchored in their bilayer substrate and need to be activated, potentially by local changes in substrate conformation or accessibility (i.e., a change producing a localized "micelle-like" environment).

A unique assay system which addresses several aspects of interfacial activation uses a series of SLUVs as substrates. The short-chain PC in a micellar matrix is an excellent substrate for phospholipases. How its rate of hydrolysis is affected when it is incorporated in a bilayer matrix of long-chain phospholipids, which are normally poor substrates, can shed light on the reasons for the kinetic differences of micelles vs. bilayers. In SLUVs, the short-chain PC is the preferred substrate for both cobra venom PLA_2 (~10-fold higher hydrolysis rates for the short-chain species over the long-chain component) and *B. cereus* PLC (~100-fold higher rates). In fact, hydrolysis of the bilayer anchored short-chain PC proceeds at rates comparable to those for pure micelles.[79,101] This is illustrated for PLC in Table 1 where V_{max} is comparable for pure diheptanoyl-PC micelles, dimyristoyl-PC in Triton® X-100 mixed micelles, and diheptanoyl-PC/dipalmitoyl-PC SLUVs.

The long-chain PC in SLUVs can be treated as an inhibitor (PLC) or poor substrate (PLA_2). Therefore, SLUVs provide a good substrate as a minor component organized in a bilayer matrix where the matrix molecules are essentially nonsubstrates. For PLC, the apparent K_m value extracted from diheptanoyl-PC/dipalmitoyl-PC SLUVs can be further reduced by treating the long-chain lipid as a competitive inhibitor; in this case the K_m for diheptanoyl-PC is 0.02 m*M* (comparable to the value for pure short-chain PC micelles) with the K_i for dipalmitoyl-PC as 1.1 m*M* (slightly tighter than the binding affinity of the enzyme for long-chain PC in Triton X-100

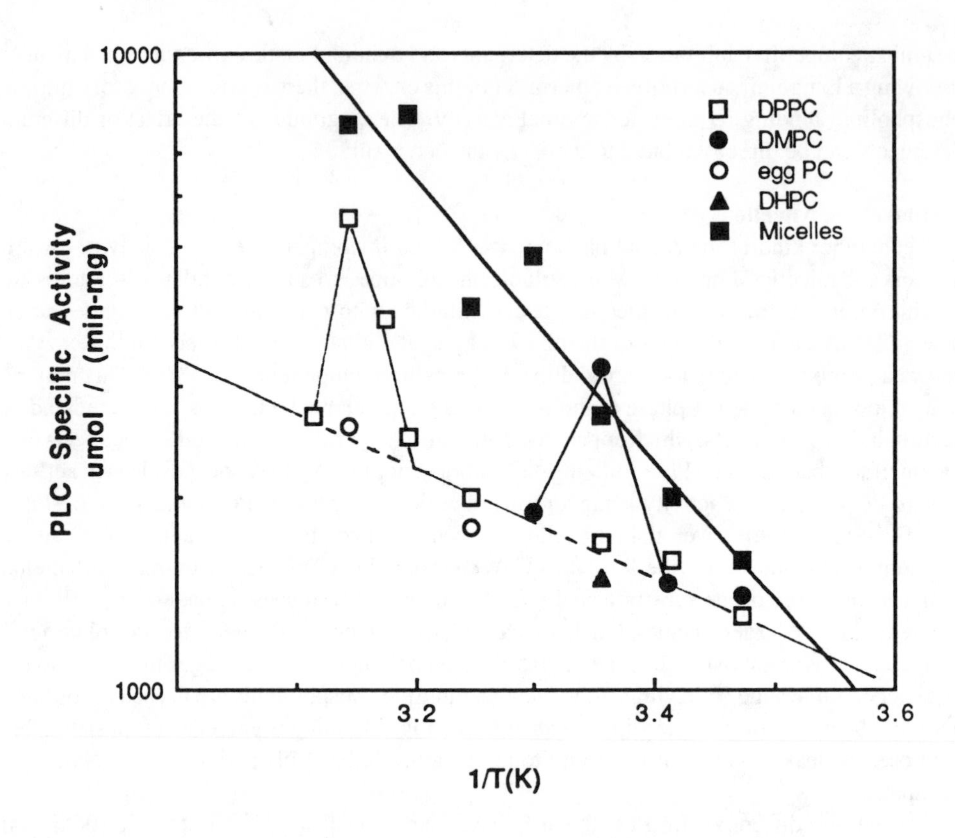

FIGURE 7. Arrhenius plot for phospholipase C *(B. cereus)* activity toward diheptanoyl-PC (5 m*M*) in micelles ([■]) and in SLUVs prepared with different long-chain PCs (20 m*M*): □, dipalmitoyl-PC; ●, dimyristoyl-PC; ○, egg PC; and ▲, dihexadecyl-PC. Note the increase in activity toward diheptanoyl-PC in vesicles (to values almost as high as those for pure diheptanoyl-PC micelles) around the gel-to-liquid crystalline transition of the long-chain PC.

mixed micelles). For PLA_2 action on SLUVs, both apparent K_m and V_{max} values show greater ranges. In a gel state matrix diheptanoyl-PC is hydrolyzed with micelle-like kinetic parameters; in a liquid-crystalline matrix, the short-chain PC as a substrate becomes comparable in both V_{max} and K_m to the long-chain component.[101]

Both enzymes also show an anomalous increase in specific activity toward short-chain PC around the phase transition temperature of the long-chain phospholipid.[101] This is illustrated for PLC in Figure 7, where activities toward diheptanoyl-PC in a wide variety of different saturated and unsaturated long-chain PC SLUVs are shown as a function of assay temperature. For comparison, the temperature dependence of PLC toward pure diheptanoyl-PC micelles is also shown. The increased PLC activity toward SLUVs around the long-chain PC T_m becomes comparable to that for pure micelles. Since the short-chain PC does not exhibit any phase transition and no residual micelles are detected in SLUV preparations around T_m, the activation must reflect fluctuations in head group area or vertical motions of the short-chain PC caused by surrounding long-chain PC molecules as they melt from a gel to a liquid crystalline state. In SLUVs the phosphocholine moiety of the short-chain species has greater motional freedom and is not strongly interacting with neighboring head groups, and the short-chain PCs are dynamically "uncoupled" or "isolated" from long-chain neighbors. Either or both of these could facilitate interactions with the phospholipases.

The results with SLUV kinetics can be used to evaluate two major differences in theories for interfacial activation of phospholipases: (1) the interface facilitates desorption of enzyme from the phospholipid (or product) at the surface, or (2) the interface facilitates adsorption of the enzyme to the surface phospholipid prior to catalysis. With SLUVs, the K_m is low for "good" substrates, i.e., a short-chain PC, and high for poor substrates (long-chain PC). The good substrate molecule has greater flexibility, mobility, and only weakly interacts with neighboring molecules. What is critical to activity is productive binding, not a nonspecific interfacial affinity, and this requires breaking phospholipid-phospholipid interactions in an aggregate and subsequently inserting hydrophobic products back into the bilayer. Thus, SLUV kinetics are more consistent with enzyme adsorption as the key step. The increase in surface "defects" that Jain and Zakim refer to in explaining pancreatic phospholipase A_2 activity in bilayers[97] is really an uncoupling or isolation of one or more phospholipids from the rest of the matrix so they are accessible for productive binding to the enzyme. More complete kinetic treatments of pancreatic PLA_2 action toward vesicles are given elsewhere.[58,98]

IV. REGULATION OF PHOSPHATIDYLCHOLINE CATABOLISM BY-PRODUCTS

A. EFFECTS OF FATTY ACIDS

One of the products of the PLA_2 reaction, fatty acid, is negatively charged as well as hydrophobic. These characteristics affect its behavior in a bilayer and can indirectly, through modulation of bilayer properties, as well as directly, through binding to the phospholipase, affect enzyme activity. The pK_a of fatty acids in bilayers is raised from 5 (monomer or micelle value) to around 7. Having this value around physiological pH makes the fatty acid a very effective uncoupler of pH gradients across cell membranes. The protonated form, which is neutral, can flip across the bilayer and release its proton inside the cell. In fact, in model studies, fatty acids show behavior consistent with a rapid transbilayer flip-flop rate,[102] which must be considerably more rapid than PC or other PC degradation products. Because the fatty acid can chelate metal ions, it can also increase membrane permeability to ions. In sufficient quantities, it is apparent that the fatty acids could be detrimental to cell metabolism since they can alter proton and ion concentrations.[103] One way for the cell to regulate the production of fatty acids is to have them act as negative feedback inhibitors for PLA_2. This could be considered for the PLA_2 from the P388D macrophage-like cell line where arachidonic acid is a potent inhibitor,[104,105] while this fatty acid does affect the lysoPLA significantly.[11]

In model PC systems, fatty acids can lead to increased activity of PLA_2 as well as a pronounced inhibition — the direction of the effect depends on the source of the enzyme and the physical characteristics of the assay system. For example, kinetic studies with the pancreatic PLA_2 indicate this fatty acid can activate the enzyme by promoting binding to the bilayer surface[106] (although this may be irrelevant physiologically since the biological substrate of this species is a bile salt/PC mixed micelle). The fatty acid does not act as a cofactor, but is theorized to lead to organizational defects in the bilayer that are critical to this enzyme binding to PC in the bilayer.[97,106] The extracellular venom enzymes do not show such fatty acid activation with bilayer substrates. With either class of phospholipase, no activation occurs toward short-chain PC micelles or phospholipid/detergent mixed micelle systems with added fatty acid. In fact, unsaturated fatty acid (e.g., oleic acid) inhibits PLA_2 hydrolysis of PC in a variety of assay systems, presumably by acting as a competitive inhibitor.[107]

B. EFFECTS OF LYSOPHOSPHATIDYLCHOLINE

LysoPC forms micelles in solution and is one of the products of PC catabolism that will destabilize the bilayer.[12] This molecule when added to cell suspensions is lytic: to avoid damage, cells have an efficient lysoPLA pathway to clear this compound and keep its concentration at safe levels. Cells can accomplish this in three ways. First, cells have high levels of lysoPLA; its activities are usually 10 to 100 times greater than the PLA_2 that produces the lysophospholipid.

Thus, the cells should be able to clear rapidly any lysoPC formed. Second, there are reacylation enzymes that convert lysoPC back into PC.[3,4] Third, the lysoPC may inhibit PLA_2. While in detergent-mixed micelles, added lysoPC is a mediocre inhibitor of PLA_2 activity toward PC, in real membranes there is one suggestion that this species may be a much more potent inhibitor of membrane-bound PLA_2,[108] and the inhibition may have a physiological role.

Membrane-bound PLA_2 from rat liver mitochondria is initially inhibited by a lipophilic species which has been identified as monolysocardiolipin.[108] Roughly one lysocardiolipid molecule causes detectable inhibition in the presence of 5000 substrate phospholipids. Given the relatively large ratio of phospholipid to lysophospholipid, it is probable that the inhibition results from a direct interaction of the lyso species with the enzyme, as opposed to an effect of the inhibitor on the physical state of the substrate. Such an inhibition could have great importance *in vivo* as a control mechanism. Assuming that cardiolipin degradation products are effective inhibitors when the PLA_2 is acting against bilayer lipids, these species could severely limit degradation of the inner mitochondrial membrane by associating with the enzyme. This negative feedback could represent an important safety valve for the cell. In the event that the normal lysoPC catabolic pathways are blocked, any slight increase in the levels of lysoPC would shut down the PLA_2 enzyme producing it.

C. EFFECTS OF DIGLYCERIDE

The PLC reaction produces a hydrophobic diglyceride (DG) and a water-soluble phosphomonoester. The DG is considerably more hydrophobic than substrate PC molecules. Up to 20 wt% DG can be solubilized in egg PC bilayers forming a single lamellar (L) phase.[109] Above 30 wt% DG, a transition to a hexagonal (H) phase occurs. For PE only 4 wt% DG suffices to promote the transition from a lamellar to a hexagonal phase.[109] Between the pure L and pure H regions both phases coexist and a second lamellar phase is detected. Both mammalian PLA_2 (from rat intestinal mucosa)[110,111] and PLC (from human platelets[110,111] and sheep seminal vesicles[28]) enzymes show activation by DG in phospholipid bilayers. The activation is most pronounced for the amounts of DG (>20 wt%) which lead to a mixture of phases. Once the pure hexagonal phase is produced, enzyme activity is dramatically inhibited. In these studies, the added product affects enzyme activity by altering the physical characteristics of the substrate. Physical studies of the effect of DG on phospholipid bilayers show that the presence of DG appears to immobilize neighboring PC molecules[112] while increasing the area per head group.[109] It may, in fact, be this latter effect which is so important for the attack of phospholipases on bilayer phospholipids. The effect of DG on substrate does not always lead to activation of phospholipases. For example, DG has no effect on PLC activity in endothelial cells;[113] such results are hard to interpret without detailed physical studies of the substrate because the existence of multiple phases depends on the exact lipid mixture.

The conformation of the DG can be affected by proximity of charged molecules;[114] this in turn can affect the activity of phospholipases. Understanding these kinetic effects is best done with micellar rather than bilayer systems to avoid the problems of multiple phases. As an example of this, the α-CH splitting in the NMR spectra for dihexanoylglyceride shows pronounced changes in the presence of deoxycholate or CTAB, but not Triton® X-100 micelles. The spectral changes detected may indicate preferential interactions of the nonionic DG with certain charged phospholipids. Evidence suggesting that removal of this product from the *B. cereus* PLC is rate limiting is derived from kinetic studies of short-chain lecithin micelles.[88] Small amounts (5 to 25 mol%) of charged, but not zwitterionic or nonionic, detergents lead to a 50 to 100 % increase in V_{max} for this enzyme. It appears to be an interfacial phenomenon since monomeric short-chain PC is not affected when similar concentrations of detergent (kept below the cmc) are added. Physical studies of detergent/DG complexes imply a strong interaction when the detergent is charged. This is consistent with the DG binding tightly to the PLC and being removed as new and tighter detergent-DG interactions form.

This type of product inhibition, which can be relieved by a negatively charged lipid or polypeptide, may well exist for membrane forms of the enzyme. It may play some role in the protein kinase C activation. Instead of detergents, if specific proteins or peptides interact with the DG and accelerate the off reaction from PLC, they may make this species accessible to the kinase.

D. WATER-SOLUBLE PHOSPHATE ESTERS

The soluble phosphate esters which are degradation products of the various phospholipase reactions will not affect substrate aggregate structure per se, but can alter phospholipase activity directly by acting as allosteric effectors or competitive inhibitors or indirectly by perturbing subsequent catabolic steps. Neither GPC nor phosphocholine bind appreciably to PLA_2 or PLC; a K_D value for binding to lysoPLA has not been determined. Therefore, it is unlikely significant product inhibition occurs via binding to the phospholipases. One way that these water-soluble products may affect phospholipase activity is by altering Ca^{2+} concentrations. GPC and phosphocholine possess phosphate moieties that can effectively chelate Ca^{2+} ions. This behavior could inhibit phospholipase activities since PLA_2, and several of the membrane-bound PI-specific PLC species require Ca^{2+} for activity.

The terminal water-soluble catabolic product choline (also ethanolamine) is an inhibitor of GPC phosphodiesterase[115] and will affect PC degradation at that level. These water-soluble species and their phosphorylated counterparts may also be involved in regulating PC biosynthesis by binding to enzymes in that pathway.

An excellent way of looking at the balance of complex interactions involving the water-soluble phosphates could be provided by ^{31}P NMR spectroscopy. It has been established that in many diseased states there is an increase in the phosphomonoester and phosphodiester content of tissues, specifically phosphocholine and phosphoethanolamine.[44,45] These may represent end products from phospholipid turnover rather than increases in pools for PC synthesis. Why the steady-state levels are increased is unknown. Trying to inhibit phospholipase activity specifically by perfusing such a system with specific enzyme modulators may lead to a good understanding of how phospholipid degradation is regulated *in vivo*.

V. SUMMARY

Phosphatidylcholine catabolism by phospholipases can be understood in systems involving purified enzymes acting on well-characterized substrates. The methods established for studying these surface-active enzymes are reasonable and have provided many insights. In particular, the influence of the interface on the observed rates of water-soluble enzymes has been addressed. A general model that unifies kinetics of phospholipases on monomers, micelles, and bilayers can be qualitatively described as follows. Any matrix with substantial lipid-lipid interactions will be a poor substrate for water-soluble phospholipases. Micelles (either pure short-chain PC or detergent-mixed micelles) are good substrates because intermolecular interactions are relatively weak. The enzyme when bound to one molecule has access to several other molecules (presumably in favorable conformations), if more than one PC is necessary for efficient hydrolysis. Also, the micelle matrix is more efficient than water at solubilizing hydrophobic products and, hence, removing them from the enzyme active site.

In more specific terms, PLA_2 shows lower relative activity toward monomers than PLC, possibly indicating a need for dehydrated, multiple phospholipids. PLC only shows a two- to threefold difference in activity toward monomers and micelles which could easily be accounted for by postulating (1) that hydrophobic DG removal is rate-limiting and the presence of micelles facilitates this, or (2) that the small conformational change observed upon PC micellization facilitates productive substrate binding to PLC. Bilayers of long-chain phospholipids are poor substrates for phospholipases because of strong lipid-lipid intermolecular interactions. While phospholipases can bind to the surface, in some cases with an apparent K_m comparable to values for mixed micelles,[51] this interaction cannot lead to catalysis unless the enzyme can "isolate" or "uncouple" the substrate molecule from its neighbors in the more tightly packed bilayer. PLC works very poorly toward bilayer PC. In this case productive binding of the enzyme to the substrate may be difficult since the area/head group has decreased and PC-PC interactions may be stronger than PLC-PC interactions. DG product release back into the bilayer may also be more difficult than with a micellar system.

The concept of PC accessibility, which involves area of the polar head group per molecule or motional uncoupling of the substrate PC from other neighboring PC molecules, may extrapolate directly to membrane-bound phospholipases. To make a phospholipid a good substrate for a membrane-bound phospholipase, one needs to uncouple the molecule from its neighbors in the bilayer. This may be facilitated by proteins, peptides, or other lipids. Inhibition of membrane-bound phospholipase activity may occur by certain products of phospholipase action, although generalizations and, indeed, the importance of this cannot be assessed at this stage. Perhaps the most intriguing area left to be addressed is the mechanism of regulation *in vivo*. Techniques such as ^{31}P NMR spectroscopy may contribute much needed information to this aspect of understanding phospholipid catabolism.

ACKNOWLEDGMENTS

We would like to thank Raymond Deems for critical reading of this manuscript. M. F. R. would like to thank Prof. John D. Baldeschwieler for his hospitality at the California Institute of Technology during the time this review was written. Financial support was provided by National Institutes of Health grants GM 26762 (M. F. R.) and GM 20501 (E. A. D.), and National Science Foundation grant DMB 88-17392 (E. A. D.).

REFERENCES

1. **Dennis, E. A.,** Phospholipases, in *The Enzymes, Third Edition, Lipid Enzymology*, Vol. 16, Boyer, P., Ed., Academic Press, New York, 1983, 307.
2. **Waite, M.,** Phospholipases, in *Biochemistry of Lipids and Membranes,* Vance, D. E. and Vance, J. E., Eds., Benjamin/Cummings, Menlo Park, CA, 1985, 299.
3. **Dennis, E. A.,** The regulation of eicosanoid production: role of phospholipases and inhibitors, *Biol Technology,* 5, 1294, 1987.
4. **Choy, P. and Arthur, G.,** Phosphatidylcholine biosynthesis from lysophosphatidylcholine, in *Phosphatidylcholine Metabolism*, Vance, D., Ed., CRC Press, Boca Raton, FL, 1989, chap. 6.
5. **Kanfer, J.**, Phospholipase D and the base exchange enzyme, in *Phosphatidylcholine Metabolism*, Vance, D., Ed., CRC Press, Boca Raton, FL, 1989, chap. 5.
6. **Dennis, E. A.**, Phospholipase A_2 mechanism: inhibition and role in arachidonic acid release, *Drug Dev. Res.*, 10, 205, 1987.
7. **Verheij, H. M., Slotboom, A. J., and de Haas, G. H.**, *Reviews of Physiology, Biochemistry, and Pharmacology*, Vol. 91, Springer-Verlag, Heidelberg, 1981, 91.
8. **Van Kuijk, F. J. G. M., Sevanian, A., Handelman, G. J., and Dratz, E. A.**, A new role for phospholipase A_2: protection of membranes from lipid peroxidation damage, *Trends Biochem. Sci.*, 12, 31, 1987.
9. **Jarvis, A. A., Cain, C., and Dennis, E. A.**, Purification and characterization of a lysophospholipase from human amnionic membranes, *J. Biol. Chem.*, 259, 15188, 1984.
10. **Dennis, E. A., Hazlett, T. L., Deems, R. A., Ross, M. I., and Ulevitch, R. J.**, Phospholipases in the macrophage, in *Prostaglandins, Leukotrienes, and Lipoxins, IV International Washington Spring Symposium,* Bailey, J. M., Ed., Plenum Press, New York, 1985, 213.
11. **Zhang, Y. and Dennis, E. A.**, Purification of a lysophospholipase from a macrophage-like cell line $P388D_1$, *J. Biol. Chem.*, 263, in press.

12. **Stafford, R. E. and Dennis, E. A.**, Lysophospholipids as biosurfactants, *Colloids Surf.*, 30, 47, 1988.
13. **Chien, K. R., Reeves, J. P., Buja, L. M., Bonte, F., Parkey, R. W., and Willerson, J. T.**, Phospholipid alterations in canine ischemic myocardium, *Circ. Res.*, 48, 711, 1981.
14. **Corr, P. B., Snyder, D. W., Lee, B. I., Gross, R. W., Keim, C. R., and Sobel, B. E.**, Pathophysiological concentrations of lysophosphatides and the slow response, *Am. J. Physiol.*, 243, H187, 1982.
15. **Fink, K. L. and Gross, R. W.**, Modulation of canine myocardial sarcolemmal membrane fluidity by amphiphilic compounds, *Circ. Res.*, 55, 585, 1984.
16. **Baine, W. B.**, Cytolytic and phospholipase C activity in Legionella species, *J. Gen. Microbiol.*, 131, 1383, 1985.
17. **Rebecchi, M. J. and Rosen, O. M.**, Purification of a phosphoinositide-specific phospholipase C from bovine brain, *J. Biol. Chem.*, 262, 12526, 1987.
18. **Ryu, S. H., Cho, K. S., Lee, K.-Y., Suh, P.-G., and Rhee, S. G.**, Purification and characterization of two immunologically distinct phosphoinositide-specific phospholipase C from bovine brain, *J. Biol. Chem.*, 262, 12511, 1987.
19. **Lee, K.-Y., Ryu, S. H., Suh, P.-G., Choi, W. C., and Rhee, S. G.**, Phospholipase C associated with particulate fractions of bovine brain, *Proc. Natl. Acad. Sci. U.S.A.*, 84, 5540, 1987.
20. **Ryu, S. H., Suh, P.-G., Cho, K. S., Lee, K.-Y., and Rhee, S. G.**, Bovine brain cytosol contains three immunologically distinct forms of inositolphospholipid-specific phospholipase C, *Proc. Natl. Acad. Sci. U.S.A.*, 84, 6649, 1987.
21. **Ebstein, R. P., Bennett, E. R., Stressman, J., and Lerer, B.**, Isoelectric focusing of human platelet phospholipase C: evidence for multimolecular forms, *Life Sci.*, 40, 161, 1987.
22. **Moscat, J., Herrero, C., Garcia-Barreno, P., and Municio, A. M.**, Phospholipase C — diglyceride lipase is a major pathway for arachidonic acid release in macrophages, *Biochem. Biophys. Res. Commun.*, 141, 367, 1986.
23. **Carter, H. R. and Smith, A. D.**, A comparison of the properties of species of phospholipase C isolated from porcine lymphocytes, *Biochem. Soc. Trans.*, 14, 1159, 1986.
24. **Melin, P. M., Sundler, R., and Jergil, B.**, Phospholipase C in rat liver plasma membranes. Phosphoinositide specificity and regulation by guanine nucleotides and calcium, *FEBS Lett.*, 198, 85, 1986.
25. **Shakir, K. M. and Eil, C.**, Phospholipase C activity in cultured human skin fibroblasts, *Enzyme*, 37, 189, 1987.
26. **Zahler, P., Reist, M., Pilarska, M., and Rosenheck, K.**, Phospholipase C and diacylglycerol lipase activities associated with plasma membranes of chromaffin cells isolated from bovine adrenal medulla, *Biochim. Biophys. Acta*, 877, 372, 1986.
27. **Low, M. G. and Weglicki, W. B.**, Resolution of myocardial phospholipase C into several forms with distinct properties, *Biochem. J.*, 215, 325, 1983.
28. **Hoffman, S. L. and Majerus, P. W.**, Modulation of phosphatidylinositol-specific phospholipase C activity by phospholipid interactions, *J. Biol. Chem.*, 257, 14359, 1982.
29. **Bennett, C. F. and Crooke, S. T.**, Purification and characterization of a phosphoinositide-specific phospholipase C from guinea pig uterus, *J. Biol. Chem.*, 262, 13789, 1987.
30. **Nishihira, J. and Ishibashi, T.**, A phospholipase C with a high specificity for platelet activating factor in rabbit liver light mitochondria, *Lipids*, 21, 780, 1986.
31. **Daniel, L. W., Small, G. W., and Schmitt, J. D.**, Alkyl-linked diglycerides inhibit protein kinase C activation by diacylglycerides, *Biochem. Biophys. Res. Commun.*, 151, 291, 1988.
32. **Davitz, M. A., Hereld, D., Shak, S., Krakow, J., Englund, P. T., and Nussenzweig, V.**, A glycan-phosphatidylinositol-specific phospholipase D in human serum, *Science*, 238, 81, 1987.
33. **Fox, J. A., Duszenko, M., Ferguson, M. A. J., Low, M. G., and Cross, G. A. M.**, Purification and characterization of a novel glycan-phosphatidylinositol-specific phospholipase C from *Trypanosoma brucei*, *J. Biol. Chem.*, 261, 15767, 1986.
34. **Wolf, R. A. and Gross, R. W.**, Identification of neutral active phospholipase C which hydrolyzes choline glycerophospholipids and plasmalogen selective phospholipase A_2 in canine myocardium, *J. Biol. Chem.*, 260, 7295, 1985.
35. **Bomalski, J. S. and Clark, M. A.**, The effect of sn-2 fatty acid substitution on phospholipase C enzyme activities, *Biochem. J.*, 244, 497, 1987.
36. **Pattinson, N. R.**, Identification of a phosphatidylcholine active phospholipase C in human gallbladder bile, *Biochem. Biophys. Res. Commun.*, 150, 890, 1988.
37. **Matsuzawa, Y. and Hostetler, K. Y.**, Properties of phospholipase C isolated from rat liver lysosomes, *J. Biol. Chem.*, 255, 646, 1980.
38. **Eisen, D., Bartolf, M., and Franson, R. C.**, Inhibition of lysosomal phospholipases C and A in rabbit alveolar macrophages, polymorphonuclear leukocytes and rat liver by sodium bisulfite, *Biochim. Biophys. Acta*, 793, 106, 1984.

39. **Huterer, S. and Wherrett, J. R.**, Degradation of lysophosphatidylcholine by lysosomes. Stimulation of lysophospholipase C by taurocholate and deficiency in Niemann-Pick fibroblasts, *Biochim. Biophys. Acta,* 794, 1, 1984.
40. **Gamache, D. A., Fawzy, A. A., and Franson, R. C.**, Preferential hydrolysis of peroxidized phospholipid by lysosomal phospholipase C, *Biochim. Biophys. Acta,* 958, 116, 1988.
41. **Meerson, F. Z., Kagan, V. E., Kozlov, Y. P., Belkina, L. M., and Arkhipenko, Y. V.**, *Basic Res. Cardiol.*, 77, 465, 1982.
42. **Burt, C. T., Pluskal, M. G., and Sreter, R. A.**, Generation of phosphodiesters during fast-to-slow muscle transformation. A ^{31}P NMR study, *Biochim. Biophys. Acta,* 721, 492, 1982.
43. **Billadello, J. J., Gard, J. K., Ackerman, J. J. H., and Gross, R. W.**, Determination of intact-tissue glycerophosphorylcholine levels by quantitative ^{31}P nuclear magnetic resonance spectroscopy and correlation with spectrophotometric quantification, *Anal. Biochem.*, 144, 269, 1985.
44. **Maris, J. M., Evans, A. E., McLaughin, A. C., D'Angio, G. J., Bolinger, L., Manos, H., and Chance, B.**, ^{31}P nuclear magnetic resonance spectroscopic investigation of human neuroblastoma in situ, *N. Engl. J. Med.*, 312, 1500, 1985.
45. **Daly, P. F., Lyon, R. C., Faustino, P. J., and Cohen, J. S.**, Phospholipid metabolism in cancer cells monitored by ^{31}P NMR spectroscopy, *J. Biol. Chem.*, 262, 14875, 1987.
46. **Dijkstra, B. W., Kalk, K. H., Hol, W. G. J., and Drenth, J.**, Structure of bovine pancreatic phospholipase A_2 at 1.7 A resolution, *J. Mol. Biol.*, 147, 97, 1981.
47. **Brunie, S., Bolin, J., Gewirth, D., and Sigler, P. B.**, The refined crystal structure of dimeric phospholipase A2 at 2.5 A, *J. Biol. Chem.*, 260, 9742, 1985.
48. **Scott, D. L., Achari, A., Zajac, M., and Sigler, P. B.**, Crystallization and preliminary diffraction studies of the Lys-49 phospholipase A_2 from *Agkistrodon piscivorus piscivorus, J. Biol. Chem.,* 261, 1233,1986.
49. **Van den Bergh, C. J., Bekkers, C. A. P. A., De Geus, P., Verheij, H. M., and De Haas, G. H.**, Secretion of biologically active porcine prophospholipase A_2 by *Saccharomyces cerevisiae, Eur. J. Biochem .,* 170, 241, 1987.
50. **Roberts, M. F., Deems, R. A., and Dennis, E. A.**, Spectral perturbations of the histidine and tryptophan in cobra venom phospholipase A_2 upon metal ion and mixed micelle binding, *J. Biol. Chem.*, 252, 6011, 1977.
51. **DeBose, C. D. and Roberts, M. F.**, The interaction of dialkyl ether lecithins with phospholipase A_2 *(Naja naja naja):* composition of the interface modulates calcium and lecithin binding by the enzyme, *J. Biol. Chem.*, 258, 6327, 1983.
52. **Menashe, M., Romero, G., Biltonen, R. L., and Lichtenberg, D.**, Hydrolysis of dipalmitoylphosphatidylcholine small unilamellar vesicles by porcine pancreatic phospholipase A_2, *J. Biol. Chem.*, 261, 5328, 1986.
53. **Roberts, M. F., Adamich, M., Robson, R. J., and Dennis, E. A.**, Phospholipid activation and specificity reversal of cobra venom phospholipase A_2. I. Lipid-lipid or lipid-enzyme interaction, *Biochemistry,* 18, 3301, 1979.
54. **Adamich, M., Roberts, M. F., and Dennis, E. A.**, Phospholipid activation and specificity reversal of cobra venom phospholipase A_2. II. Characterization of the phospholipid-enzyme interaction, *Biochemistry,* 18, 3308, 1979.
55. **Pluckthun, A. and Dennis, E. A.**, The role of monomeric activators in cobra venom phospholipase A_2 action, *Biochemistry,* 21, 1750, 1982.
56. **Dennis, E. A. and Pluckthun, A.**, Mechanism of interaction of phospholipase A_2 with phospholipid substrates and activators, in *Enzymes of Lipid Metabolism II,* Freysz, L., Dreyfus, H., Massarelli, R., and Gatt, S., Eds., Plenum Press, New York, 1986, 121.
57. **Hendrickson, H. S. and Dennis, E. A.**, Kinetic analysis of the dual phospholipid model for phospholipase A_2 action, *J. Biol. Chem.,* 259, 5734, 1984.
58. **Roberts, M. F., Deems, R. A., and Dennis, E. A.**, Dual role of interfacial phospholipid in phospholipase A_2 catalysis, *Proc. Natl. Acad. Sci. U.S.A.,* 74, 1950, 1977.
59. **DeBose, C. D., Burns, R. A., Jr., Donovan, J. M., and Roberts, M. F.**, Methyl branching in short-chain lecithins: are both chains important for effective phospholipase A_2 activity?, *Biochemistry*, 24, 1298, 1985.
60. **Roberts, M. F., Otnaess, A.-B., Kensil, C. A., and Dennis, E. A.**, The specificity of phospholipase A_2 and phospholipase C in a mixed micellar system, *J. Biol. Chem.*, 253, 1252, 1978.
61. **Sundler, R., Alberts, A. W., and Vagelos, P. R.**, Enzymatic properties of phosphatidylinositol inositol-phosphohydrolase from *Bacillus cereus, J. Biol. Chem.*, 253, 4175, 1978.
62. **Little, C. and Otnaess, A.-B.**, The metal ion dependence of phospholipase C from *Bacillus cereus, Biochim. Biophys. Acta,* 391, 326, 1975.
63. **El-Sayed, M. Y. and Roberts, M. F.**, Lanthanide derivatives of phospholipase C from *Bacillus cereus, Biochim. Biophys. Acta,* 744, 291, 1983.
64. **El-Sayed, M. Y.**, Interaction of phospholipase C *(Bacillus cereus)* with Amphipathic Molecules: A Model for Protein Lipid Interactions, Ph.D. dissertation, Massachusetts Institute of Technology, Cambridge, 1984.

65. **Clark, M. A., Shorr, R. G. L., and Bomalski, I. S.**, Antibodies prepared to *Bacillus cereus* phospholipase C crossreact with a phosphatidylcholine preferring phospholipase C in mammalian cells, *Biochem. Biophys. Res. Commun.*, 140, 114, 1986.
66. **El-Sayed, M. Y., DeBose, C. D., Coury, L. A., and Roberts, M. F.**, Sensitivity of phospholipase C (Bacillus cereus) activity to lecithin structural modifications, *Biochim. Biophys. Acta*, 837, 325, 1985.
67. **Tausk, R. J. M., Esch, J. van, Karmiggelt, J., Voordouw, G., and Overbeek, J. Th. G.**, Physical chemical studies of short-chain lecithin homologues. I. Micellar weights of dihexanoyl- and diheptanoyllecithin, *Biophys. Chem.*, 1, 184, 1974.
68. **Lin, T.-L., Chen, S.-H., and Roberts, M. F.**, Thermodynamic analyses of the growth and structure of asymmetric linear short-chain lecithin micelles based on small angle neutron scattering data, *J. Am. Chem. Soc.*, 109, 2321, 1987.
69. **Burns, R. A., Jr. and Roberts, M. F.**, ^{13}C NMR studies of short-chain lecithins: motional and conformational characteristics of micellar and monomeric phospholipid, *Biochemistry*, 19, 3100, 1980.
70. **Burns, R. A., Jr., Roberts, M. F., Dluhy, R. A., and Mendelsohn, R.**, Monomer-to-micelle transition of dihexanoyl phosphatidylcholine: ^{13}C NMR and Raman studies, *J. Am. Chem. Soc.*, 104, 430, 1982.
71. **Lin, T.-L., Chen, S.-H., Gabriel, N. E., and Roberts, M. F.**, Use of small angle neutron scattering to determine the structure and interaction of dihexanoylphosphatidylcholine micelles, *J. Am. Chem. Soc.*, 108, 3499, 1986.
72. **Robson, R. J. and Dennis, E. A.**, Characterization of mixed micelles of phospholipids of various classes and a synthetic, homogeneous analogue of the nonionic detergent Triton X-100 containing nine oxyethylene groups, *Biochim. Biophys. Acta*, 508, 513, 1978.
73. **Robson, R. J. and Dennis, E. A.**, Micelles of nonionic detergents and mixed micelles with phospholipids, *Acc. Chem. Res.*, 16, 251, 1983.
74. **Burns, R. A., Jr., Stark, R. E., Vidusek, D. A., and Roberts, M. F.**, Dependence of phosphatidylcholine ^{31}P relaxation times and $^{31}P\{^{1}H\}$ NOE distribution on aggregate structure, *Biochemistry*, 22, 5084, 1983.
75. **Kensil, C. A. and Dennis, E. A.**, Alkaline hydrolysis of phospholipids in model membranes and the dependence on their state of aggregation, *Biochemistry*, 20, 6079, 1981.
76. **Gabriel, N. E. and Roberts, M. F.**, Spontaneous formation of stable unilamellar vesicles, *Biochemistry*, 23, 4011, 1984.
77. **Gabriel, N. E. and Roberts, M. F.**, Short-chain lecithins: how to form unilamellar vesicles from micelles and phospholipid multibilayers, *Colloids Surf.*, 30, 113, 1988.
78. **Gabriel, N. E. and Roberts, M. F.**, Interaction of short-chain lecithin with long-chain phospholipids: characterization of vesicles which form spontaneously, *Biochemistry*, 25, 2812, 1986.
79. **Gabriel, N. E. and Roberts, M. F.**, Short-chain lecithin/long-chain phospholipid unilamellar vesicles: asymmetry, dynamics, and enzymatic hydrolysis of the short-chain component, *Biochemistry*, 26, 2432, 1987.
80. **Hazlett, T. L.**, Ligand Binding and Ligand-Induced Aggregation of Phospholipase A_2 from *Naja naja naja* (Indian Cobra) Venom, Ph.D. dissertation, University of California at San Diego, La Jolla, 1986.
81. **Roberts, M. F., Bothner-By, A. A., and Dennis, E. A.**, Magnetic nonequivalence within the fatty acyl chains of phospholipids in membrane models: ^{1}H NMR studies of the α-methylene groups, *Biochemistry*, 17, 935, 1978.
82. **Thuren, T.**, A model for the molecular mechanism of interfacial activation of phospholipase A_2 supporting the substrate theory, *FEBS Lett.*, 229, 95, 1988.
83. **Barlow, P. N., Vidal, J.-C., Lister, M. D., Hancock, A. J., and Sigler, P. B.**, Synthesis and some properties of constrained short-chain phosphatidylcholine analogues: (+)- and (-)-(1,3/2)-1-*O*-(phosphocholine)-2,3-*O*-dihexanoylcyclopentane-1,2,3-triol, *Chem. Phys. Lipids*, 46, 157, 1988.
84. **Barlow, P. N., Lister, M. D., Sigler, P. B., and Dennis, E. A.**, Probing the role of substrate conformation in phospholipase A_2 action on aggregated phospholipids using constrained phosphatidylcholine analogues, *J. Biol. Chem.*, 263, 12954, 1988.
85. **Hazlett, T. L. and Dennis, E. A.**, Aggregation studies on fluorescein coupled cobra venom phospholipase A_2, *Biochemistry*, 24, 6152, 1985.
86. **Hazlett, T. L. and Dennis, E. A.**, Effect of phospholipid on fluorescence polarization and lifetimes of a fluorescein-labeled phospholipase A_2, *Biochim. Biophys. Acta*, 958, 172, 1988.
87. **Hazlett, T. L. and Dennis, E. A.**, Lipid-induced aggregation of phospholipase A_2: sucrose density gradient ultracentrifugation and crosslinking studies, *Biochim. Biophys. Acta*, 961, 22, 1988.
88. **El-Sayed, M. Y. and Roberts, M. F.**, Charged detergents enhance the activity of phospholipase C *(Bacillus cereus)* toward micellar short-chain lecithins, *Biochim. Biophys. Acta*, 831, 325, 1985.
89. **Lombardo, D., Fanni, T., Pluckthun, A., and Dennis, E. A.**, Rate determining step in phospholipase A_2 mechanism: ^{18}O isotope exchange examined by ^{13}C NMR, *J. Biol. Chem.*, 261, 11663, 1986.
90. **Dennis, E. A.**, Phospholipase A_2 activity towards phosphatidylcholine in mixed micelles: surface dilution kinetics and the effect of thermotropic phase transitions, *Arch. Biochem. Biophys.*, 158, 485, 1973.

91. **Deems, R.A., Eaton, B. R., and Dennis, E. A.**, Kinetic analysis of phospholipase A_2 activity toward mixed micelles and its implications for the study of lipolytic enzymes, *J. Biol. Chem.*, 250, 9013, 1975.
92. **Eaton, B. R. and Dennis, E. A.**, Analysis of phospholipase C (*Bacillus cereus*) action toward mixed micelles of phospholipid and surfactant, *Arch. Biochem. Biophys.*, 176, 604, 1976.
93. **Roberts, M. F., Deems, R.A., and Dennis, E. A.**, Dual role of interfacial phospholipid in phospholipase A_2 catalysis, *Proc. Natl. Acad. Sci. U.S.A.*, 74, 1950, 1977.
94. **Hendrickson, H. S. and Dennis, E. A.**, Analysis of the kinetics of phospholipid activation of cobra venom phospholipase A_2, *J. Biol. Chem.*, 259, 5740, 1984.
95. **Burns, R. A., Jr., El-Sayed, M. Y., and Roberts, M. F.**, Kinetic model for surface active enzymes based on the Langmuir adsorption isotherm: phospholipase C *(Bacillus cereus)* activity toward dimyristoylphosphatidylcholine/detergent mixed micelles, *Proc. Nat. Acad. Sci. U.S.A.*, 79, 4902, 1982.
96. **Wilschut, J. C., Regts, J., Westenberg, H., and Scherphof, G.**, Action of phospholipases A_2 on phosphatidylcholine bilayers. Effects of the phase transition, bilayer curvature and structural defects, *Biochim. Biophys. Acta,* 508, 185, 1978.
97. **Jain, M. K. and Zakim, D.**, The spontaneous incorporation of proteins into preformed bilayers, *Biochim. Biophys. Acta,* 906, 33, 1987.
98. **Lichtenberg, D., Romero, G., Menashe, M., and Biltonen, R. L.**, Hydrolysis of dipalmitoylphosphatidylcholine large unilamellar vesicles by porcine pancreatic phospholipase A_2, *J. Biol. Chem.*, 261, 5334, 1986.
99. **Kensil, C. A. and Dennis, E. A.**, Action of cobra venom phospholipase A_2 on the gel and liquid crystalline states of dimyristoyl and dipalmitoyl phosphatidylcholine vesicles, *J. Biol. Chem.*, 254, 5843, 1979.
100. **DeBose, C. D.**, Interaction of Phospholipase A_2 with Substrates and Substrate Analogues: A Model for Surface Active Enzymes, Ph.D. dissertation, Massachusetts Institute of Technology, Cambridge, 1984.
101. **Gabriel, N. E., Agman, N. V., and Roberts, M. F.**, Enzymatic hydrolysis of short-chain lecithin/long-chain phospholipid unilamellar vesicles: sensitivity of phospholipases to matrix phase state, *Biochemistry,* 26, 7409, 1987.
102. **Hamilton J. A. and Cistola, D. P.**, Transfer of oleic acid between albumin and phospholipid vesicles, *Proc. Natl. Acad. Sci. U. S. A.*, 83, 82, 1986.
103. **Herrero, A. A., Gomez, R. F., Snedecor, B., Tolman, C. J., and Roberts, M. F.**, Growth inhibition of *Clostridium thermocellum* by carboxylic acids: a mechanism based on uncoupling by weak acids, *Appl. Microbiol. Biotechnol.*, 22, 53, 1985.
104. **Ulevitch, R. J., Sano, M., Watanabe, Y., Lister, M. D., Deems, R. A., and Dennis, E. A.**, Solubilization and characterization of a membrane-bound phospholipase A_2 from the P388D_1 macrophage-like cell line, *J. Biol. Chem.*, 263, 3079, 1988.
105. **Lister, M. D., Deems, R. A., Watanabe, Y., Ulevitch, R. J., and Dennis, E. A.**, Kinetic analysis of the Ca^{2+} dependent, membrane-bound, macrophage phospholipase A_2 and the effects of arachidonic acid, *J. Biol. Chem.*, 263, 7506, 1988.
106. **Jain, M. K., Edmond, M. R., Verheij, H. M., Apitz-Castro, R., Dijkman, R., and De Haas, G. H.**, Interaction of phospholipase A_2 and phospholipid bilayers, *Biochim. Biophys. Acta,* 688, 341, 1982.
107. **Pluckthun, A. and Dennis, E. A.**, Activation, aggregation, and product inhibition of cobra venom phospholipase A_2 and comparison with other phospholipases, *J. Biol. Chem.*, 260, 11099, 1985.
108. **Reers, M. and Pfeiffer, D. R.**, Inhibition of mitochondrial phospholipase A_2 by mono- and dilysocardiolipin, *Biochemistry,* 26, 8038, 1987.
109. **Das, S. and Rand, R. P.**, Modification by diacylglycerol of the structure and interaction of various phospholipid bilayer memnbranes, *Biochemistry*, 25, 2882, 1986.
110. **Dawson, R. M. C., Hemington, N. L., and Irvine, R. F.**, Diacylglycerol potentiates phospholipase attack upon phospholipid bilayers: possible connection with cell stimulation, *Biochem. Biophys. Res. Commun.*, 117, 196, 1983.
111. **Dawson, R. M. C., Irvine, R. F., Bray, J., and Quinn, P. J.**, Long-chain unsaturated diacylglycerols cause a perturbation in the structure of phospholipid bilayers rendering them susceptible to phospholipase attack, *Biochem. Biophys. Res. Commun.*, 125, 836, 1984.
112. **DeBoeck, H. and Zidovetzki, R.**, Nuclear magnetic resonance study of mixed diacylglycerol/phosphatidylcholine bilayers, *Biophys. J.*, 53, 495a, 1988.
113. **Martin, T. W., Wysolmerski, R. B., and Lagunoff, D.**, Phosphatidylcholine metabolism in endothelial cells: evidence for phospholipase A_2 and a novel calcium-independent phospholipase C, *Biochim. Biophys. Acta,* 917, 296, 1987.
114. **Roberts, M. F.**, submitted.
115. **Baldwin, J. J. and Cornatzer, W. E.**, Rat kidney glycerylphosphorylcholine diesterase, *Biochim. Biophys. Acta,* 164, 195, 1968.
116. **Little, C.**, Phospholipase C from *Bacillus cereus.* Action on some artificial lecithins, *Acta Chem. Scand.*, B31, 267, 1977.

Chapter 9

METABOLISM OF THE ETHER ANALOGUES OF PHOSPHATIDYLCHOLINE, INCLUDING PLATELET ACTIVATING FACTOR

Fred Snyder, Ten-ching Lee, and Merle L. Blank

TABLE OF CONTENTS

I. INTRODUCTION

The alkyl and alk-1-enyl (plasmalogen) types of ether-linked phospholipids are present in mammalian cells almost exclusively as structural analogues of phosphatidylcholine and phosphatidylethanolamine. With few exceptions, the alkyl type is found predominantly as choline phosphoglycerolipids, whereas the plasmalogens occur mainly as ethanolamine phosphoglycerolipids. The terms plasm*anyl* and plasm*enyl* designate the alkylacyl and alk-1-enylacyl types of glycerophospholipids, respectively. These names were originally recommended in 1976 by the Working Group on Lipid Nomenclature for the IUPAC-IUB Commission on Biochemical Nomenclature[1] in order to provide a simple term for the "diradylglycerophospho-" radicals that would be comparable to "phosphatidyl", which designates the "1,2-diacyl-*sn*-glycero-3-phospho" radical. Thus, plasmanylcholine and plasmenylcholine can be interchanged with 1-alkyl-2-acyl-*sn*-glycero-3-phosphocholine (alkylacyl-GPC) and 1-alk-1′-enyl-2-acyl-*sn*-glycero-3-phosphocholine (alkenylacyl-GPC), respectively (Figure 1).

Although this chapter is devoted to the ether-linked phosphoglycerolipids that contain choline, it is noteworthy that elucidation of the first molecular structures of ether-linked phosphoglycerolipids was for those containing ethanolamine. The chemical structures of plasmenylethanolamine (see Reference 2 for review) and plasmanylethanolamine[3] were first reported in the late 1950s. However, it soon became apparent that plasmanylcholines also existed.[4-7] In the early 1960s plasmanylcholines were reported to be present in ox heart,[8] swine[9] and bovine[10] red bone marrow, and human serum.[11] The occurrence of ether-linked phosphoglycerolipids in a wide variety of tissues has been reported in many subsequent investigations and has been reviewed in depth.[12-16]

Perhaps the most significant discovery and advancement in studies of the ether-linked analogues of phosphatidylcholine was the elucidation of the chemical structure of platelet activating factor (PAF) which was identified in 1979 as 1-alkyl-2-acetyl-*sn*-glycero-3-phosphocholine (Figure 1).[17-19] The potent and diverse biological activities associated with PAF as well as the ubiquitous tissue distribution of enzymes that can synthesize PAF make this phospholipid unique among cellular mediators yet described. Although its position in the hierarchy of other mediators and/or factors involved in cellular signal transduction events is not yet established, it is clear that PAF can modulate the biological responses induced by other mediators, e. g., via products of the phosphatidylinositol cycle and eicosanoid metabolism, as well as by directly influencing numerous other metabolic events.[20-24]

The function of ether lipids in cellular processes is still poorly understood. In fact, no function has been established for plasmanylcholine, plasmenylcholine, or their ethanolamine counterparts, except that plasmanylcholines can serve as a precursor of PAF. Nevertheless, even the biological function of the potent bioactive molecule of PAF, other than its villain role as an autocoid involved in hypersensitivity reactions, is not clear. Plasmanylcholine and plasmenylcholine are membrane components, but how they function in this capacity is unknown. In model membranes the replacement of ester bonds with ether bonds in glycerolipids can cause significant alterations in membrane properties, such as changes in permeability to ions, surface potential, phase transition temperatures, and their interactions with cholesterol and proteins.[25] Thus, the presence of ether lipids in biological membranes could have profound effects on cellular functions. Although there is considerable literature about the possible function of ether-linked phospholipids, we have chosen not to cover the subject in this chapter since it is purely speculative at this time.

Enzymes involved in the metabolism of PAF, PAF precursors, and other choline-containing ether phospholipids, including their regulatory controls, will be the emphasis of this chapter. Nevertheless, we have also discussed the distribution and composition of both the *O*-alkyl and *O*-alk-1-enyl analogues of phosphatidylcholine in various mammalian tissues since this

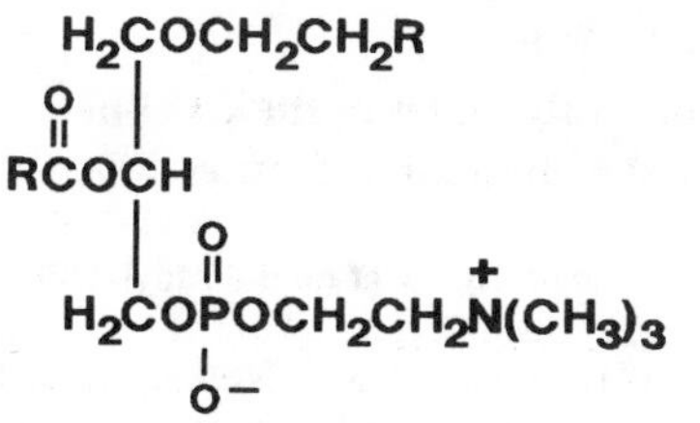

$H_2COCH_2CH_2R$
$RC(=O)OCH$
$H_2COP(=O)(O^-)OCH_2CH_2\overset{+}{N}(CH_3)_3$

1-alkyl-2-acyl-sn-glycero-3-phosphocholine
(plasmanylcholine)

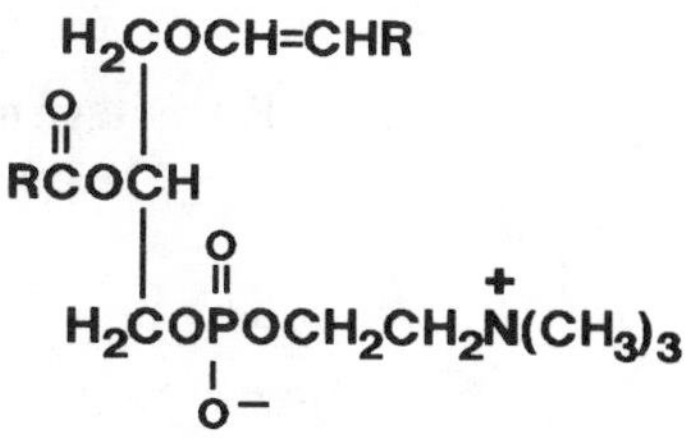

$H_2COCH{=}CHR$
$RC(=O)OCH$
$H_2COP(=O)(O^-)OCH_2CH_2\overset{+}{N}(CH_3)_3$

1-alk-1′-enyl-2-acyl-sn-glycero-3-phosphocholine
(plasmenylcholine; choline plasmalogen)

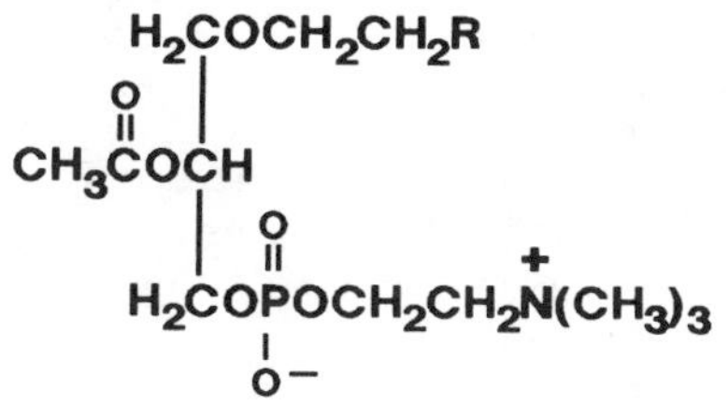

$H_2COCH_2CH_2R$
$CH_3C(=O)OCH$
$H_2COP(=O)(O^-)OCH_2CH_2\overset{+}{N}(CH_3)_3$

1-alkyl-2-acetyl-sn-glycero-3-phosphocholine
(platelet activating factor; PAF)

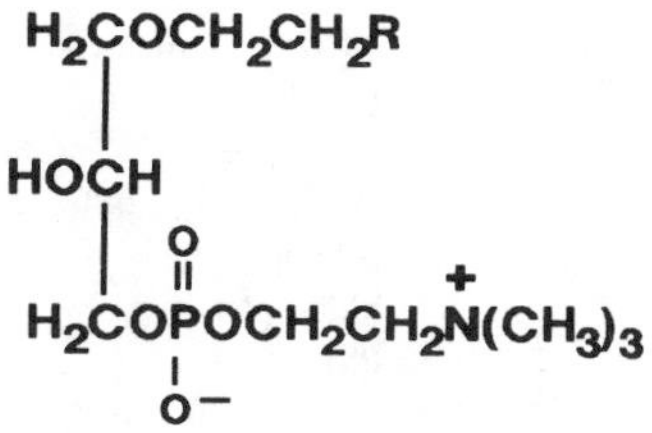

$H_2COCH_2CH_2R$
$HOCH$
$H_2COP(=O)(O^-)OCH_2CH_2\overset{+}{N}(CH_3)_3$

1-alkyl-2-lyso-sn-glycero-3-phosphocholine
(lysoplasmanylcholine)

FIGURE 1. Chemical structures of key ether-linked phospholipids that contain choline.

information is relevant to their metabolism. Although the metabolism of the ethanolamine-containing ether-linked phospholipids is clearly related to their choline-containing counterparts, we have purposely ignored the ethanolamine type (except when specifically pertinent) in order to keep within the overall theme of this book on phosphatidylcholine.

II. TISSUE DISTRIBUTION AND COMPOSITION

Much of the interest in the ether composition of choline phosphoglycerolipids was prompted by the discovery that plasmanylcholine can serve as an inactive storage precursor of PAF (see Section III). Therefore, until recently, it was not unusual to read scientific articles that completely ignored the possible presence of ether-linked lipids in the "phosphatidylcholine" fraction from mammalian tissues. Excellent detailed reviews of the ether-lipid content and composition of many tissues were written by Horrocks in 1972[26] and Horrocks and Sharma in 1982.[16] In 1987, the ether composition of ethanolamine and choline phosphoglycerolipids from several mammalian tissues was extensively reviewed by Sugiura and Waku.[27] Because of the exhaustive detail and recent nature of this latest review,[27] this section will merely provide a synopsis of conclusions about the tissue distribution and composition of the choline-type ether-linked phospholipids based on the earlier surveys of the literature.

A. PLASMANYLCHOLINES

Most mammalian tissues contain at least small amounts of plasmanylcholine, with some cells having levels as high as 50% of the choline glycerophospholipids (Table 1). The lowest levels of plasmanylcholine are found in the livers and erythrocytes of all species, in rat hearts, and in the plasma of humans and rats. Lymphocytes, macrophages, and polymorphonuclear leukocytes (PMNs) have relatively high amounts of plasmanylcholine. These cells and platelets have

TABLE 1
Percentage of Ether Lipid Subclasses in the Choline Phosphoglycerides of Mammalian Tissues

Tissue	Percentage of total diradyl–GPC	
	Plasmanylcholine	Plasmenylcholine
Liver	0—2.4	0—0.14
Erythrocytes	1.9—3.9	0—0.6
Brain	1.0—5.6	Tr—4.4
Lung	2.1—5.9	Tr—2.5
Plasma	0.9—5.5	0—0.2
Kidney	Tr—6.2	0—3.7
Testes	5.6—9.8	0.4—2.6
Platelets	4.5—18.0	1.4—8.8
Heart (rat)	0.6—1.4	1.7—3.8
Heart (species other than rat)	1.6—5.2	33—52
Lymphocytes	10.2—23.2	2.2—3.2
Macrophages	13.5—35.2	2.3—5.6
PMNs[a]	16.4—50.2	Tr—9.4

Note: A range of values for different species and experiments are summarized from the recent detailed review by Sugiura and Waku.[27] Tr = trace.

[a] Polymorphonuclear leukocytes.

received particular attention in their responses to PAF and the metabolic events that follow. It should be noted that differences between animal species account for some of the wide ranges of values listed in Table 1. For example, PMNs from guinea pigs contain 16.4% of the total choline glycerophospholipids as plasmanylcholine, compared to 44.1 to 50.2% in PMNs from humans and rabbits.

There is some evidence to suggest that the subclasses of diradyl-GPC may have an asymmetric distribution within cellular membranes. For example, it was reported that plasmanylcholine is located in the inner leaflet of plasma membranes from Krebs II ascites cells, while the phosphatidylcholine and plasmenylcholine subclasses were localized in the external leaflet.[28,29] Although these results are from tumor cells, it would not be surprising if similar asymmetric distributions of subclasses of choline glycerophospholipids also occur in normal cells.

In most tissues the *sn*-1 position of plasmanylcholine is comprised primarily of 16:0, 18:0, and 18:1 alkyl ether moieties.[26,27] However, there are situations where this fraction from certain tissues may contain relatively high amounts of other alkyl chains. Rabbit peritoneal PMNs, for example, were found to possess 16 and 9% of 20:0 and 22:0 alkyl chains, respectively, in addition to the normally encountered 16:0, 18:0, and 18:1 moieties.[30]

In contrast to the somewhat predictable composition of the *sn*-1 alkyl chains, the *sn*-2 acyl groups of plasmanylcholine appear to be more variable, as summarized in Table 2.[30-35] Analysis of the arachidonic acid content at the *sn*-2 position (Table 2) was made because it has been proposed that the alkylarachidonoyl species of plasmanylcholine is the major precursor of PAF. These arachidonate-containing species ranged from a high of 43.7% in human platelets to a low of 1.3% in rabbit PMNs. Because of the differences observed between animal species and differences seen among different tissues of the same species, it is difficult to generalize about the composition of *sn*-2 acyl groups of plasmanylcholine. Nevertheless, there does appear to be some tendency for the percentage of arachidonoyl species to be higher in the plasmanylcholine subclass than in the phosphatidylcholine subclass of the same cells or tissue, although even this

TABLE 2
Content of Arachidonate at the *sn*–2 Position of Phosphatidylcholine and Plasmanylcholine

Tissue: species	Percentage of *sn*–2–acyl groups as 20:4	
	Phosphatidyl-choline	Plasmanyl-choline
Polymorphonuclear leukocytes		
Rabbit[30]	2.7	1.3
Rat[31]	21.6[a]	28.3
Human[32]	4.4[a]	10.4
Platelets		
Rabbit[33]	12.6	20.9
Rat[34]	30.0	26.4
Human[35]	23.2	43.7

[a] We have multiplied the values given in the references (*sn*–1 plus *sn*–2 positions) by two with the assumption that all of the archidonate is located at the *sn*–2 position.

trend shows some exceptions (Table 2). Therefore, if analytical data is not available for the system under investigation, it would be prudent to determine quantitatively the content and composition of each subclass.

B. PLASMENYLCHOLINE (CHOLINE PLASMALOGENS)

Most mammalian tissues contain only minimal levels of plasmenylcholine, except for heart tissue of some species that can contain as much as 52% of the total diradyl-GPC as plasmalogens (Table 1). In contrast, rat, mouse, and hamster hearts have only small amounts of plasmenylcholine (Table 1 and Reference 27). Another tissue that has significant amounts of plasmenylcholine is the preputial gland from adult rats.[36,37] We have recently applied new methodology[38] to the quantitative analysis of diradyl-GPC subclasses of adult, male rat preputial glands and found a distribution of 47.4, 7.2, and 45.3% as alk-1-enylacyl-, alkylacyl-, and diacyl-GPC, respectively. These results are in general agreement with the earlier semiquantitative data of Hack and Helmy.[36,37]

Like the *sn*-1 position of the plasmanylcholine, the 16:0, 18:0, and 18:1 alk-1-enyl groups also normally predominate at the *sn*-1 position of this subclass.[16,26,27] Although information is somewhat more limited, most of the generalizations made about the composition of the *sn*-2 acyl groups of plasmanylcholine, particularly with regard to arachidonate, appear to apply equally well for the *sn*-2 acyl moieties of plasmenylcholine.

III. METABOLISM

A. BACKGROUND

The end products in the biosynthesis of choline-containing ether-linked glycerophospholipids are plasmanylcholine, plasmenylcholine, and PAF. In this section the intermediary metabolic steps in the biosynthetic pathways leading to the formation of these products will be described. It will become clear that the plasmanylcholine and PAF can be derived from a common intermediate, 1-alkyl-2-lyso-Gro(glycerol)P, since the lyso position of this phospholipid can be either acylated by a long-chain acyl-CoA acyltransferase or acetylated by an acetyl-CoA acetyltransferase. Thus, this lyso intermediate ether analogue of phosphatidic acid is a crucial branch point in the formation of both plasmanylcholine and PAF. Plasmanylcholine itself is also a known source of PAF via a remodeling mechanism involving sequential steps catalyzed by a phospholipase A_2 and an acetyltransferase; the latter enzyme activity is distinctly different from the acetyltransferase in the *de novo* route of PAF biosynthesis. Also the inactivation of PAF by acetylhydrolase can lead to the formation of a highly polyunsaturated species of plasmanylcholine.

In contrast to the alkyl lipids that contain phosphocholine, only limited knowledge is available about the origin of the alk-1-enyl moiety of plasmenylcholine. However, lysoplasmenylcholine participates in acylation and transacylation reactions in a manner similar to that of lysoplasmanylcholine.

Although beyond the scope of this review, it is worthwhile to mention briefly how the alkyl bond in glycerolipids is formed. Alkyldihydroxyacetone-P (DHAP) synthase (EC 2.5.1.26) is known to catalyze a unique exchange reaction (Reaction I, Figure 2) that involves the replacement of the acyl group of acyl-DHAP by a long-chain fatty alcohol (see earlier reviews[39,40] of this subject). Hajra and Bishop[41] have shown that alkyl-DHAP synthase is located in peroxisomes in a variety of tissues, but studies with microsomes from Ehrlich ascites cells (devoid of peroxisomes) indicate that alkyl-DHAP synthase is not just restricted to peroxisomes.[42-44] In fact, Raberi et al.[45] have recently reported a bimodal distribution of alkyl-DHAP synthase between peroxisomal and microsomal fractions isolated from rat liver. Experiments with a partially purified preparation of alkyl-DHAP synthase from Ehrlich ascites cells support a ping-pong-type enzymatic mechanism for this unusual exchange reaction; thus, an enzyme-DHAP complex has been postulated as the key intermediate.[42-44] Once alkyl-DHAP is formed, the ketone group is then reduced[39,40] by an NADPH-dependent oxidoreductase (EC 1.1.1.101) to produce 1-alkyl-2-lyso-*sn*-glycero-3-phosphate (alkyllyso-GroP) (Reaction II, Figure 2), the branch point intermediate that is converted to the more complex ether-linked phospholipids discussed in the next section.

B. BIOSYNTHETIC ENZYMES

1. Biosynthesis of Plasmanylcholine

a. De Novo Route from Alkyllyso-GroP

Acylation of alkyllyso-GroP via an acyl-CoA acyltransferase (EC 2.3.1.63) produces plasmanic acid, the alkylacyl analogue of phosphatidic acid (Reaction V, Figure 2). Alkyllyso-GroP:acyl-CoA acyltransferase activity is present in several normal[46] as well as tumor[47] tissues and is primarily associated with the microsomal subcellular fraction. This acyltransferase is apparently a different enzyme than the acyl-CoA acyltransferase that is responsible for the acylation of acyllyso-GroP.[46] After hydrolysis of the phosphate group[39,40] from alkylacyl-GroP by a phosphohydrolase (Reaction VI, Figure 2), the resulting alkylacyl-Gro can be converted to plasmanylcholine by a dithiothreitol(DTT)-sensitive CDP-choline:alkylacyl-Gro cholinephosphotransferase enzyme (EC 2.7.8.2) (Reaction VII, Figure 2). Although cholinephosphotransferase and ethanolaminephosphotransferase are probably different enzymes,[48] available evidence suggests that the same, or a very similar cholinephosphotransferase, is involved in the conversion of both diacyl-Gro and alkylacyl-Gro to the respective diradyl-GPC.[49-51] As shown in the biosynthesis of phosphatidylcholine,[52] the cholinephosphotransferase is probably not a rate-limiting step in the formation of plasmanylcholine.[51] Studies of cholinephosphotransferase using alkylacyl-Gro as a substrate suggest that the cholinephosphotransferase reaction is reversible.[53,54]

b. De Novo Route from Alkylglycerols

Alkylglycerols that originate from the diet or from catabolic enzymes acting on endogenous ether-linked glycerolipids can also serve as precursors of plasmanylcholine. In this situation[55,56] the alkylglycerols are first phosphorylated by a microsomal phosphotransferase (EC 2.7.1.93) that requires ATP/Mg^{+2}. Phosphorylation of alkylglycerols is stereoselective for the 1-alkyl-*sn*-

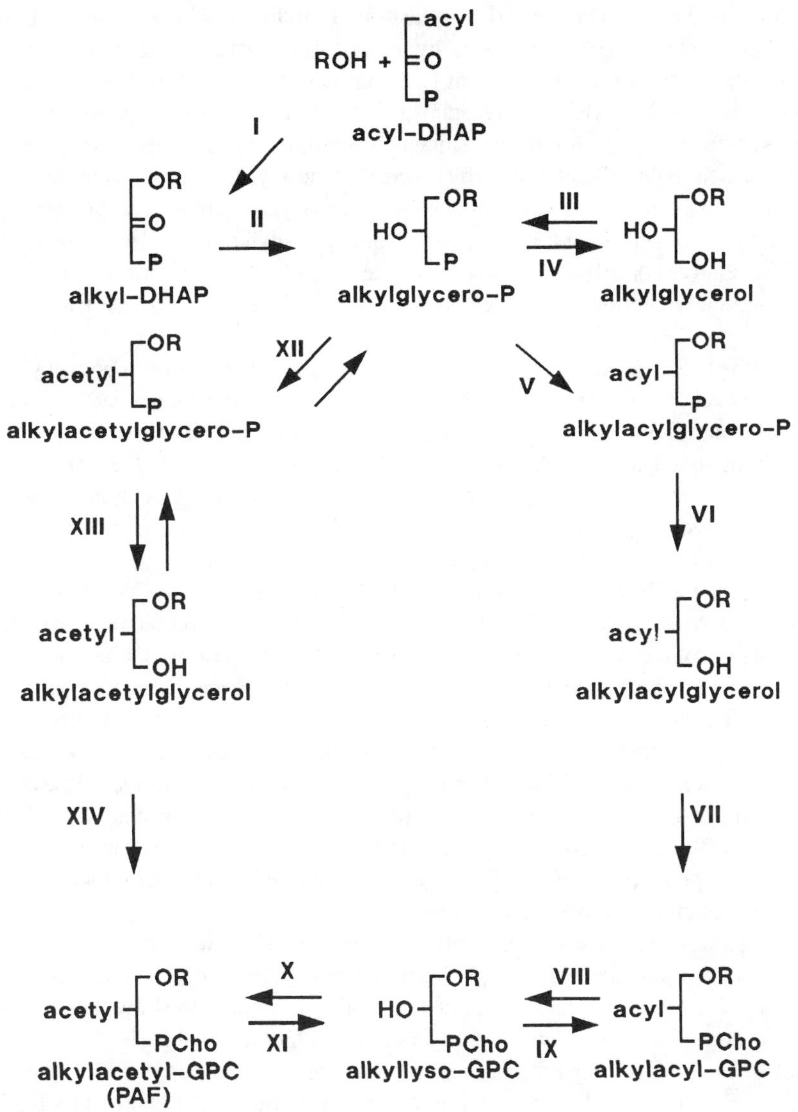

ROH + acyl-DHAP (acyl, O, P)

I

alkyl-DHAP (OR, O, P)

II

alkylglycero-P (HO, OR, P)

III

IV

alkylglycerol (HO, OR, OH)

XII

V

alkylacetylglycero-P (acetyl, OR, P)

alkylacylglycero-P (acyl, OR, P)

XIII

VI

alkylacetylglycerol (acetyl, OR, OH)

alkylacylglycerol (acyl, OR, OH)

XIV

VII

X

XI

VIII

IX

alkylacetyl-GPC (PAF) (acetyl, OR, PCho)

alkyllyso-GPC (HO, OR, PCho)

alkylacyl-GPC (acyl, OR, PCho)

FIGURE 2. Metabolic pathway for the biosynthesis of *O*-alkyl phospholipids that contain choline. The Roman numerals refer to the following enzymes: I, alkyl-DHAP synthase (EC 2.5.1.26); II, NADPH:alkyl-DHAP oxidoreductase (EC 1.1.1.101); III, ATP:1-alkyl-*sn*-glycerol phosphotransferase (EC 2.7.1.93); IV, alkyl-*sn*-glycero-3-P phosphohydrolase; V, acyl-CoA:1-alkyl-2-lyso-*sn*-glycero-3-P acyltransferase (2.3.1.63); VI, 1-alkyl-2-acyl-*sn*-glycero-3-P phosphohydrolase; VII, CDP-choline:1-alkyl-2-acyl-*sn*-glycerol DTT-sensitive cholinephosphotransferase (EC 2.7.8.2); VIII, phospholipase A_2; IX, polyenoic acid transacylase; X, acetyl CoA:1-alkyl-2-lyso-*sn*-glycero-3-phosphocholine acetyltransferase (EC 2.3.1.67); XI, 1-alkyl-2-acetyl-*sn*-glycero-3-phosphocholine acetylhydrolase (EC 3.1.1.48); XII, acetyl-CoA:1-alkyl-2-lyso-*sn*-glycero-3-P acetyltransferase; XIII, 1-alkyl-2-acetyl-*sn*-glycero-3-P phosphohydrolase; and XIV, CDP-choline:1-alkyl-2-acetyl-*sn*-glycerol DTT-insensitive cholinephosphotransferase.

glycerol structure.[56] The alkyllyso-GroP formed in this reaction can then enter the *de novo* pathway for the biosynthesis of plasmanylcholine. The contribution of this alternative pathway for the biosynthesis of plasmanylcholine from alkylglycerols in relation to the *de novo* synthesis starting from alkyl-DHAP has not been evaluated.

2. Acylation and Transacylation of Lysoplasmanylcholine and Lysoplasmenylcholine

Lysoplasmanylcholine can be acylated by at least three different mechanisms to produce plasmanylcholine. One route of acylation is via an acyl-CoA acyltransferase originally described by Lands[57] for the acylation of lysophosphatidylcholines. A second pathway is by a CoA-dependent transfer of acyl groups from an intact phospholipid to an acceptor lysophosphatide,[58] a reaction that is thought to be catalyzed by reversal of the acyl-CoA acyltransferase reaction.[59] The third mechanism involves a CoA-independent transacylation of the lysophospholipid with acyl groups from intact phospholipids, where phosphatidylcholine appears to be the acyl source.[60] The relative contribution from each of these mechanisms in controlling the composition of the plasmanylcholine pool depends on the tissue and specific experimental conditions employed.

Studies using cell-free membrane fractions with exogenous additions of both acyl-CoA and radyllyso-GPC acceptors, have shown that the acyl-CoA acyltransferase system can acylate both lysoplasmanylcholine[61-65] and lysoplasmenylcholine.[66] However, lysophosphatidylcholine functioned as a better acceptor molecule than either of the two ether-linked phospholipid substrates in these experiments.[61-63,65,66] The acyl-CoA acyltransferase showed selectivity for exogenously added polyunsaturated acyl-CoA derivatives in the acylation of both lysoplasmanylcholine[61-63] and lysoplasmenylcholine.[66]

Other experiments have been conducted with the addition of only CoA (CoA-dependent transacylase), CoA plus ATP/Mg^{2+} (acyl-CoA acyltransferase), or no cofactors (CoA-independent transacylase) to microsomal preparations along with exogenous lysoplasmanylcholine to assess the relative activities and/or acyl selectivities of the three distinct acylation mechanisms.[64,67-71] All three types of reactions for the acylation of lysoplasmanylcholine were found to be present in membrane preparations from rat platelets,[68] Fischer tumors,[70] rat brains,[71] and rabbit alveolar macrophages.[64,69] However, only the CoA-independent transacylase contributed to the acylation of lysoplasmanylcholine in membranes from human platelets.[67] There was general agreement in all of these reports that the CoA-independent transacylase enzyme demonstrated the greatest selectivity for polyunsaturated fatty acids (particularly arachidonic acid) in the acylation of lysoplasmanylcholine.

Some experiments have also been described where exogenous preparations of both the diacyl donor phospholipid and the lysophospholipid acceptor were added to cell-free membrane preparations.[64,71-74] Using this technique, Flesch et al.[72] were able to show that lysoplasmenylcholine can be acylated in the presence of CoA by arachidonic acid derived from phosphatidylinositol in membrane preparations derived from murine thymocytes and bone marrow macrophages. Sugiura et al.[64] found that microsomes from rabbit alveolar macrophages could catalyze the transfer of arachidonate from exogenously added phosphatidylcholine to lysoplasmenylcholine in the absence of added CoA and that the addition of CoA increased this transfer by about 40%. These workers also observed that the transfer of arachidonate from exogenous phosphatidylinositol to lysoplasmanylcholine did require the addition of CoA.[64] In a similar manner, the CoA-independent transfer of arachidonic acid[73] and docosahexaenoic acid[74] from exogenous phosphatidylcholine to lysoplasmanylcholine by microsomes from rabbit alveolar macrophages was demonstrated. The CoA-independent transfer of palmitate, stearate, oleate, linoleate, and linolenate from exogenous phosphatidylcholine to lysoplasmanylcholine was very low compared to the transfer of arachidonate, docosahexaenoate, eicosapentaenoate, and docosatetraenoate.[64] In these experiments[64,73,74] only a slight increase in transfer of fatty acids from phosphatidylcholine to lysoplasmanylcholine was noted after the addition of CoA to the incubations. However, addition of CoA stimulated the transfer of arachidonate from phosphatidylcholine to lysophosphatidylcholine by a factor of about three.[64]

Studies of PAF inactivation in intact cells also implicate the importance of the CoA-independent transacylase in the formation of species of plasmanylcholine highly enriched in polyenoic acids. Such experiments have shown that arachiodonic acid is the major fatty acid found at the *sn*-2 position of plasmanylcholine when PAF and/or lysoplasmanylcholine is metabolized by human neutrophils,[75] rabbit platelets,[76] and rabbit alveolar macrophages.[64,69] Exceptions to acylation with arachidonate were noted for intact rabbit neutrophils[77] and Ehrlich ascites cells[78] where the major acyl groups of plasmanylcholine were linoleate and docosahexenoate, respectively.

3. Biosynthesis of Platelet Activating Factor (PAF)

a. Remodeling Route

In response to inflammatory agents, some of the plasmanylcholine of membranes is remodeled at the *sn*-2 position through the combined actions of a phospholipase A_2 and an acetyl-CoA dependent acetyltransferase[79] to produce PAF. A considerable number of studies[20,24] have focused on this acetyltransferase, whereas information about the phospholipase A_2 involved is sparse at the present time.

Our understanding[20,24] of the acetyltransferase in the remodeling pathway reveals that (1) it is readily activated by inflammatory stimuli, (2) activation occurs by phosphorylation of the enzyme, (3) the dephosphorylated form of the enzyme has no activity, and (4) it is distinctly different in its properties from those of the acetyltransferase in the *de novo* pathway and the long-chain acyltransferases involved in other aspects of lipid metabolism (see Section III, B).

It is apparent from a number of reports (see Section IV on regulation) that the activation of acetyltransferase in the remodeling route is activated by phosphorylation of a serine residue of the enzyme and that inactivation of the enzyme requires a dephosphorylation step. However, as will be noted later it is not entirely clear as to what actual mechanism(s) is responsible for this activation-inactivation sequence in different cells.

b. De Novo Route

As mentioned earlier in Section III.A, 1-alkyl-2-lyso-GroP is an important branch point in the biosynthesis of more complex ether-linked phospholipids. With the choline-containing ether lipids, this lyso intermediate can be converted by a *de novo* route directly to PAF by reaction steps catalyzed by an acetyltransferase (Figure 2, Reaction XII), a phosphohydrolase (Figure 2, Reaction XIII), and a DTT-insensitive cholinephosphotransferase (Figure 2, Reaction XIV). All enzymes in the *de novo* route are membrane bound[80-82] and have not yet been purified. However, characterization of their properties[80-82] have clearly demonstrated that they are different from their counterparts in either the remodeling pathway of PAF biosynthesis or those involved in the *de novo* biosynthesis of phosphatidylcholine (see Table 3). Rate limiting steps in this route are the acetyltransferase[80] and cytidylyltransferase;[83] these enzymes are discussed in regard to their role as regulatory controls in Section IV.

Recent experiments with HL-60 cells[84] have revealed that the 1-alkyl-2-acetyl-Gro formed in the *de novo* route of PAF biosynthesis can also serve as a substrate for a novel acyl-CoA acyltransferase that differs from the one that acylates long-chain diradylglycerols. Thus, the alkylacetylglycerols can be diverted either into PAF or into a neutral lipid pool (1-alkyl-2-acetyl-3-acyl-*sn*-glycerols) that can serve as a stored form of acetylated lipids and a potential precursor of PAF.

4. Biosynthesis of Plasmenylcholine

Formation of the alk-1-enyl linkage in glycerophospholipids has been well documented in studies of the biosynthesis of plasmenylethanolamine.[14-16,40] The alk-1-enyl bond in plasmenylethanolamine is formed from plasmanylethanolamine by the action of a microsomal cytochrome b_5-dependent desaturase.[85-88] This mixed function oxygenase has been comprehensively described in an earlier review.[40] However, the plasmanylethanolamine Δ-1 desaturase system does

TABLE 3
Comparison of Properties of Enzymes Catalyzing Similar Types of Reactions in the Metabolism of PAF and Phosphatidylcholine

	Properties			
Enzymes	**pH optimum**	**Temperature sensitivity**	**Detergent sensitivity**	**Substrate specificity with respect to *sn*–1 alklyl chain length**
Acetyltransferase				
Remodeling (PAF)	6.9 [79]	More	ND[a]	18:0 > 16:0
De novo (PAF)	8.4 [80]	Less	ND	18:0 = 16:0
Phosphohydrolase[81]				
De novo (PAF)	7.4	More	More	18:0 = 16:0
De novo (phosphatidylcholine)	6.2	Less	Less	NA[b]
Cholinephosphotransferase[82]				
De novo (PAF) (DTT–insensitive)	8.0	Less	More	16:0 > 18:0
De novo (phosphatidylcholine) (DTT–sensitive)	8.0—9.0	More	Less	NA

[a] ND = not determined.
[b] NA = not applicable.

not convert plasmanylcholine to plasmenylcholine. Presently, the biosynthesis of choline plasmalogens is poorly understood.

Several potential metabolic routes that could contribute to the biosynthesis of plasmenylcholine are

1. A Ca^{2+}-dependent base exchange reaction
2. Methylation of plasmenylethanolamine
3. Phospholipase C-dependent cleavage of plasmenylethanolamine and subsequent condensation of the alk-1-enylacyl-Gro with CDP-choline by a cholinephosphotransferase
4. A CMP-dependent reverse reaction involving plasmenylethanolamine catalyzed by ethanolaminephosphotransferase coupled with cholinephosphotransferase activity
5. Sequential actions of phospholipase D- and phosphatidate phosphohydrolase-catalyzed hydrolysis of plasmenylethanolamine to produce alk-1-enylacyl-Gro which is then converted to choline plasmalogens by the action of cholinephosphotransferase

or

6. Introduction of an alk-1-enyl bond by a desaturase that uses alkylacyl-Gro as a substrate along with the subsequent addition of the phosphocholine group via cholinephosphotransferase

Enzymatic methylation of plasmenylethanolamine to form plasmenylcholine has been reported in rabbit myocardial membranes.[89] However, this activity amounts to only 0.4% of the alk-1-enylacyl-Gro:CDP-choline cholinephosphotransferase activity (0.065 pmol/min·mg protein for methylation vs. 18 pmol/min·mg protein for the cholinephosphotransferase).[90] In addition, plasmenylcholine contained less than 0.3% of the radioactivity found in plasmenylethanolamine when isolated guinea pig hearts were perfused with [1-^{3}H]ethanolamine for 60 and 120 min.[91] Therefore, it would appear that the methylation of plasmenylethanolamine probably does not contribute significantly to the biosynthesis of plasmenylcholine.

TABLE 4
Substrate Specificity of Phospholipase A_2 in Various Cells and Tissues Toward Subclasses of Diradyl–GPC

Cells or tissues	Phosphatidyl-choline	Plasmanyl-choline	Plasmenyl-choline
Rat epididymal fat cells[93]	Same	Same (1)	Lower (0.7)
Rat brain mitochondrial fraction[94]	High	Medium (0.6)	Low (0.5)
Acetone powder of human cerebral cortex[94]	Higher	Same (0.5)	Same (0.5)
Canine myocardial cytosolic fraction[95]	Low	N.D.	High (3—5)
Sheep platelet cytosolic fraction[96]			
Peak I	Same	Same	Same
Peak II	Low	N.D.	High (100)
Human platelets[97]	Same	Same (1)	N.D.
Human neutrophil cytosolic fractions[98]	Same	Same (1)	N.D.

Note: N.D. = not determined. Numbers in parenthesis represent relative enzyme activity using either plasmanylcholine or plasmenylcholine as a substrate in relation to that of phosphatidylcholine as the substrate.

For the likelihood of pathways 3 to 6 to operate in a given tissue, it is necessary to demonstrate the existence of alk-1-enylacyl-Gro:CDP-choline cholinephosphotransferase and a sufficient amount of a free alk-1-enylacyl-Gro pool. The presence of alk-1-enylacyl-Gro:CDP-choline cholinephosphotransferase activity in rat liver was first documented by Kiyasu and Kennedy[92] and further substantiated in ox heart by Poulos et al.[93] Recently, the activity of this enzyme was also reported in guinea pig heart[91] and rabbit myocardial microsomes.[90] The pool size of the alk-1-enylacyl-Gro was found to be 17 nmol/g in guinea pig hearts[91] and 0.46 μg/g (≈0.7 pmol/g) in rabbit hearts;[90] only hexadec-1′-enylacyl-Gro was detected. Nevertheless, it remains uncertain as to which of the enzymes listed above for steps 3 to 6 is responsible for the production of alk-1-enylacyl-Gro and which of the metabolic routes could account for the formation of plasmenylcholine.

C. CATABOLIC ENZYMES

1. Phospholipases

a. Phospholipase A_2

Studies on the susceptibility of ether analogues of phosphatidylcholine to phospholipase A_2 in mammalian cells gained momentum after ample evidence accumulated to suggest this enzyme plays an important role in the stimulus-induced production of PAF. However, a unique phospholipase A_2 that is responsible for the specific hydrolysis of the *sn*-2 fatty acyl group of plasmanylcholine has not yet been identified. The substrate specificity of phospholipase A_2 toward diradylglycerophosphocholine in various cells and tissues is listed in Table 4.[94-99] Most phospholipase A_2 preparations hydrolyze phosphatidylcholine and plasmanylcholine at equal or similar rates and plasmenylcholine at slower rates[90,93-95] except for a specific plasmenylcholine phospholipase A_2 present in the cytosolic fractions from both canine heart[96] and sheep platelets.[97] The latter hydrolyze plasmenylcholine at rates much higher (5 to 100-fold) than phosphatidylcholine.

Partially purified phospholipase A_2 from the cytosolic fraction of canine myocardium (obtained by tandem DEAE-Sephacel-hydroxyapatite chromatography) did not require Ca^{2+} for activity, exhibited a V_{max} and K_m of 5 nmol/mg·h and 7 μM, respectively, for plasmenylcholine and 1 nmol/mg·h and 3 μM, respectively, for phosphatidylcholine; the pH optimum for both substrates was between 6 and 7.[96]

A plasmenylcholine specific phospholipase A_2 from sheep platelet cytosol that could be fully activated by physiological concentrations of calcium ions has been purified to homogeneity.[97] The physical state of this enzyme is a dimeric form with an apparent molecular mass of 58 kDa, which appears to consist of multiple isoforms.

The microsomal, mitochondrial, and soluble fractions of guinea pig heart also contain a phospholipase A_2 activity that hydrolyzes plasmenylcholine. The activities in these fractions had an optimal pH of 8.5, were enhanced by Ca^{2+}, and had a K_m of 5 μM with alk-1-enyloleoyl-GPC as the substrate.[100] Similar to the phospholipase A_2 in rat epididymal fat cells[94] and rat brain mitochondrial fraction,[95] the guinea pig heart phospholipase A_2 had an order of preference for plasmenylcholine with different acyl moieties at the *sn*-2 position that corresponded to alk-1-enyloleoyl-GPC > alk-1-enyllinoleoyl-GPC > alk-1-enylarachidonoyl-GPC > alk-1-enylstearoyl-GPC.

The phospholipase A_2 in human platelets had a biphasic-dependence for Ca^{2+} (10 μM and 2 mM), was optimal at pH 9 to 10, and increased in activity by the addition of albumin and diacylglycerols into the substrate dispersion.[98] The phospholipase A_2 activity in the cytosol of human neutrophils showed an absolute requirement for Ca^{2+} that was maximal at 10 mM, an optimal pH of 8.0, and a 3.5-fold increase in activity when neutrophils were stimulated with ionophore A23187.[99]

b. Phospholipase C

Recent studies have described a phospholipase C from two different mammalian sources that could hydrolyze ether-containing choline glycerophospholipids.[96,101] A partially purified phospholipase C from canine myocardial cytosol with an apparent molecular weight of 29,000 Da and a pI of 7.4 utilizes both plasmenylcholine (K_m = 20 μM) and phosphatidylcholine (K_m = 14 μM) equally well as substrates. This enzyme hydrolyzed phosphatidylethanolamine to a considerably lesser extent, but it did not cleave sphingomyelin or phosphatidylinositol.[96] The myocardial enzyme was easily distinguished from a lysosomal and bacterial phospholipase C. It had a maximum activity between pH 7 and 8, was inhibited by EDTA and Zn^{2+}, and was activated by Ca^{2+} in contrast to the lysosomal phospholipase C which had an acidic pH optimum and did not require divalent cations and was not inhibited by EDTA.[102] The bacterial phospholipase C was activated by zinc ions instead of calcium and had a different pH profile, molecular weight, and isoelectric point than the myocardial phospholipase C.[103-105]

A second phospholipase C which catalyzes the hydrolysis of PAF to alkylacetyl-Gro and phosphocholine was purified to near homogeneity from rabbit liver lysosomes.[101] This enzyme consisted of two forms, having pIs of 4.7 and 5.8 with apparent molecular weights of 33,000 and 75,000, a pH optima of 8.2 and 8.5, and an apparent K_ms of 55.6 and 45.5 μM for PAF, respectively. Since this enzyme activity had an alkaline pH optimum and was also completely inhibited by 1 mM EDTA and returned to the original level upon addition of Ca^{2+}, it obviously differs from the rat liver lysosomal phospholipase C as previously described by Matsuzawa and Hostetler.[102] The PAF-related enzyme from rabbit liver lysosomes hydrolyzed all other phospholipids (i.e., phosphatidyl-choline, -ethanolamine, -serine, and -inositol; sphingomyelin, phosphatidic acid; and lysoplasmanylcholine) at about 20 to 30% of the rate of PAF. Therefore, this phospholipase C[101] appears to be relatively specific for ether-linked lipids.

c. Lysophospholipase D

Wykle et al.[106-108] were the first to describe a Mg^{2+}-dependent lysophospholipase Da that is specific for ether lipids; it catalyzes the hydrolysis of choline from lysoplasmanylcholine to form alkyllysoglycerophosphate in the microsomal fraction of rat liver and other tissues. This enzyme is unique in that it also hydrolyzes ethanolamine from lysoplasmanylethanolamine or lysoplasmenylethanolamine,[106,107] but is not active toward the corresponding acyl analogues, i.e., lysophosphatidylcholine and lysophosphatidylethanolamine.[106] When the *sn*-2 position of the lysoplasmanylcholine and lysoplasmanylethanolamine was acylated, only minimal hydrolysis of the base group occurred.[106] However, Ca^{2+}-dependent lysophospholipase D with optimal activity at pH 8.4 was recently identified in the microsomal fraction of rabbit kidney medulla.[109] Thus, the divalent metal ion requirements for expression of maximum lysophospholipase D activities can apparently differ among animal species. The Ca^{2+}-dependent lysophospholipase D was predominant in rabbit tissues, whereas the Mg^{2+}-dependent enzyme was the major activity found in rat tissues.[109] In addition, Tokumura et al.[110] reported the existence of still another type of lysophospholipase D in rat plasma that is not specific for ether-lipids; both lysophosphatidylcholine and lysoplasmanylcholine could be converted to their respective radyllysoglycerophosphates after incubation with rat plasma for 48 h at 37° C.

2. Pte·H_4-Dependent Alkyl Monooxygenase

A Pte·H_4-dependent alkyl monooxygenase catalyzes the oxidative cleavage of the *O*-alkyl ether linkage of glycerolipids in mammalian tissues.[40] However, neither plasmanylcholine nor PAF can serve directly as a substrate for this enzyme. The acetate group has to be hydrolyzed from PAF by an acetylhydrolase[111,112] and the acyl group must be hydrolyzed from plasmanylcholine[113] by a phospholipase A_2; the resulting product in both situations, lysoplasmanylcholine, can then be cleaved by the microsomal alkyl monooxygenase to form a fatty aldehyde and glycero-3-phosphocholine. Alternatively, through the combined actions of lysophospholipase D and a phosphohydrolase, lysoplasmanylcholine can be further metabolized to alkylglycerols. The alkylglycerols are also degraded by the alkyl monooxygenase to generate free glycerol and fatty aldehydes. Results on the Pte·H_4 requirement, tissue distribution, and responses to thermal inactivation and inhibitors indicate that the same enzyme is responsible for the hydrolysis of both lysoplasmanylcholine and alkylglycerol,[112] but alkylglycerol is the preferable substrate.[112-114] Details of the properties of the Pte·H_4-dependent alkyl monooxygenase have been reviewed.[40]

Alkyl monooxygenase from rat liver microsomes has been solubilized with 2% Triton X-100[115] and purified to a near homogeneous state using affinity chromatography on chimyl alcohol-Sepharose 4B columns.[116] The purified enzyme has an apparent molecular weight of 45,000 Da, exhibits a pH optimum of 8.5, and requires reduced glutathione and phospholipids for the full expression of enzyme activity.

3. Lysoplasmenylcholine (Ethanolamine) Hydrolase or Lysoplasmalogenase

Based on enzymatic studies, it appears that the major route for plasmenycholine catabolism in brain involves the initial hydrolysis of the alk-1-enyl moiety by a plasmalogenase and the subsequent removal of the acyl moiety by a lysophospholipase.[40] However, the mammalian heart does not have the ability to initiate the direct hydrolysis of the alk-1-enyl bond in plasmenylcholine; thus, the majority of plasmenylcholine in cardiac tissue is first deacylated by phospholipase A_2 to form lysoplasmenylcholine which is then further metabolized by a lysoplasmalogenase.[100]

Warner and Lands[117] first demonstrated that lysoplasmenylcholine hydrolase in rat liver microsomes catalyzes the hydrolysis of lysoplasmenylcholine (choline lysoplasmalogens) to produce free aldehydes and glycerophosphocholine. This enzyme requires no cofactors[117,118] and appears to be able to use both lysoplasmenylcholine and lysoplasmenylethanolamine as substrates,[118,119] but not their acylated derivatives, i.e., plasmenylcholine or plasmenylethanolamine.[117] The activity is inhibited by chloromercuribenzoate (K_i = 50 μM)[118] and imidazole (competitively),[120] but not by derivatives of imidazole containing a negatively charged carboxyl group (i.e., histidine).[120] The partially purified enzyme is a negatively charged protein composed of multiple isoforms, has a molecular weight of 20,000 Da, and pH optimum of 7.2. K_m values for lysoplasmenylcholine and lysoplasmenylethanolamine were 18 and 55 μM, respectively.[118]

Lyosplasmenylcholine hydrolase found in the microsomal fraction of guinea pig hearts[100] has several properties that differ from the one in rat liver microsomes. It has a pH optimum of 8.5

and requires detergent for activity. Although it has no absolute requirement for cations, both Ca^{2+} and Mg^{2+} cause an enhancement of activity. D'Amato et al.[121] reported a plasmalogenase activity in an acetone-dried powder preparation from bovine brain with a substrate specificity that varies significantly from the one in liver and heart. A single enzyme in the brain may cleave both plasmenylcholine and plasmenylethanolamine.

4. PAF Acetylhydrolase

PAF is inactivated by hydrolysis of the acetate moiety via a reaction catalyzed by an acetylhydrolase.[111] Although phospholipase A_2 can also utilize PAF as a substrate, the properties of the acetylhydrolase drastically differ from those of phospholipase A_2 as shown in their different responses to calcium, magnesium, EDTA, dithiothreitol, deoxycholate, and diisopropylfluorophosphate. Also, the addition of phosphatidylcholine to PAF preparations does not influence the rates of hydrolysis by the acetylhydrolase. Acetylhydrolase activity occurs intracellularly (mainly in cytosol)[111] and in the vascular compartment.[122-125] The biochemical properties of both forms are similar, except the enzyme in plasma has a slightly higher molecular weight than the intracellular enzyme; also the activity in plasma is resistant to proteolytic hydrolysis by proteases, whereas the intracellular acetylhydrolase is readily attacked.[122]

Recently, Stafforini et al.[124] purified acetylhydrolase from human plasma 25,000-fold. Characteristics possessed by the purified enzyme were (1) molecular weight of 43,000, (2) a K_m of 13.7 μM and a V_{max} of 568 μmol/h/mg, (3) preference for the micellar form of the substrate, and (4) no calcium requirement. Approximately 30% of the activity was found with the high-density lipoproteins (HDL) and the remainder with low-density lipoproteins (LDL). The activity in the LDL fraction appears to be responsible for PAF degradation, whereas the HDL-associated enzyme represents an inactive form.[125] However, the acetylhydrolase can be transferred reversibly between these two lipoprotein fractions, but only the apoE subclass of HDL participates.[125]

IV. METABOLIC REGULATION

Regulation of the enzymatic steps that catalyze the *de novo* synthesis of plasmanylcholines from 1-alkyl-2-lyso-GroP is poorly understood. In fact, the rate-limiting reactions in this pathway have not yet been established. Presumably, cytidylyltransferase could be an important contributing factor, especially since it is known to be rate limiting in the biosynthesis of phosphatidylcholine[52] and PAF.[83] Enzymic control of the cellular levels of PAF has recently been reviewed.[126] Therefore, only recent developments will be the focus of discussion in this section.

As mentioned in Section III.B.3, PAF can be synthesized by either the remodeling or *de novo* route. Three factors in the remodeling pathway play important roles in mediating the PAF levels: (1) the activity of lysoplasmanylcholine:acetyl-CoA acetyltransferase, (2) the activity of phospholipase A_2, and (3) the pool size of plasmanylcholine, in particular, alkylarachidonoyl-GPC.

The activity of acetyltransferase is stimulated by various inflammatory stimuli and regulated by activation/inactivation through phosphorylation/dephosphorylation. With homogenates from neutrophils, Nieto et al.[127] obtained data to suggest acetyltransferase is phosphorylated by a mechanism linked to the cyclic-AMP dependent protein kinase, whereas a phospholipid-sensitive calcium-dependent kinase promotes negative signals leading to inactivation of the enzyme. The enhancement of acetyltransferase activity by the catalytic subunit of cyclic-AMP-dependent protein kinase in the presence of Mg^{2+} and ATP,[127,128] along with data showing the incorporation of ^{32}P from [γ-^{32}P]ATP into serine[129] of a single protein band of approximately 30 kDa isolated by SDS/polyacrylamide-gel electrophoresis, has provided sound support for the cAMP-dependent protein kinase in the activation step *in vitro*. However, in studies with microsomal preparations of guinea pig parotid glands, Domenech et al.[128] concluded that a calcium/calmodulin-dependent protein kinase is responsible for phosphorylating the acetyltransferase. In their experiments, protein kinase C had little effect, and although the cAMP-dependent protein kinase could also activate the acetyltransferase activity, β agonists had no effect on acetyltransferase activity even though such agonists are known to activate the cAMP-dependent protein kinase.[128] The authors suggest that compartmentalization of the cAMP-dependent kinase might explain this enigma.

Even though the cAMP-dependent protein kinase and calcium/calmodulin-dependent protein kinase can phosphorylate and activate acetyltransferase *in vitro,* the protein kinase(s) responsible for the cellular activation of acetyltransferase by different agonists has not been identified. Likewise, alkaline phosphatase can reduce acetyltransferase activity in rat spleen microsomes,[130,131] in antigen-stimulated mouse mast cell microsomes,[132] and in zymosan-stimulated neutrophil homogenates.[127] Also, protein phosphatase 2A *in vitro* decreases acetyltransferase activity in microsomes from both stimulated and unstimulated exocrine glands.[127] Yet, the specific enzyme responsible for the dephosphorylation of the acetyltransferase in intact cells remains unknown.

The suggestion that activation of phospholipase A_2-like activity for the production of PAF may require protein kinase C comes from evidence[133] that shows long-chain amines such as sphingosine, stearylamine, and palmitoylcarnitine (known inhibitors of protein kinase C) also inhibit A23187-induced PAF and LTB_4 synthesis, as well as arachidonic acid release in human neutrophils. Phorbol-12-myristate-13-acetate reverses these inhibitory processes. In addition, LTB_4 synthesis from exogenously supplied arachidonate is not inhibited by sphingosine. In dimethyl sulfoxide-induced differentiated human promyelocytic leukemic HL-60 cells, *N*-formylMet-Leu-Phe, or ionophore A23187, stimulates the production of PAF, while stimulation of the uninduced cells produce little PAF by the remodeling pathway.[134] However, based on pH dependence, the Ca^{2+} requirement, and kinetic characteristics, the activities of phospholipase A_2 and acetyltransferase in the cell extracts of uninduced cells were virtually indistinguishable from those of differentiated cells. These data led to the suggestion by Billah et al.[134] that the mere presence of phospholipase A_2 and acetyltransferase in cells and a mechanism to increase the free cytosolic Ca^{2+} concentration are not sufficient to turn on PAF synthesis, at least in certain cells. The appearance and/or disappearance of, as yet, unidentified factors or conditions that regulate the activation of phospholipase A_2 and acetyltransferase must accompany the onset of differentiation.

Peripheral or elicited PMNs from rats fed a fat-free diet for 3 to 4 months are 90% deficient in arachidonic acid. When such cells are stimulated with ionophore A23187, they produce 84% less PAF and incorporate 86% less acetate into PAF than cells with normal levels of arachidonate.[135] Reduction in PAF generation could be partially restored when arachidonate-deficient cells were incubated with arachidonate. Also, when lysoplasmanylcholine was added to incubation mixtures containing A23187-stimulated cells, the inhibition of PAF synthesis in the arachidonate-depleted cells returned to near normal levels. These studies indicate that alkylarachidonoyl-GPC and a specific phospholipase A_2 for this substrate are necessary for PAF biosynthesis in the remodeling pathway.

The importance of the *de novo* pathway (see Section III.B.3) in contributing to PAF production has been strengthened by recent findings[136] that show PAF generated by chick retinas upon stimulation with neurotransmitters, such as acetylcholine and dopamine, was due to an increase in the DTT-insensitive cholinephosphotransferase activity, but not the alkyllyso-GPC:acetyl-CoA acetyltransferase activity. Furthermore, phorbol-12-myristate-13-acetate also enhances PAF biosynthesis through the *de novo* route.[137] The regulation step under these conditions was not identified, but the activity of DTT-insensitive cholinephosphotransferase was not increased. Recently, Blank et al.[83] demonstrated that treatment of Ehrlich ascites cells with 2 m*M* oleic acid caused a >10-fold increase in the formation of PAF from alkylacetyl-Gro.

Under these conditions, oleic acid induced a translocation of CTP:phosphocholine cytidylyltransferase from the cytosol to cellular membranes with a concomitant increase in activity (32%). Thus, the activity of cytidylyltransferase can regulate PAF production[83] in the same manner that it does the biosynthesis of phosphatidylcholine[52] (see Chapter 3).

Cellular levels of PAF can also be influenced by catabolic enzymes. In most mammalian cells, PAF is deacetylated by an acetylhydrolase (see Section III.C.4) to form lysoplasmanylcholine; this lyso intermediate is then acylated by the CoA-independent transacylase (Section III.B.2) to produce plasmanylcholine. Activated peritoneal macrophages from mice injected with bacilli Calmette-Guerin, trehalose dimycolate, or streptococci C_{74} synthesize two to three times less PAF in response to a zymosan challenge than resident macrophages.[138] Under these conditions it appears that increased catabolism, but not impaired synthesis is responsible for the reduced formation of PAF. Along the same line, Touqui et al.[139] noted a marked and prolonged inhibition of PAF catabolism when rabbit platelets were incubated with labeled PAF and ionophore A23187 (or ionomycin) that increases the Ca^{2+} level. In these experiments the deacetylation of PAF *(in vivo)* and reacylation of lysoplasmanylcholine, both *in vivo* and *in vitro*, were inhibited by Ca^{2+}.

Acetylhydrolase appears to be of great importance in regulating the levels of PAF, both within cells and in extracellular fluids. In hypertensive rats[122] and hypertensive white males,[140] the acetylhydrolase activity is significantly decreased, which is consistent with the lower quantity of circulating PAF found in a hypertensive animal model.[141] The activity of acetylhydrolase also is modulated when physiological conditions are altered. For example, the activity of this enzyme in plasma decreases during pregnancy and lactation, thus indicating that the maternal plasma plays a role in controlling the PAF of fetal origin that comes in contact with the placenta.[142]

V. NEW FRONTIERS

There are many unsolved problems in the ether-lipid field as this decade nears its end. However, space limitations prevent a thorough discussion of these existing challenges. Therefore, we have chosen to emphasize what we consider frontier areas in the form of a series of questions that will require extensive research efforts if adequate answers are to be forthcoming.

Important questions still awaiting answers with regard to the biochemistry and functional role of the choline-containing ether-linked glycerophospholipids are

1. Can the membrane-bound enzymes leading to the biosynthesis of plasmanylcholine, plasmenylcholine, and PAF be purified for their chemical and physical characterization, as well as for the production of specific antibodies to use in probing their regulation?
2. How are plasmenylcholines synthesized?
3. What is the cellular function of plasmanylcholine and plasmenylcholine, and do they play a specific role in biological membranes?
4. Are ether lipids essential dietary components, and how does nutrition influence the content, distribution, and composition of plasmanylcholine and plasmenylcholine in tissues?
5. Why is there a high affinity of arachidonate for the choline (and ethanolamine) ether-linked phospholipids?
6. What is the interrelationship between the ether-linked phospholipids that contain choline and those that contain ethanolamine?
7. What biochemical mechanisms are responsible for the diverse biological actions of PAF?
8. How are the metabolism and levels of PAF in tissues and the vascular compartment regulated?

9. Can the PAF receptor be isolated and characterized?
10. Is there a physiological role for PAF as both an intracellular and extracellular mediator?
11. Is PAF a responsible factor in specific diseases?
12. What is the mechanism that explains why PAF production is stimulated by arachidonic acid and other polyunsaturates?
13. Is a specific phospholipase A_2 (for polyenoic acids) involved in the remodeling pathway of PAF biosynthesis?
14. Do different receptors account for the hypotensive and inflammatory responses induced by PAF?
15. What is the mechanism responsible for the antihypertensive responses induced by PAF?

ACKNOWLEDGMENT

This work was supported by the Office of Energy Research, U.S. Department of Energy (Contract No. DE-AC05-760R00033), the American Cancer Society (Grant BC-70R), and the National Heart, Lung, and Blood Institute (Grants HL-27109-08 and HL-35495-04).

REFERENCES

1. IUPAC-IUB Commission on Biochemical Nomenclature, The nomenclature of lipids, *Lipids,* 12, 455, 1976.
2. **Rapport, M.,** The discovery of plasmalogen structure, *J. Lipid Res.,* 25, 1522, 1984.
3. **Carter, H. E., Smith, D. B., and Jones, D. N.,** A new ethanolamine containing lipid from egg yolk, *J. Biol. Chem.,* 232, 681, 1958.
4. **Gray, G. M.,** The structure of the plasmalogens of ox heart, *Biochem. J.,* 70, 425, 1958.
5. **Rapport, M. M. and Franzl, R. E.,** Hydrolysis of phosphatidylcholine by lecithinase a, *J. Biol. Chem.,* 225, 851, 1957.
6. **Marinetti, G. V. and Erbland, J.,** The structure of pig heart plasmalogens, *Biochim. Biophys. Acta,* 26, 429, 1957.
7. **Uziel, M. and Hanahan, D. J.,** An enzyme-catalyzed acyl migration: a lysolecithin migratase, *J. Biol. Chem.,* 226, 789, 1957.
8. **Pietruszko, R. and Gray, G. M.,** The products of mild alkaline and mild acid hydrolysis of plasmalogens, *Biochim. Biophys. Acta,* 56, 232, 1962.
9. **Pietruszko, R.,** Lipids of red bone marrow from pig epiphysis, *Biochim. Biophys. Acta,* 64, 562, 1962.
10. **Thompson, G. A., Jr. and Hanahan, D. J.,** The nature and formation of α-glyceryl ether lipids in bovine bone marrow, *Biochemistry,* 2, 641, 1963.
11. **Renkonen, O.,** Lecithins of glycerol ether character in normal human serum, *Biochim. Biophys. Acta,* 59, 497, 1962.
12. **Snyder, F.,** The biochemistry of lipids containing ether bonds, in *Progress in the Chemistry of Fats and Other Lipids,* Vol. 10 (Part 3), Holman, R. T., Ed., Pergamon Press, Oxford, 1969.
13. **Snyder, F. and Snyder, C.,** Glycerolipids and cancer, *Prog. Biochem. Pharmacol.,* 10, 1, 1975.
14. **Snyder, F.,** *Ether Lipids: Chemistry and Biology,* Academic Press, New York, 1972.
15. **Mangold, H. K. and Paltauf, F.,** *Ether Lipids, Biochemical and Biomedical Aspects,* Academic Press, New York, 1983.
16. **Horrocks, L. A. and Sharma, M.,** Plasmalogens and *O*-alkyl glycerophospholipids, in *Phospholipids,* Hawthorne, J. N. and Ansell, G. B., Eds., Elsevier Biomedical Press, 1982, chap. 2.
17. **Demopoulos, C. A., Pinckard, R. N., and Hanahan, D. J.,** Platelet-activating factor. Evidence for 1-*O*-alkyl-2-acetyl-*sn*-glyceryl-3-phosphorycholine as the active component, *J. Biol. Chem.,* 254, 9355, 1979.
18. **Blank, M. L., Snyder, F., Byers, L. W., Brooks, B., and Muirhead, E. E.,** Antihypertensive activity of an alkyl ether analog of phosphatidylcholine, *Biochem. Biophys. Res. Commun.,* 90, 1194, 1979.

19. **Benveniste, J., Tence, M., Varenne, P., Bidault, J., Boullet, C., and Polonsky, J.,** Semi-synthese et structure purpose du facteur activant les plaquettes (P.A.F.); PAF-acether, un alkyl ether analogue de la lysophosphatidylcholine, *C. R. Acad. Sci. Paris,* 289, 1037, 1979.
20. **Snyder, F., Ed.,** *Platelet Activating Factor,* Plenum Press, New York, 1987, 471.
21. **Hanahan, D. J.,** Platelet activating factor: a biologically active phosphoglyceride, *Annu. Rev. Biochem.,* 55, 483, 1986.
22. **Braquet, P., Touque, L., Shen, T. Y., and Vargaftig, B. B.,** Perspectives in platelet-activating factor research, *Pharmacol. Rev.,* 39, 97, 1987.
23. **O'Flaherty, J. T. and Wykle, R. L.,** Biology and biochemistry of platelet-activating factor, *Clin. Lab. Med.,* 3, 619, 1983.
24. **Lee, T.-c. and Snyder, F.,** Function, metabolism, and regulation of platelet activating factor and related ether lipids, in *Phospholipids and Cellular Regulation,* Vol. 2, Kuo, J. F., Ed., CRC Press, Boca Raton, FL, 1985, chap. 1.
25. **Paltauf, F.,** Ether lipids in biological and model membranes, in *Ether Lipids: Biochemical and Biomedical Aspects,* Mangold, H. K. and Paltauf, F., Eds., Academic Press, New York, 1983, 309.
26. **Horrocks, L. A.,** Content, composition, and metabolism of mammalian and avian lipids that contain ether groups, in *Ether Lipids: Chemistry and Biology,* Snyder, F., Ed., Academic Press, New York, 1972, 177.
27. **Sugiura, T. and Waku, K.,** Composition of alkyl ether-linked phospholipids in mammalian tissues, in *Platelet Activating Factor and Related Lipid Mediators,* Snyder, F., Ed., Plenum Press, New York, 1987, 55.
28. **Record, M., El Tamer, A., Chap, H., and Douste-Blasy, L.,** Evidence for a highly asymmetric arrangement of ether- and diacyl-phospholipid subclasses in the plasma membrane of Krebs II ascites cells, *Biochim. Biophys. Acta,* 778, 449, 1984.
29. **Terce, F., Record, M., Chap, H., and Douste-Blazy, L.,** Different susceptibility of alkylacyl- versus diacyl- and alkenylacyl-phosphatidylcholine subclasses to stimulation of biosynthesis by phospholipase C, *Biochem. Biophys. Res. Commun.,* 125, 413, 1984.
30. **Mueller, H. W., O'Flaherty, J. T., and Wykle, R. L.,** Ether lipid content and fatty acid distribution in rabbit polymorphonuclear neutrophil phospholipids, *Lipids,* 17, 72, 1982.
31. **Ramesha, C. S. and Pickett, W. C.,** Fatty acid composition of diacyl, alkylacyl, and alkenylacyl phospholipids of control and arachidonate-depleted rat polymorphonuclear leukocytes, *J. Lipid Res.,* 28, 326, 1987.
32. **Mueller, H. W., O'Flaherty, J. T., Greene, D. G., Samuel, M. P., and Wykle, R. L.,** 1-*O*-Alkyl-linked glycerophospholipids of human neutrophils: distribution of arachidonate and other acyl residues in the ether-linked and diacyl species, *J. Lipid Res.,* 25, 383, 1984.
33. **Sugiura, T., Soga, N., Nitta, H., and Waku, K.,** Occurrence of alkyl ether phospholipids in rabbit platelets: composition of fatty chain profiles, *J. Biochem.,* 94, 1719, 1983.
34. **Colard, O., Breton, M., and Bereziat, G.,** Arachidonyl transfer from diacyl phosphatidylcholine to ether phospholipids in rat platelets, *Biochem. J.,* 222, 657, 1984.
35. **Mueller, H. W., Purdon, A. D., Smith, J. B., and Wykle, R. L.,** 1-*O*-Alkyl-linked phosphoglycerides of human platelets: distribution of arachidonate and other acyl residues in the ether-linked and diacyl species, *Lipids,* 18, 814, 1983.
36. **Helmy, F. M. and Hack, M. H.,** Comparative lipid biochemistry. V. The neutral lipid plasmalogen and other lipids of some specialized sebaceous glands, *Comp. Biochem. Physiol.,* 23, 329, 1967.
37. **Hack, M. H. and Helmy, F. M.,** On the plasmalogenation of myocardial choline glycerophospholipid during maturation of various vertebrates, *Comp. Biochem. Physiol.,* 89B, 111, 1988.
38. **Blank, M. L., Cress, E. A., and Snyder, F.,** Separation and quantitation of phospholipid subclasses as their diradylglycerobenzoate derivatives by normal-phase high-performance liquid chromatography, *J. Chromatogr.,* 392, 421, 1987.
39. **Hajra, A. K.,** Biosynthesis of *O*-alkylglycerol ether lipids, in *Ether Lipids Biochemical and Biomedical Aspects,* Mangold, H. K. and Paltauf, F., Eds., Academic Press, New York, 1983, 85.
40. **Snyder, F., Lee, T.-C., and Wykle, R. L.,** Ether-linked glycerolipids and their bioactive species: enzymes and metabolic regulation, in *The Enzymes of Biological Membranes,* Vol. 2, Martonosi, A. N., Ed., Plenum Press, New York, 1985, chap. 14.
41. **Hajra, A. K. and Bishop, J. E.,** Glycerolipid biosynthesis in peroxisomes via the acyl dihydroxyacetone phosphate pathway, *Ann. N.Y. Acad. Sci.,* 386, 170, 1982.
42. **Brown, A. J. and Snyder, F.,** Alkyldihydroxyacetone-P synthase solubilization, partial purification, new assay method, and evidence for a ping-pong mechanism, *J. Biol. Chem.,* 257, 8835, 1982.
43. **Brown, A. J. and Snyder, F.,** The mechanism of alkyldihydroxyacetone-P synthase, *J. Biol. Chem.,* 258, 4184, 1983.
44. **Brown, A. J., Glish, G. L., McBay, E. H., and Snyder, F.,** Alkyldihydroxyacetonephosphate synthase mechanism: ^{18}O studies of fatty acid release from acyldihydroxyacetone phosphate, *Biochemistry,* 24, 8012, 1985.

45. **Raberi, U., Volkl, A., and Debauch, H.,** Distribution of alkylglycerone-phosphate synthase in subcellular fractions of rat liver, *Biol. Chem. Hoppe-Seyler,* 367, 215, 1986.
46. **Fleming, P. J. and Hajra, A. K.,** 1-Alkyl-*sn*-glycero-3-phosphate: acyl-CoA acyltransferase in rat brain microsomes, *J. Biol. Chem.,* 252, 1663, 1977.
47. **Wykle, R. L. and Snyder, F.,** Biosynthesis of an *O*-alkyl analogue of phosphatidic acid and *O*-alkylglycerols via *O*-alkyl ketone intermediates by microsomal enzymes of Ehrlich ascites tumor, *J. Biol. Chem.,* 245, 3047, 1970.
48. **Bell, R. M. and Coleman, R. A.,** Enzymes of glycerolipid synthesis in eukaryocytes, *Annu. Rev. Biochem.,* 49, 459, 1980.
49. **Radominska-Pyrek, A., Strosznajder, J., Dabrowiecki, Z., Chojnacki, T., and Horrocks, L. A.,** Effects of free fatty acids on the enzymic synthesis of diacyl and ether types of choline and ethanolamine phosphoglycerides, *J. Lipid Res.,* 17, 657, 1976.
50. **Radominska-Pyrek, A., Strosznajder, J., Dabrowiecki, Z., Goracci, G., Chojnacki, T., and Horrocks, L. A.,** Enzymic synthesis of ether types of choline and ethanolamine phosphoglycerides by microsomal fractions from rat brain and liver, *J. Lipid Res.,* 18, 53, 1977.
51. **Lee, T.-c., Blank, M. L., Fitzgerald, V., and Snyder, F.,** Formation of alkylacyl- and diacylglycerophosphocholines via diradylglycerol cholinephosphotransferase in rat liver, *Biochim. Biophys. Acta,* 713, 479, 1982.
52. **Pelech, S. L. and Vance, D. E.,** Regulation of phosphatidylcholine biosynthesis, *Biochim. Biophys. Acta,* 779, 217, 1984.
53. **Goracci, G., Horrocks, L. A., and Porcellati, G.,** Reversibility of ethanolamine and choline phosphotransferases (EC 2.7.8.1 and EC 2.7.8.2) in rat brain microsomes with labelled alkylacylglycerols, *FEBS Lett.,* 80, 41, 1977.
54. **Goracci, G., Horrocks, L. A., and Porcellati, G.,** Studies of rat brain choline ethanolamine phosphotransferases using labeled alkylacylglycerol as substrate with evidence for reversibility of the reactions, *Adv. Exp. Med. Biol.,* 101, 269, 1978.
55. **Chae, K., Piantadosi, C., and Snyder, F.,** An alternate enzymic route for the synthesis of the alkyl analog of phosphatidic acid involving alkylglycerol, *Biochem. Biophys. Res. Commun.,* 51, 119, 1973.
56. **Rock, C. O. and Snyder, F.,** Biosynthesis of 1-alkyl-*sn*-glycero-3-phosphate via adenosine triphosphate: 1-alkyl-*sn*-glycerol phosphotransferase, *J. Biol. Chem.,* 249, 5382, 1974.
57. **Lands, W. E. M.,** Metabolism of glycerolipids. II. The enzymatic acylation of lysolecithin, *J. Biol. Chem.,* 235, 2233, 1960.
58. **Irvine, R. F. and Dawson, R. M. C.,** Transfer of arachidonic acid between phospholipids in rat liver microsomes, *Biochem. Biophys. Res. Commun.,* 91, 1399, 1979.
59. **Trotter, J. and Ferber, E.,** CoA-dependent cleavage of arachidonic acid from phosphatidylcholine and transfer to phosphatidylethanolamine in homogenates of murine thymocytes, *FEBS Lett.,* 128, 237, 1981.
60. **Kramer, R. M. and Deykin, D.,** Arachidonoyl transacylase in human platelets. Coenzyme A-independent transfer of arachidonate from phosphatidylcholine to lysoplasmenylethanolamine, *J. Biol. Chem.,* 258, 13806, 1983.
61. **Waku, K. and Nakazawa, Y.,** Regulation of the fatty acid compositon of alkyl ether phospholipid in Ehrlich ascites tumor cells, *J. Biochem.,* 82, 1779, 1977.
62. **McKean, M. L. and Silver, M. J.,** Phospholipid biosynthesis in human platelets, the acylation of lyso-platelet-activating factor, *Biochem. J.,* 225, 723, 1985.
63. **McKean, M. L., Silver, M. J., Authi, K. S., and Crawford, N.,** Formation of diacyl- and alkylacylphosphatidylcholine by the membranes of human platelets, *FEBS Lett.,* 195, 38, 1986.
64. **Sugiura, T., Masuzawa, Y., Nakagawa, Y., and Waku, K.,** Transacylation of lyso-platelet-activating factor and other lysophospholipids by macrophage microsomes, *J. Biol. Chem.,* 262, 1199, 1987.
65. **Chilton, F. H., Hadley, J. S., and Murphy, R. C.,** Incorporation of arachidonic acid into 1-acyl-2-lyso-*sn*-glycero-3-phosphocholine of the human neutrophil, *Biochim. Biophys. Acta,* 917, 48, 1987.
66. **Arthur, G. and Choy, P. C.,** Acylation of 1-alkenyl-glycerophosphocholine and 1-acyl-glycerophosphocholine in guinea pig heart, *Biochem. J.,* 236, 481, 1986.
67. **Kramer, R. M., Patton, G. M., Pritzker, C. R., and Deykin, D.,** Metabolism of platelet-activating factor in human platelets, *J. Biol. Chem.,* 259, 13316, 1984.
68. **Cornic, M., Breton, M., and Colard, O.,** Acylation of 1-alkyl- and 1-acyl-lysophospholipids by rat platelets, *Pharmacol. Res. Commun.,* 18, 43, 1986.
69. **Robinson, M., Blank, M. L., and Snyder, F.,** Acylation of lysophospholipids by rabbit alveolar macrophages, *J. Biol. Chem.,* 260, 7889, 1985.
70. **Lee, T.-c., Blank, M. L., Fitzgerald, V., and Snyder, F.,** Mechanisms of acylation for alkylysophospholipids, *Fed. Proc. Fed. Am. Soc. Exp. Biol.,* 46, 2126 (Abstr. 1168), 1987.

71. **Ojima, A., Nakagawa, Y., Sugiura, T., Masuzawa, Y., and Waku, K.,** Selective transacylation of 1-*O*-alkylglycerophosphoethanolamine by docosahexaenoate and arachidonate in rat brain microsomes, *J. Neurochem.*, 48, 1403, 1987.
72. **Flesch, I., Ecker, B., and Ferber, E.,** Acyltransferase-catalyzed cleavage of arachidonic acid from phospholipids and transfer to lysophosphatides in macrophages derived from bone marrow, *Eur. J. Biochem.*, 139, 431, 1984.
73. **Sugiura, T. and Waku, K.,** CoA-independent transfer of arachidonic acid from 1,2-diacyl-*sn*-glycero-3-phosphocholine to 1-*O*-alkyl-*sn*-glycero-3-phosphocholine (lyso platelet-activating factor) by macrophage microsomes, *Biochem. Biophys. Res. Commun.*, 127, 384, 1985.
74. **Sugiura, T., Masuzawa, Y., and Waku, K.,** Transacylation of 1-*O*-alkenyl-*sn*-glycero-3-phosphoethanolamine with docosahexaenoic acid (22:6ω3), *Biochem. Biophys. Res. Commun.*, 133, 574, 1985.
75. **Chilton, F. H., O'Flaherty, J. T., Ellis, J. M., Swendsen, C. L., and Wykle, R. L.,** Selective acylation of lyso platelet activating factor by arachidonate in human neutrophils, *J. Biol. Chem.*, 258, 7268, 1983.
76. **Malone, B., Lee, T.-c., and Snyder, F.,** Inactivation of platelet activating factor by rabbit platelets, *J. Biol. Chem.*, 260, 1531, 1985.
77. **Chilton, F. H., O'Flaherty, J. T., Ellis, J. M., Swendsen, C. L., and Wykle, R. L.,** Metabolic fate of platelet-activating factor in neutrophils, *J. Biol. Chem.*, 258, 6357, 1983.
78. **Masuzawa, Y., Okano, S., Nakagawa, Y., Ojima, A., and Waku, K.,** Selective acylation of alkyllysophospholipids by docosahexaenoic acid in Ehrlich ascites cell, *Biochim. Biophys. Acta*, 876, 80, 1986.
79. **Wykle, R. L., Malone, B., and Snyder, F.,** Enzymatic synthesis of 1-alkyl-2-acetyl-*sn*-glycero-3-phosphocholine, a hypotensive and platelet-aggregating lipid, *J. Biol. Chem.*, 255, 10256, 1980.
80. **Lee, T.-c., Malone, B., and Snyder, F.,** A new *de novo* pathway for the formation of 1-alkyl-2-acetyl-*sn*-glycerols, precursors of platelet activating factor. Biochemical characterization of 1-alkyl-2-lyso-*sn*-glycero-3-P:acetyl-CoA acetyltransferase in rat spleen, *J. Biol. Chem.*, 261, 5373, 1986.
81. **Lee, T.-c., Malone, B., and Snyder, F.,** Formation of 1-alkyl-2-acetyl-*sn*-glycerols via *de novo* biosynthetic pathway for platelet activating factor. Characterization of 1-alkyl-2-acetyl-*sn*-glycero-3-phosphate phosphohydrolase in rat spleen, *J. Biol. Chem.*, 263, 1755, 1988.
82. **Woodard, D. S., Lee, T.-c., and Snyder, F.,** The final step in the *de novo* biosynthesis of platelet activating factor. Properties of a unique CDP-choline:1-alkyl-2-acetyl-*sn*-glycerol cholinephosphotransferase in microsomes from the renal inner medulla of rats, *J. Biol. Chem.*, 262, 2520, 1987.
83. **Blank, M. L., Lee, Y. J., Cress, E. A., and Snyder, F.,** Stimulation of the *de novo* pathway for the biosynthesis of platelet activating factor (PAF) via cytidylyltransferase activation in cells with minimal endogenous PAF production, *J. Biol. Chem.*, 263, 5656, 1988.
84. **Kawasaki, T. and Snyder, F.,** Synthesis of a novel acetylated neutral lipid related to platelet activating factor by acyl-CoA:1-*O*-alkyl-2-acetyl-*sn*-glycerol acyltransferase in HL-60 cells, *J. Biol. Chem.*, 263, 2593, 1988.
85. **Wykle, R. L., Blank, M. L., Malone, B., and Snyder, F.,** Evidence for a mixed-function oxidase in the biosynthesis of ethanolamine plasmalogens from 1-alkyl-2-acyl-*sn*-glycero-3-phosphorylethanolamine, *J. Biol. Chem.*, 247, 5442, 1972.
86. **Wykle, R. L. and Schremmer Lockmiller, J. M.,** The biosynthesis of plasmalogens by rat brains: involvement of the microsomal electron transport system, *Biochim. Biophys. Acta*, 380, 291, 1975.
87. **Paltauf, F. and Holasek, A.,** Enzymatic synthesis of plasmalogens. Characterization of the 1-*O*-alkyl-2-acyl-*sn*-glycero-3-phosphorylethanolamine desaturase from mucosa of hamster small intestine, *J. Biol. Chem.*, 248, 1609, 1973.
88. **Paltauf, F., Prough, R. A., Masters, B. S. S., and Johnston, J. M.,** Evidence for the participation of cytochrome b_5 in plasmalogen biosynthesis, *J. Biol. Chem.*, 249, 2661, 1974.
89. **Mogelson, S., Wilson, G. E., Jr., and Sobel, B. E.,** Ethanolamine plasmalogen methylation by rabbit myocardial membranes, *Biochim. Biophys. Acta*, 666, 205, 1981.
90. **Ford, D. A. and Gross, R. W.,** Identification of endogenous 1-*O*-alk-1′-enyl-2-acyl-*sn*-glycerol in myocardium and its effective utilization by choline phosphotransferase, *J. Biol. Chem.*, 263, 2644, 1988.
91. **Wientzek, M., Man, R. Y. K., and Choy, P. C.,** Choline glycerophospholipid biosynthesis in the guinea pig heart, *Biochem. Cell Biol.*, 65, 860, 1987.
92. **Kiyasu, J. Y. and Kennedy, E. P.,** The enzymatic synthesis of plasmalogens, *J. Biol. Chem.*, 235, 2590, 1960.
93. **Poulos, A., Hughes, B. P., and Cumings, J. N.,** The biosynthesis of choline plasmalogen by ox heart, *Biochim. Biophys. Acta*, 152, 627, 1968.
94. **Woelk, H.,** The action of phospholipase A_2 in isolated fat cells on specifically labeled 2-acyl-1-alk-1′-enyl and 2-acyl-1-alkyl-*sn*-glycero-3-phosphocholine, *Biochem. Biophys. Res. Commun.*, 59, 1278, 1974.
95. **Woelk, H., Goracci, G., and Porcellati, G.,** The action of brain phospholipases A_2 on purified, specifically labeled 1,2-diacyl-, 2-acyl-1-alk-1′-enyl- and 2-acyl-1-alkyl-*sn*-glycero-3-phosphorylcholine, *Physiol. Chem.*, 355, 75, 1974.
96. **Wolf, R. and Gross, R. W.,** Identification of neutral active phospholipase C which hydrolyzes choline glycerophospholipids and plasmalogen selective phospholipase A_2 in canine myocardium, *J. Biol. Chem.*, 260, 7295, 1985.

97. **Loeb, L. A. and Gross, R. H.**, Identification and purification of sheep platelet phospholipase A_2 isoforms, *J. Biol. Chem.*, 261, 10467, 1986.
98. **Kramer, R. M., Jakubowski, J. A., and Deykin, D.**, Hydrolysis of 1-alkyl-2-arachidonoyl-*sn*-glycero-3-phosphocholine, a common precursor of platelet-activating factor and eicosanoids by human platelets phospholipase A_2, *Biochim. Biophys. Acta*, 959, 269, 1988.
99. **Alonso, F., Henson, P. M., and Leslie, C. C.**, A cytosolic phospholipase in human neutrophils that hydrolyzes arachidonoyl-containing phosphatidylcholine, *Biochim. Biophys. Acta*, 878, 273, 1986.
100. **Arthur, G., Page, L., Mock, T., and Choy, P. C.**, The catabolism of plasmenylcholine in the guinea pig heart, *Biochem. J.*, 236, 475, 1986.
101. **Nishihira, J. and Ishibashi, T.**, A phospholipse C with a high specificity for platelet-activating factor in rabbit liver light mitochondria, *Lipids*, 21, 780, 1986.
102. **Matsuzawa, Y. and Hostetler, K. Y.**, Properties of phospholipase C isolated from rat liver lysosomes, *J. Biol. Chem.*, 255, 646, 1980.
103. **Otanaess, A. B., Prydz, H., Bjorklid, E., and Berre, A.**, Phospholipase C from bacillus cereus and its use in studies of tissue thromboplastin, *Eur. J. Biochem.*, 27, 238, 1982.
104. **Zwaal, R. F. A., Roelofsen, B., Comfurius, P., and van Deenen, L. L. M.**, Complete purification and some properties of phospholipase C from Bacillus cereus, *Biochim. Biophys. Acta*, 233, 474, 1971.
105. **Takahashi, T., Sugahara, T., and Ohsaka, A.**, Purification of Clostridium perfringes phospholipase (α-toxin) by affinity chromatography on agrose-linked egg-yolk lipoprotein, *Biochim. Biophys. Acta*, 351, 155, 1974.
106. **Wykle, R. L. and Schremmer, J. M.**, A lysophospholipase D pathway in the metabolism of ether-linked lipids in brain microsomes, *J. Biol. Chem.*, 249, 1742, 1974.
107. **Wykle, R. L., Kraemer, W. F., and Schremmer, J. M.**, Studies of lysophospholipase D of rat liver and other tissues, *Arch. Biochem. Biophys.*, 184, 149, 1977.
108. **Wykle, R. L., Kraemer, W. F., and Schremmer, J. M.**, Specificity of lysophospholipase D., *Biochim. Biophys. Acta*, 619, 58, 1980.
109. **Kawasaki, T. and Snyder, F.**, The metabolism of lyso-platelet-activating factor (1-*O*-alkyl-2-lyso-*sn*-glycero-3-phosphocholine) by a calcium-dependent lysophospholipase D in rabbit kidney medulla, *Biochim. Biophys. Acta*, 920, 85, 1987.
110. **Tokumura, A., Harada, K., Fukuzawa, K., and Tsukatani, H.**, Involvement of lysophospholipase D in the production of lysophosphatidic acid in rat plasma, *Biochim. Biophys. Acta*, 875, 31, 1986.
111. **Blank, M. L., Lee, T.-c., Fitzgerald, V., and Snyder, F.**, A specific acetylhydrolase for 1-alkyl-2-acetyl-*sn*-glycero-3-phosphocholine (a hypotensive and platelet-activating lipid), *J. Biol. Chem.*, 256, 175, 1981.
112. **Lee, T.-c., Blank, M. L., Fitzgerald, V., and Snyder, F.**, Substrate specificity in the biocleavage of the *O*-alkyl-bond: 1-alkyl-2-acetyl-*sn*-glycero-3-phosphocholine (a hypotensive and platelet-activating lipid) and its metabolites, *Arch. Biochem. Biophys.*, 208, 353, 1981.
113. **Snyder, F., Malone, B., and Piantadosi, C.**, Tetrahydropteridine-dependent cleavage enzyme for *O*-alkyl lipids: substrate specificity, *Biochim. Biophys. Acta*, 316, 259, 1973.
114. **Hoffman, D. R., Hoffman, L. H., and Snyder, F.**, Cytotoxicity and metabolism of alkyl phospholipid analogues in neoplastic cells, *Cancer Res.*, 46, 5803, 1986.
115. **Ishibashi, T. and Imai, Y.**, Solubilization and partial characterization of alkylglycerol monooxygenase from rat liver microsomes, *Eur. J. Biochem.*, 132, 23, 1983.
116. **Ishibashi, T. and Imai, Y.**, Affinity purification of alkylglycerol monooxygenase from rat liver microsome by chimyl alcohol-Sepharose 4B column chromatography, *J. Lipid Res.*, 26, 393, 1985.
117. **Warner, H. R. and Lands, W. E. M.**, The metabolism of plasmalogen: enzymatic hydrolysis of the vinyl ether, *J. Biol. Chem.*, 236, 2404, 1961.
118. **Alexander, M. S. J., Ebata, H., Mills, J. S., Hirashima, Y., and Horrocks, L. A.**, Plasmalogen hydrolysis in liver: partial purificaiton of alkenyl hydrolase, *FASEB J.*, 2, A1367, 1988.
119. **Gunawan, J. and Debuch, H.**, Liberation of free aldehyde from 1-(1-alkeynl)-*sn*-glycero-3-phosphoethanolamine (lysoplasmalogen) by rat liver microsomes, *Hoppe Seylers Z. Physiol. Chem.*, 362, 445, 1981.
120. **Ellingson, J. S. and Lands, W. E. M.**, Phospholipid reactivation of plasmalogen metabolism, *Lipids*, 3, 111, 1968.
121. **D'Amto, R. A., Horrocks, L. A., and Richardson, K. E.**, Kinetic properties of plasmalogenase from bovine brain, *J. Neurochem.*, 24, 1251, 1975.
122. **Blank, M. L., Hall, M. N., Cress, E. A., and Snyder, F.**, Inactivation of 1-alkyl-2-acetyl-*sn*-glycero-3-phosphocholine by a plasma acetylhydrolase: higher activities in hypertensive rats, *Biochem. Biophys. Res. Commun.*, 113, 666, 1983.
123. **Farr, R. S., Wardlow, M. L., Cox, C. P., Meng, K. E., and Greene, D. E.**, Human serum acid-labile factor is an acylhydrolase that inactivates platelet-activating factor, *Fed. Proc., Fed. Am. Soc. Exp. Biol.*, 42, 3120, 1983.
124. **Stafforini, D. M., Prescott, S. M., and McIntyre, T. M.**, Human plasma platelet-activating factor acetylyhydrolase. Purification and properties, *J. Biol. Chem.*, 262, 4223, 1987.

125. **Stafforini, D. M., McIntyre, T. M., Carter, M. E., and Prescott, S. M.,** Human plasma platelet-activating factor acetylydrolase. Association with lipoprotein particles and role in the degradation of platelet-activating factor, *J. Biol. Chem.,* 262, 4215, 1987.
126. **Lee, T.-c.,** Enzymatic control of the cellular levels of platelet activating factor, in *Platelet Activating Factor and Related Mediators,* Snyder, F., Ed., Plenum Press, New York, 1987, 115.
127. **Nieto, M. L., Velasco, S., and Sanchez Crespo, M.,** Modulation of acetyl-CoA:1-alkyl-2-lyso-*sn*-glycero-3-phosphocholine (lyso-PAF) acetyltransferase in human polymorphonuclears. The role of cyclic AMP-dependent and phospholipid sensitive calcium-dependent protein kinases, *J. Biol. Chem.,* 263, 4607, 1988.
128. **Domenech, C., Machado-De Domenech, E., and Soling, H. D.,** Regulation of acetyl-CoA:1-alkyl-*sn*-glycero-3-phosphocholine O^2-acetyltransferase (lyso-PAF-acetyltransferase) in exocrine glands. Evidence for an activation via phosphorylation by calcium/calmodulin-dependent protein kinase, *J. Biol. Chem.,* 262, 5671, 1987.
129. **Gomez-Cambronero, J., Mato, J. M., Vivanco, F., and Sanchez-Crespo, M.,** Phosphorylation of partially purified 1-*O*-alkyl-2-lyso-*sn*-glycero-3-phosphocholine:acetyl-CoA acetyltransferase from rat spleen, *Biochem. J.,* 245, 893, 1987.
130. **Lenihan, D. J. and Lee, T.-c.,** Regulation of platelet activating factor synthesis: modulation of 1-alkyl-2-*sn*-glycero-3-phosphocholine:acetyl-CoA acetyltransferase by phosphorylation and dephosphorylation in rat spleen microsomes, *Biochem. Biophys. Res. Commun.,* 120, 834, 1984.
131. **Gomez-Cambronero, J., Velasco, S., Mato, J. M., and Sanchez-Crespo, M.,** Modulation of lyso-platelet activating factor:acetyl-CoA acetyltransferase from rat splenic microsomes. The role of cyclic AMP-dependent protein kinase, *Biochim. Biophys. Acta,* 845, 516, 1985.
132. **Ninio, E., Joly, F., Heiblot, C., Bessou, G., Mencia-Huerta, J. M., and Benveniste, J.,** Biosynthesis of PAF-acether. IX. Role for a phosphorylation-dependent activation of acetyltransferase in antigen-stimulated mouse mast cells, *J. Immunol.,* 139, 154, 1987.
133. **McIntyre, T. M., Reinhold, S. L., Prescott, S. M., and Zimmerman, G. A.,** Protein kinase C activity appears to be required for the synthesis of platelet-activating factor and leukotriene B_4 by human neutrophils, *J. Biol. Chem.,* 263, 15370, 1987.
134. **Billah, M., Eckel, S., Myers, R., and Siegel, M.,** Metabolism of platelet-activating factor (1-*O*-alkyl-2-acetyl-*sn*-glycero-3-phosphocholine) by human promyelocytic leukemic HL-60 cells, *J. Biol. Chem.,* 261, 5824, 1986.
135. **Ramesha, C. S. and Pickett, W. C.,** Platelet-activating factor and leukotriene biosynthesis is inhibited in polymorphonuclear leukocytes depleted of arachidonic acid, *J. Biol. Chem.,* 261, 7592, 1986.
136. **Bussolino, F., Gremo, F., Teea, C., Pescarmona, G. P., and Camussi, G.,** Production of platelet-activating factor by chick retina, *J. Biol. Chem.,* 261, 16502, 1986.
137. **Nieto, M. L., Velasco, S., and Sanchez Crespo, M.,** Biosynthesis of platelet-activating factor in human polymorphonuclear leukocytes. Involvement of the cholinephosphotransferase pathway response to the phorbol esters, *J. Biol. Chem.,* 263, 2217, 1988.
138. **Roubin, R., Dulioust, A., Haye-Legrand, I., Ninio, E., and Benveniste, J.,** Biosynthesis of PAF-acether. VIII. Impairment of PAF-acether production in activated macrophages does not depend upon acetyltransferase activity, *J. Immunol.,* 36, 1796, 1986.
139. **Touqui, L., Shaw, A. M., Dumatey, C., Jacquemin, C., and Vargaftig, B. B.,** The role of Ca^{2+} in regulating the catabolism of PAF-acether (1-*O*-alkyl-2-acetyl-*sn*-glycero-3-phosphocholine) by rabbit platelets, *Biochem. J.,* 241, 55, 1987.
140. **Crook, J. E., Mroczkowski, P. J., Cress, E. A., Blank, M. L., and Snyder, F.,** Serum platelet activating factor acetylhydrolase activity in white and black essential hypertensive patients, *Circulation Supplement, Abstracts from the 59th Scientific Sessions,* 74, (4):II-329, American Heart Association, Dallas, October 1986.
141. **McGowan, H. M., Vandongen, R., Kelly, L. D., and Hill, K. J.,** Increased levels of platelet activating factor (1-*O*-alkyl-2-acetylglycerophosphocholine) in blood after reversal of renal clip hypertension in the rat, *Clin. Sci.,* 74, 393, 1988.
142. **Maki, N., Hoffman, D. R., and Johnston, J. M.,** Platelet activity in maternal, fetal, and newborn rabbit plasma during pregnancy and lactation, *Proc. Natl. Acad. Sci.,* 85, 728, 1988.

Chapter 10

PHOSPHATIDYLCHOLINE METABOLISM IN *SACCHAROMYCES CEREVISIAE*

George M. Carman

TABLE OF CONTENTS

I. INTRODUCTION

Phosphatidylcholine (PC) is the major phospholipid found in eukaryotic membranes.[1,2] In the unicellular eukaryote, the yeast *Saccharomyces cerevisiae,* PC accounts for approximately 40 to 50% of the total membrane phospholipids.[2] As in animal cells,[1,3] PC is synthesized in *S. cerevisiae* by one of two alternative pathways.[2] The *de novo* pathway for PC synthesis culminates with the three sequential methylation reactions of phosphatidylethanolamine (PE). Alternatively, PC is synthesized by a series of reactions culminating with the reaction between CDP-choline and diacylglycerol (DG). The regulation of PC biosynthesis in *S. cerevisiae* is complex and involves a number of biosynthetic reactions important to overall phospholipid biosynthesis. Many of the enzymes in the *de novo* PC biosynthetic pathway are regulated by the availability of water-soluble phospholipid precursors, the cellular growth phase of cells, the membrane phospholipid composition, and by the competition of enzymes found at branch points in the pathway for common substrates. Biochemical, genetic, and molecular studies have shown that in *S. cerevisiae* the *de novo* biosynthesis of PC by the reaction sequence phosphatidate (PA) → CDP-diacylglycerol (CDP-DG) → phosphatidylserine (PS) → PE → phosphatidylmonomethylethanolamine (PMME) → phosphatidyldimethylethanolamine (PDME) → PC is coordinately regulated to the synthesis of phosphatidylinositol (PI).

Over the past few years a number of the phospholipid biosynthetic enzymes from *S. cerevisiae* have been purified to homogeneity and characterized, several structural and regulatory gene mutants have been isolated and characterized, and genes have been cloned by the genetic complementation of mutants. From this, a clearer understanding of PC metabolism in *S. cerevisiae* is emerging.

II. PHOSPHATIDYLCHOLINE BIOSYNTHETIC PATHWAYS

Phospholipid biosynthesis in *S. cerevisiae*[2] is carried out by enzymes common to higher eukaryotic organisms.[1,3] One exception is that in *S. cerevisiae,* PS is synthesized from CDP-DG and serine,[4] whereas in mammals PS is synthesized via an exchange reaction between PE and serine.[5] The pathways for the *de novo* and auxiliary routes for the synthesis of PC in *S. cerevisiae* were identified by Lester and co-workers[6,7] and are shown in Figure 1. Reactions leading to the formation of PI are also shown in Figure 1 since they are involved in the coordinate regulation of PC biosynthesis. Other reactions involved in phospholipid biosynthesis, including the formation of phosphatidylglycerol, cardiolipin, and the polyphosphoinositides, are not shown in Figure 1 and are described in the review of phospholipid biosynthesis by Henry.[2]

III. PURIFIED PHOSPHOLIPID BIOSYNTHETIC ENZYMES

Most of the enzymes of phospholipid metabolism have not been purified from any eukaryotic source. The enzymes of phospholipid metabolism from *S. cerevisiae* that have been purified and characterized are listed in Table 1. In higher eukaryotic cells, only choline kinase (Chapter 2), cholinephosphate cytidylyltransferase (Chapter 3), and PE-*N*-methyltransferase (Chapter 7) have been purified.

A. CDP-DIACYLGLYCEROL SYNTHASE

CDP-DG synthase (Figure 1, Reaction 2) is associated with the mitochondrial and microsomal fractions of *S. cerevisiae.*[8] About 75% of the total CDP-DG synthase activity is associated with the mitochondrial fraction.[9,10] The enzyme has been purified from the mitochondrial fraction by Triton X-100 solubilization of mitochondrial membranes, affinity chromatography with CDP-DG-Sepharose, and hydroxyapatite chromatography.[10] The major purification of the

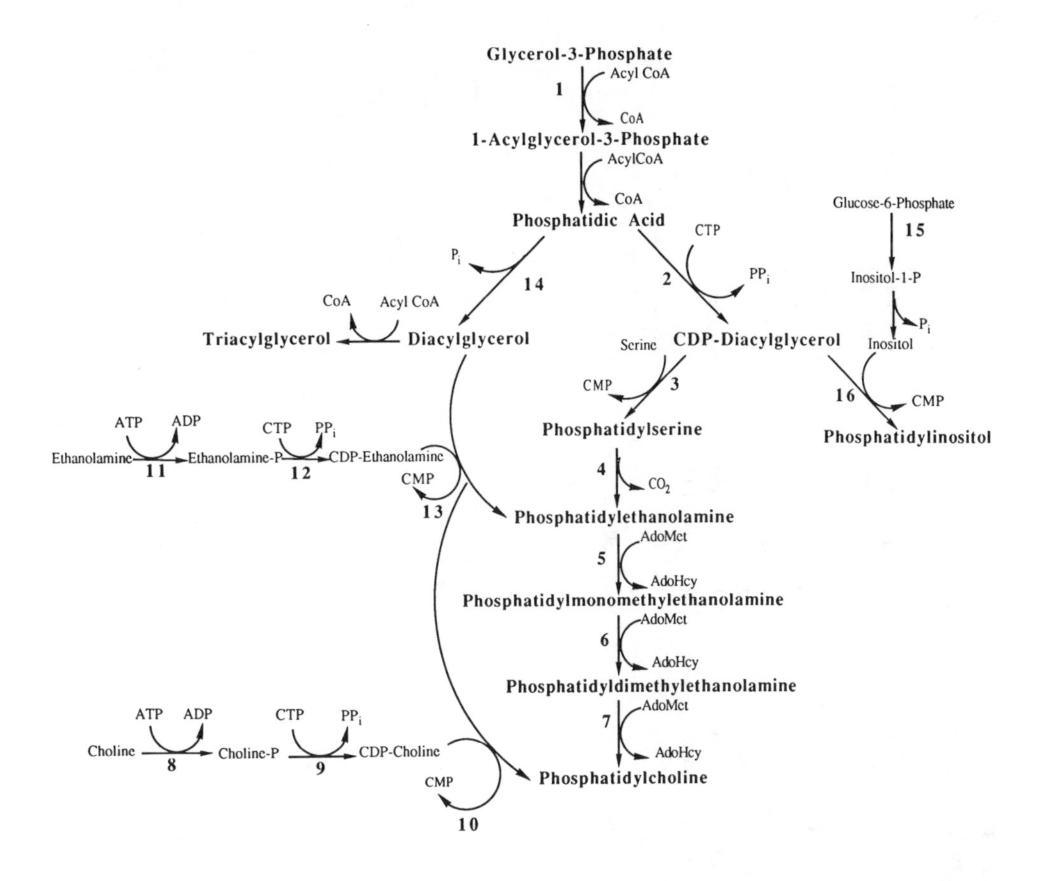

FIGURE 1. Phosphatidylcholine biosynthetic pathways in *S. cerevisiae*. The indicated reactions are catalyzed by the following enzymes: (1) glycerol–3–phosphate acyltransferase; (2) CDP–diacylglycerol synthase; (3) phosphatidylserine synthase; (4) phosphatidylserine decarboxylase; (5) phosphatidylethanolamine methyltransferase; (6 and 7) phospholipid methyltransferase; (8) choline kinase; (9) cholinephosphate cytidylyltransferase; (10) cholinephosphotransferase; (11) ethanolamine kinase; (12) ethanolaminephosphate cytidylyltransferase; (13) ethanolaminephosphotransferase; (14) phosphatidate phosphatase; (15) inositol–1–phosphate synthase; (16) phosphatidylinositol synthase.

enzyme is achieved by the CDP-DG-Sepharose chromatography step. Effective binding of the enzyme to the affinity resin occurs because the CDP-DG synthase reaction is favored in the reverse direction.[10] The enzyme has been purified 2309-fold over the cell extract with an activity yield of 31%.[10] Radiation inactivation analysis of the mitochondrial-associated and purified enzyme suggests that a functional CDP-DG synthase has a target size of 114,000.[10] Based on the results of sodium dodecyl sulfate(SDS)-polyacrylamide gel electrophoresis, the native enzyme appears to be a dimer composed of identical subunits of 56,000.[10] Maximum CDP-DG synthase activity is dependent on magnesium ions and Triton X-100 at the pH optimum of 6.5.[10] The activation energy for the reaction is 9 kcal/mol, the enzyme is labile above 30°C, and thioreactive agents inhibit activity.[10] In a mixed micelle substrate of Triton X-100 and PA, the K_m values for CTP and PA are 1 m*M* and 0.5 m*M*, respectively, and the V_{max} is 4.7 μmol/min/mg.[10] The results of kinetic experiments[9,10] and the ability of the enzyme to catalyze a variety of isotopic exchange reactions[10] suggest that CDP-DG synthase catalyzes a sequential Bi Bi reaction. CDP-DG synthase binds to CTP prior to PA and PPi is released prior to CDP-DG in the reaction sequence.[10] dCTP is both a substrate[10] and a competitive inhibitor for the enzyme.[9,10]

TABLE 1
Phospholipid Biosynthetic Enzymes Purified from *S. cerevisiae*

Enzyme	Subcellular localization	Purification (fold)	Molecular weight	Activity requirements	Kinetic constants	Reaction mechanism	Ref.
CDP–DG synthase	Mitochondria	2,309	114,000[a]	pH optimum 6.5, Mg^{2+},	K_{CTP}, 1 m*M*	Sequential	8, 10
(EC 2.7.7.41)	Microsomes		56,000[b]	Triton® X–100	K_{PA}, 0.5 m*M*		
PS synthase	Mitochondria	5,000	23,000[b]	pH optimum 8.0, Mn^{2+}	K_{CDP-DG}, 83μ*M*	Sequential	4, 8, 11,
(EC 2.7.8.8)	Microsomes			or Mg^{2+}, Triton® X–100	K_{serine}, 0.83 m*M*		14, 15
PI synthase	Mitochondria	3,300	34,000[b]	pH optimum 8.0, Mn^{2+}	K_{CDP-DG}, 66μ*M*	Sequential	8, 11, 14,
(EC 2.7.8.11)	Microsomes			or Mg^{2+}, Triton® X–100	$K_{inositol}$, 0.21 m*M*		18,19
IP synthase	Cytosol	500	240,000[a]	pH optimum 7.0, NAD,	K_{NAD}, 8 μ*M*	Unknown	20
(EC 5.5.1.4)			62,000[b]	NH_4^+	$K_{glucose-6-p}$, 1.2 m*M*		
PA phosphatase	Cytosol	600	75,000[a]	pH optimum 7—8, Mg^{2+}	K_{PA}, 50 μ*M*	Unknown	21

[a] Native molecular weight.
[b] Subunit molecular weight.

B. PHOSPHATIDYLSERINE SYNTHASE

PS synthase (Figure 1, Reaction 3) is associated with the mitochondrial and microsomal fractions of *S. cerevisiae*.[8] The enzyme has been purified from wild-type cells as well as from a strain bearing the structural gene (*CHO1*) for PS synthase on a hybrid plasmid which directs the overproduction of the enzyme.[4] The purification procedure includes the solubilization of microsomal membranes with Triton X-100, affinity chromatography with CDP-DG-Sepharose, and ion-exchange chromatography with DEAE-cellulose.[4] Overall PS synthase has been purified about 5000-fold relative to the cell extract of wild-type cells with activity yields of 42 to 70%.[4] The subunit molecular weight of the purified enzyme is 23,000 as determined by SDS-polyacrylamide gel electrophoresis[4] and the electroblotting of PS synthase activity.[11] When PS synthase is partially purified in the presence of the protease inhibitor PMSF, an $M_r = 30{,}000$ peptide as well as the $M_r = 23{,}000$ peptide is isolated.[12] The $M_r = 23{,}000$ peptide is a proteolytic cleavage product of the Mr = 30,000 peptide of PS synthase.[12] However, only the $M_r = 23{,}000$ peptide appears to have PS synthase activity.[12] The predicted subunit molecular weight of PS synthase based on the DNA sequence of the PS synthase structural gene is $M_r = 30{,}804$.[12,13]

Maximum PS synthase activity is dependent on either manganese or magnesium ions for activity and Triton X-100 at the pH optimum of 8.0.[4] In a mixed-micelle substrate of Triton X-100 and CDP-DG, the apparent K_m values for serine and CDP-DG are 0.83 m*M* and 83 μ*M*, respectively, and the V_{max} is 4 μmol/min/mg.[4,14] The enzyme is labile above 40°C and is sensitive to thioreactive agents.[4] PS synthase utilizes dCDP-DG as well as CDP-DG as a substrate.[4] Based on the results of kinetic experiments,[14] isotopic exchange reactions between substrates and products,[4] and a stereochemical analysis of the reaction using ^{31}P NMR,[15] PS synthase catalyzes a Bi Bi reaction mechanism where the enzyme binds to CDP-DG before serine and PS is released prior to CMP in the reaction sequence.

Phospholipid vesicles have become an accepted model of biological membranes which are attractive for studying membrane-associated proteins.[16] To gain insight into the modulation of PS synthase activity by phospholipids (see Section V.B), the purified enzyme has been reconstituted into unilamellar phospholipid vesicles containing its substrate CDP-DG.[17] Since PS synthase is purified in the presence of Triton X-100,[4] it is necessary to reconstitute the enzyme into vesicles directly from mixed micelles containing pure enzyme, phospholipid, and Triton X-100. Reconstitution of the enzyme is achieved by removing detergent from an octyl glucoside/phospholipid/Triton X-100/enzyme mixed micelle by Sephadex G-50 superfine chromatography.[17,18] In this procedure, the vesicles containing PS synthase elute near the void volume of the column and are well separated from the detergent molecules.[17,18] The average size of the vesicles containing reconstituted PS synthase is 90 nm in diameter. Each vesicle contains about 1 molecule of PS synthase which is reconstituted asymmetrically with 80 to 90% of its active site facing outward.[17] The enzymatic properties of reconstituted PS synthase assayed in the absence of detergent[17] are similar to those determined in the presence of detergent.[4] Therefore, these properties have not been changed by reconstituting the enzyme.

C. PHOSPHATIDYLINOSITOL SYNTHASE

PI synthase (Figure 1, Reaction 16) is associated with the mitochondrial and microsomal fractions of *S. cerevisiae*.[8] PI synthase has been purified 3300-fold from the microsomal fraction over the cell extract with a yield of 60%.[19] Successful purification of PI synthase requires the solubilization of the enzyme from microsomal membranes with Triton X-100, CDP-DG-Sepharose affinity chromatography, and chromatofocusing chromatography.[19] The purified enzyme has a subunit molecular weight of 34,000 as determined by SDS-polyacrylamide gel electrophoresis[19] and the electroblotting of PI synthase activity.[11] Maximum PI synthase activity is dependent on manganese or magnesium ions and Triton X-100 for activity at the pH optimum of 8.0.[19] The K_m values for inositol and CDP-DG are 0.21 m*M* and 66 μ*M*, respectively, as determined in a Triton X-100-CDP-DG mixed-micelle system.[14] The enzyme is labile above

60°C and is inactivated by thioreactive agents.[19] The activation energy for the reaction is 35 kcal/mol.[19] Based on the results of kinetic experiments[14] and isotopic exchange reactions,[18] PI synthase catalyzes a Bi Bi sequential reaction mechanism. PI synthase binds to CDP-DG before inositol, and PI is released prior to CMP in the reaction sequence.[18]

Purified PI synthase has been reconstituted into unilamellar phospholipid vesicles containing its substrate CDP-DG by the same procedure used for the reconstitution of PS synthase.[18] The average size of the vesicles containing PI synthase is 40 nm and each 40-nm vesicle contains about 1 molecule of enzyme.[18] PI synthase is reconstituted asymmetrically with about 90% of its active site facing outward.[18] The basic enzymatic properties of reconstituted PI synthase[18] are similar to that of the enzyme measured in the Triton X-100-CDP-DG mixed-micelle assay system.[19] The reconstituted system has been used to study the modulation of PI synthase activity by phospholipids (see Section V.B) in a system that more closely mimics the *in vivo* environment of an intact membrane.[18]

D. INOSITOL-1-PHOSPHATE SYNTHASE

Inositol-1-phosphate (IP) synthase (Figure 1, Reaction 15) is associated with the cytosolic fraction of *S. cerevisiae* and has been purified 500-fold relative to the activity in cell extracts with yields of 20 to 40%.[20] The purification procedure includes streptomycin sulfate precipitation and ammonium sulfate fractionation, followed by chromatography with DEAE-cellulose, hexyl-agarose, Bio-Gel A-0.5 m, and DEAE-cellulose.[20] The native molecular weight of the purified enzyme is 240,000 as determined by gel filtration.[20] The enzyme is a tetramer composed of four identical subunits with a molecular weight of 62,000 as determined by SDS-polyacrylamide gel electrophoresis.[20] The pH optimum of purified IP synthase is 7.0, and activity is dependent on NAD and ammonium ions.[20] The glucose-6-phosphate analogue, 2-deoxyglucose-6-phosphate is an effective inhibitor of the enzyme.[20] The K_m values for glucose-6-phosphate and NAD are 1.18 mM and 8 μM, respectively.[20]

E. PHOSPHATIDATE PHOSPHATASE

PA phosphatase (Figure 1, Reaction 14) is associated with the cytosolic and membrane fractions of *S. cerevisiae*.[21] The cytosolic form of the enzyme has been partially purified 600-fold by ammonium sulfate and polyethylene glycol fractionations followed by column chromatography on DEAE-Sepharose, Sephadex G-100, and Blue-Sepharose.[22] The native molecular weight for the enzyme is appoximately 75,000 as determined by gel filtration chromatography.[22] PA phosphatase is dependent on magnesium ions, and the pH optimum for the reaction is between 7 and 8.[22] The enzyme is specific for PA with a K_m of 50 μM.[22]

IV. MUTANTS OF PHOSPHOLIPID BIOSYNTHESIS

Mutants with defects in structural and regulatory genes in the PC biosynthetic pathways have been identified. Table 2 provides a list of these mutants and their properties.

A. GLYCEROL-3-PHOSPHATE ACYLTRANSFERASE MUTANTS

Mutants defective in glycerol-3-phosphate acyltransferase *(gat)* (Figure 1, Reaction 1) have been identified[23] using the colony autoradiographic screen technique[24] developed for yeast phospholipid biosynthetic enzymes.[23,25] These mutants do not have an obvious *in vivo* phenotype. Glycerol-3-phosphate acyltransferase activity is reduced over 70-fold in some mutants. The enzyme shows alterations in the K_m for glycerol-3-phosphate and the V_{max} for the reaction.[23] The isolation of these glycerol-3-phosphate acyltransferase mutants has confirmed that dihydroxyacetone phosphate acyltransferase activity in *S. cerevisiae* is a second activity of the glycerol-3-phosphate acyltransferase.[23]

TABLE 2
Mutants of Phospholipid Biosynthesis in *S. cerevisiae*

Mutant	Phenotype and biochemical defect	Gene cloned	Ref.
gat	No phenotype, defective in glycerol-3-phosphate acyltransferase activity	No	23
cdg1	Excretes inositol, reduced CDP-DG synthase activity	No	26
cho1/pss	Ethanolamine/choline auxotrophs, defective in PS synthase activity	Yes	12, 13, 27—30, 34
pis	Requires high concentration of inositol, defective in PI synthase activity	Yes	36, 37
pem1	Choline auxotroph, defective in PE methyltransferase activity	Yes	39, 43
cho2	No phenotype, defective in PE methyltransferase activity	No	41
pem2	Choline auxotroph, defective in phospholipid methyltransferase activity	Yes	40, 43
opi3	Excretes inositol, defective in phospholipid methyltransferase activity	No	42
cki	No phenotype, defective in choline kinase activity	No	44, 45
cct	No phenotype, defective in CDP-choline synthase activity	Yes	46, 47
cpt1	No phenotype, defective in cholinephosphotransferase activity	Yes	45, 48
ept1	No phenotype, defective in ethanolaminephosphotransferase activity	Yes	74
ino1	Inositol auxotroph, defective in IP synthase activity	Yes	20, 49, 50
ino2	Inositol auxotrophs, lacks IP synthase activity, pleiotropic defects in PC *de novo* pathway enzymes	No	20, 32, 49, 54, 56
ino4	Inositol auxotrophs, lacks IP synthase activity, pleiotropic defects in PC biosynthetic *de novo* pathway enzymes	Yes	20, 32, 49, 54, 56—58
opi1	Excretes inositol, overproduces IP synthase, pleiotropic defects in PC *de novo* biosynthetic pathway enzymes	No	56, 59—63

B. CDP-DIACYLGLYCEROL SYNTHASE MUTANT

The *cdg1* mutant is isolated on the basis of its inositol excretion phenotype and exhibits pleiotropic deficiencies in phospholipid biosynthesis.[26] Genetic analysis of the *cdg1* mutant has confirmed that a defect in a single gene is responsible for the *cdg1* phenotype. CDP-DG synthase activity in mutant haploid cells is less than 25% of the wild-type derepressed level. Biochemical and immunoblot analyses have shown that the defect in CDP-DG activity in mutant *cdg1* is due to a reduced level of the CDP-DG synthase $M_r = 56,000$ subunit rather than an alteration in the enzymatic properties of the enzyme. This defect results in a reduced rate of CDP-DG synthesis, elevated PA content, and alterations in overall phospholipid synthesis. Unlike wild-type cells, CDP-DG synthase in mutant cells is not regulated in response to water-soluble phospholipid precursors. The *cdg1* lesion also causes constitutive expression of IP synthase and elevated PS synthase. PI synthase is not affected in the *cdg1* mutant. Whether or not the *cdg1* mutant has a structural gene defect for CDP-DG synthase or a regulatory gene defect that affects CDP-DG synthase is not clear.

C. PHOSPHATIDYLSERINE SYNTHASE MUTANTS

The PS synthase mutants *(cho1* and *pss)* are defective in PS synthase activity and synthesize little or no PS unless supplemented with high levels of serine.[27-30] These mutants are believed to be PS synthase K_m mutants.[29] When not supplemented with serine, the PS synthase mutants require ethanolamine or choline for growth.[27-30] Ethanolamine and choline are converted to CDP-ethanolamine and CDP-choline which react with DG to form PE and PC, respectively.[31] Cells supplemented with only ethanolamine would synthesize PE from DG and CDP-ethanolamine as catalyzed by ethanolaminephosphotransferase.[31] The cells supplemented with ethanolamine still synthesize PC by the sequential methylation of PE. Cells supplemented with only choline synthesize PC from DG and CDP-choline as catalyzed by cholinephosphotransferase.[31] Null *cho1* allele PS synthase mutants have been constructed by disruption of the *CHO1* gene.[32] These null *cho1* mutants lack *CHO1* RNA and PS synthase activity and only grow when supplemented with ethanolamine or choline.[32]

When starved for ethanolamine or choline, the *cho1* mutants show a balanced rapid decline in PE, PC, and PI biosynthesis.[33] IP synthase which is regulated in response to exogenous inositol in wild-type cells,[20] is not regulated in starved *cho1* cells.[33]

The structural gene encoding for PS synthase *(CHO1/PSS)* has been cloned by genetic complementation of PS synthase mutants using a yeast genomic library constructed on the plasmid YEp13.[13,34] The plasmid YEp13 is capable of autonomous replication in *S. cerevisiae* and *Escherichia coli*. This plasmid can be selected for in *S. cerevisiae* strains of genotype *leu2*$^-$ because it bears the yeast *LEU2* gene.[35] YEp13 can be selected in *E. coli* by ampicillin resistance because it contains the plasmid pBR322.[35] Strains carrying the PS synthase gene on the YEp13 hybrid plasmid have a four- to eightfold amplification of PS synthase activity.[13,34] This is due to the high copy number of the YEp13 plasmid.[35] The availability of strain YEp*CHO1* which overproduces PS synthase[34] has facilitated the acquisition of larger amounts of purified enzyme.[4]

The structural gene for PS synthase has been sequenced and contains an open reading frame capable of encoding 276 amino acid residues with a calculated molecular weight of 30,804.[12,13] The predicted amino acid sequence contains four potential glycosylation sites and one putative phosphorylation site.[13]

D. PHOSPHATIDYLINOSITOL SYNTHASE MUTANT

The *pis* mutant is defective in PI synthase and requires a high concentration of inositol for growth.[36] PI synthase activity in the mutant has an apparent K_m for inositol that is over 200 times that of the enzyme in wild-type cells.[36] The PI content of the *pis* mutant is markedly lower than that of wild-type cells.[36]

The structural gene *(PIS)* encoding PI synthase has been cloned by genetic complementation of the PI synthase K_m mutant using a yeast genomic library inserted into the YEp13 shuttle vector.[37] PI synthase activity is elevated eightfold in the transformed strain and has a normal K_m for inositol.[37] Although PI synthase activity is elevated in the transformed strain, the cellular level of PI is similar to that found in wild-type cells.[37] The *PIS* gene has been sequenced and the *PIS* coding frame is capable of encoding 220 amino acid residues with a calculated molecular weight of 24,823.[38] There are two potential glycosylation sites on the predicted amino acid sequence.[38] The discrepancy of the inferred molecular weight of the enzyme (M_r = 24,823) and that (M_r = 34,000) of the purified enzyme[19] could be the result of PI synthase being glycosylated.

Cells which carry a disrupted copy of the *PIS* gene do not synthesize PI and are found in an arrested state with a characteristic terminal (death) phenotype.[38] These studies suggest that PI synthase is essential for progression of the yeast cell cycle.[38]

E. PHOSPHOLIPID *N*-METHYLTRANSFERASES MUTANTS

The results of genetic[39-42] and molecular studies[43] have shown that in *S. cerevisiae* two enzymes catalyze the reaction sequence PE $\rightarrow$ PMME $\rightarrow$ PDME $\rightarrow$ PC. As suggested by Kodaki and Yamashita,[43] the enzyme catalyzing the first methylation reaction (PE → PMME) and the enzyme catalyzing the last two methylation reactions (PMME → PDME → PC) will be referred to as PE methyltransferase and phospholipid methyltransferase, respectively.

1. Phosphatidylethanolamine Methyltransferase Mutants

The *pem1* mutant is auxotrophic for choline and is defective in PE methyltransferase activity (Figure 1, Reaction 5).[39] The *cho2* mutant is also defective in the first methylation reaction; however, unlike the *pem1* mutant, it is not a choline auxotroph.[41] The *cho2* mutant accumulates PE and overproduces IP synthase.[41] When the *cho2* mutant is supplemented with monomethylethanolamine, PC biosynthesis is normal and the regulation of IP synthase is restored.[41] The second two methylation steps are normal in the *pem1*[39] and *cho2*[41] mutants.

The *PEM1* gene has been cloned by genetic complementation of the PE methyltransferase mutants.[43] The nucleotide sequence of the *PEM1* gene encodes 869 amino acid residues with a calculated molecular weight of 101,202.[43]

2. Phospholipid Methyltransferase Mutants

The *pem2* mutant is auxotrophic for choline and is defective in the phospholipid methyltransferase (Figure 1, Reactions 6 and 7).[40] The *opi3* mutant is also defective in the synthesis of PC via the last two methylation reactions but does not require choline for growth.[42] The *opi3* mutant accumulates PMME and PDME and contains only 2 to 3% PC compared to 40 to 50% PC in wild-type cells.[42] Like the *cho2* mutant, the *opi3* mutant is constitutive for and overproduces IP synthase.[42] The *pem2*[40] and *opi3*[42] mutants are normal for PE methyltransferase activity.

The *PEM2* gene has been cloned by genetic complementation of the *pem2* mutant.[43] The *PEM2* gene has been sequenced and encodes 206 amino acid residues with a calculated molecular weight of 23,150.[43] The *PEM2* gene has sequence homology with portions of the *PEM1* gene, suggesting that the *PEM1* and *PEM2* genes evolved by gene duplication.[43] The relationship between *PEM1* and *CHO2* and between *PEM2* and *OPI3* is unclear.

F. CHOLINE KINASE MUTANTS

The *cki* mutants[44,45] are defective in choline kinase activity (Figure 1, Reaction 8). Two *cki* mutants were isolated from a mutagenized conditional choline auxotroph[39] by selecting colonies which no longer grow on medium containing inositol.[44] Two other *cki* mutants were isolated by selecting choline-resistant revertants from a choline-sensitive mutant[45]

G. CHOLINEPHOSPHATE CYTIDYLYLTRANSFERASE MUTANT

The cholinephosphate cytidylyltransferase (Figure 1, Reaction 9) temperature sensitive mutant *(cct)* has negligible cholinephosphate cytidylyltransferase activity compared to wild-type cells.[46] The *cct* mutant does not exhibit a growth phenotype.[46] The activity of the other enzymes in the CDP-choline-based pathway (i.e., choline kinase and cholinephosphotransferase, Figure 1, Reactions 8 and 10) are normal.[46] Studies with this mutant have shown that ethanolaminephosphate cytidylyltransferase (Figure 1, Reaction 12) and cholinephosphate cytidylyltransferase activities are encoded by different genes.[46] Furthermore, studies with this mutant have shown that the CDP-choline-based pathway for PC biosynthesis is not necessary, providing PC is synthesized via the PE methylation pathway.[46]

The structural gene encoding for cholinephosphate cytidylyltransferase *(CCT)* has been isolated by the genetic complementation of mutant *cct* using a *S. cerevisiae* genomic library on the plasmid YEp13.[47] Strains carrying the *CCT* gene on a multicopy plasmid overproduce cholinephosphate cytidylyltransferase activity about 100-fold.[47] Although enzyme overproduction results in increased CDP-choline synthesis, the overall rate of PC synthesis in transformed cells is normal.[47]

The cloned DNA containing the *CCT* gene has been subcloned into a 2.5-kb fragment and was used to determine the nucleotide sequence of the DNA.[47] A single open reading frame of 1.2 kb is contained in the sequence which is capable of encoding a protein with 424 amino acid residues with a predicted molecular weight of 49,379.[47] There are ten potential N-linked glycosylation sites and 11 possible phosphorylation sites of protein kinase C within the predicted amino acid sequence.[47] Northern blot analysis of wild-type mRNA has shown that the cloned DNA segment is transcribed in *S. cerevisiae* and the length of the transcript is consistent with the putative translation product.[47]

H. CHOLINEPHOSPHOTRANSFERASE MUTANTS

The *cpt1* mutants are defective in cholinephosphotransferase activity (Figure 1, Reaction 10).[45,48] Studies with the *cpt1* mutants have confirmed that cholinephosphotransferase and ethanolaminephosphotransferase in *S. cerevisiae* are encoded by different genes.[48]

The *CPT1* gene has been cloned by genetic complementation of mutant *cpt1* using a yeast genomic library on plasmid YEp13.[48] The *CPT1* gene has been sequenced and contains an open reading frame capable of encoding 407 amino acids with a predicted molecular weight of 46,000.[74] Transformed strains with the *CPT1*-bearing plasmid have a fivefold overproduction of cholinephosphotransferase activity and wild-type enzyme kinetic properties.[48] *CPT1* gene disruption mutants have been constructed which have fivefold reduced cholinephosphotransferase activity compared to wild-type cells.[48] This residual cholinephosphotransferase activity is due to ethanolaminephosphotransferase activity utilizing CDP-choline as a substrate.[74] The results of studies with the *CPT1* gene disruption mutants indicate that cholinephosphotransferase activity is not essential for growth.[48]

I. ETHANOLAMINEPHOSPHOTRANSFERASE MUTANTS

The *ept1* mutants were identified by colony autoradiography and are defective in ethanolaminephosphotransferase (Figure 1, Reaction 13) activity.[74] The *EPT1* gene has been cloned by genetic complementation of an *ept1* mutant and the complete nucleotide sequence determined.[74] The *EPT1* sequence contains an open reading frame encoding 391 amino acids with an inferred molecular weight of 44,525. The *EPT1* gene product shows 54% amino acid sequence homology with the *CPT1* gene product. Strains containing the *EPT1* gene on multicopy plasmids overproduce ethanolaminephosphotransferase activity 22- to 33-fold. Studies with *EPT1* gene disruption mutants have shown that ethanolaminephosphotransferase activity is not essential for growth.[74]

The aminoalcoholphosphotransferase activities of the *CPT1* and *EPT1* gene products have been independently studied using *cpt1* and *ept1* disruption mutants.[74] The *EPT1* gene product utilizes CDP-ethanolamine, CDP-monomethylethanolamine, CDP-dimethylethanolamine, and CDP-choline whereas the *CPT1* gene product uses CDP-choline and CDP-dimethylethanolamine. Together, the *CPT1* and *EPT1* gene products account for all the monomethylethanolaminephosphotransferase and dimethylethanolaminephosphotransferase activities in yeast, precluding the existence of distinct enzymes catalyzing these reactions.

J. INOSITOL-1-PHOSPHATE SYNTHASE MUTANTS

The *ino1* mutant is auxotrophic for inositol[49] due to a defect in the structural gene for IP synthase.[20] The *ino1* mutants have 60- to 240-fold lower IP synthase activity compared to wild-type cells.[50] However, *ino1* mutants do produce normal amounts of the IP synthase $M_r = 62,000$ subunit.[20] When *ino1* mutants are deprived of inositol, cells undergo changes in the metabolism of lipids, carbohydrates, proteins, and nucleic acids, resulting in a loss of cell viability known as inositol-less death.[51,52] Starved *ino1* cells accumulate PA and CDP-DG,[51] which are precursors to PC as well as PI synthesis.

The *INO1* gene has been cloned by genetic complementation of an *ino1* mutant using a genomic library on plasmid YEp13.[53] The transformed strain containing the *INO1* gene is prototrophic for inositol and contains IP synthase.[53]

K. REGULATORY MUTANTS

The *INO2* and *INO4* genes are regulatory genes whose wild-type gene product is required for the expression of the *INO1* gene product IP synthase.[54,55] The *ino2* and *ino4* mutants are inositol auxotrophs[20,49] and are unable to derepress IP synthase, despite the presence of a nonmutated copy of its structural gene *INO1*.[20,50,53] The *ino2* and *ino4* mutants also have reduced constitutive levels of CDP-DG synthase,[56] PS synthase,[32,56] and the PE methyltransferase pathway enzymes.[54]

The *INO4* gene has been isolated by genetic complementation of an *ino4* mutant using a yeast DNA genomic library on the vector YEp13.[57] The *INO4* gene has been sequenced, and the predicted *INO4* protein has a molecular weight of 17,390.[58] The *INO4* protein may be a DNA-binding protein which affects the regulation of IP synthase as well as the other enzymes in the *de novo* route of PC biosynthesis.[58]

The *opi1* mutant excretes inositol[59] and is constitutively elevated for IP synthase.[60] The *opi1* mutant also has constitutively elevated levels of CDP-DG synthase,[56,61] PS synthase,[56,62,63] and the PE methyltransferase pathway enzymes.[62]

V. REGULATION OF PHOSPHATIDYLCHOLINE BIOSYNTHESIS

The phospholipid compositions of wild-type and mutant strains of *S. cerevisiae* may vary dramatically depending on culture conditions.[33,42,51,52,56,62,64] However, it appears that the average charge of the membrane phospholipids remains relatively constant.[51] It has been suggested that a mechanism exists that allows wide fluctuations in the amount of phospholipids.[51] The cell compensates for these changes by causing parallel fluctuations in lipids of another charge, thereby regulating the total phospholipid charge.[51] Biochemical and genetic evidence suggests that the biosynthesis of PC by the primary and auxiliary biosynthetic routes is coordinately regulated to the synthesis of PI. The enzymes in the biosynthetic pathways are regulated at the genetic level as well as by modulation of enzyme activity.

A. GENETIC REGULATION OF ENZYMES

The regulation of PC biosynthesis by the water-soluble phospholipid precursor choline was first demonstrated by Lester and co-workers.[6,65,66] It is now known that several enzymes of the *de novo* route of PC biosynthesis are regulated by the availability of phospholipid precursors. A summary of the regulation of these enzyme activities is presented in Table 3. CDP-DG synthase,[61] PS synthase,[62,63,67] PS decarboxylase (Figure 1, Reaction 4),[68] and the PE methyltransferase pathway enzymes[6,40] are repressed when wild-type exponential phase cells are grown in medium containing inositol plus choline. CDP-DG synthase[56,61] and PS synthase[56,63] activities are also repressed when wild-type exponential phase cells are grown in media containing inositol plus ethanolamine and inositol plus serine. The PE methyltransferase pathway enzymes are also reduced when cells are supplemented with inositol plus monomethylethanolamine and inositol plus dimethylethanolamine.[6] Water-soluble phospholipid precursors have no effect on CDP-DG synthase,[56,61] PS synthase,[56,62,63] and the PE methyltransferase pathway enzymes[40,62] in the absence of inositol. The addition of inositol alone to the growth medium causes a partial repression of CDP-DG synthase,[56,61] PS synthase,[56,62,63] and the PE methyltransferase pathway enzymes.[40] The repression of these enzyme activities is not due to the presence of water-soluble effector molecules.[40,56,61,63] Furthermore, the water-soluble phospholipid precursors serine, ethanolamine, and choline have no effect on the activities of purified preparations of CDP-DG synthase[56] and PS synthase,[56,63] indicating that the regulation

TABLE 3
Regulation of Phospholipid Biosynthetic Enzymes of Wild-Type Cells in Response to Water-Soluble Phospholipid Precursors

Enzyme	Relative activity (%) from cells grown in the presence of								Ref.
	None	Inositol	Serine	Inositol + serine	Ethanolamine	Inositol + ethanolamine	Choline	Inositol + choline	
CDP–DG synthase	100	60	99	56	104	40	100	44	56, 61
PS synthase	100	62	92	33	100	39	109	29	56, 62, 63, 67
PS decarboxylase	—	100	—	—	—	—	—	34	68
PE methyltransferase pathway enzymes	100	68	—	—	—	—	90	27	6, 40
IP synthase	100	2	—	—	—	—	100	2	20, 54, 69
PI synthase	100	94	100	100	97	100	97	90	18, 56, 62
PA phosphatase	100	200	95	210	105	180	105	180	70

of these enzymes by precursors does not occur as a direct result of modulation of enzyme activity.

The availability of antibodies to CDP-DG synthase[10] and PS synthase[63] has permitted experiments to determine if the regulation occurs at the level of enzyme formation. Under the growth conditions where a reduction in enzyme activity occurs, there is a corresponding reduction of the CDP-DG synthase (M_r = 56,000)[56] and the PS synthase (M_r = 23,000)[56,63] subunits as determined by immunoblot analysis.

The orginal 4.5-kb *CHO1* clone 34 was subcloned, and the smallest fragment able to complement the *cho1* mutation is 2.8 kb.[32] Studies using this subclone as a probe have detected a single transcript of 1.2 kb on Northern and slot blots.[32] The transcript is regulated in a fashion consistent with the regulation of PS synthase activity.[32] Thus, the regulation of PS synthase by water-soluble phospholipid precursors occurs at the level of transcription.

CDP-DG synthase,[56,61] PS synthase,[32,56,61-63] and the PE methyltransferase pathway enzymes[54,62] are not regulated in the regulatory mutants *ino2, ino4,* and *opi1* when grown in the presence of water-soluble phospholipid precursors.

Cytosolic-associated IP synthase activity and its M_r = 62,000 subunit are repressed in wild-type exponential phase cells by inositol and derepressed in the absence of inositol.[20] RNA blot hybridization studies using the cloned *INO1* DNA have detected two RNA species of 1.8 and 0.6 kb.[69] The 1.8-kb transcript as well as IP synthase activity are repressed in wild-type cells grown in the presence of inositol.[69] The addition of choline to inositol-containing medium results in a further repression of the IP synthase 1.8-kb transcript.[69] The 0.6-kb IP synthase transcript is also regulated in response to these precursors, but to a lesser degree.[69]

The levels of IP synthase activity,[20] M_r = 62,000 subunit,[54] and 1.8-kb transcript[69] are constitutively reduced in the *ino2* and *ino4* regulatory mutants. On the other hand, IP synthase activity,[60] subunit,[60,62] and 1.8-kb transcript[69] levels are overproduced and not regulated in response to inositol and inositol plus choline in the *opi1* mutant.

Unlike IP synthase and the enzymes leading to the formation of PC in the *de novo* biosynthetic route, PI synthase activity[18,56,62] and M_r = 34,000 subunit[18,56] levels are not regulated in exponential phase wild-type cells in response to water-soluble phospholipid precursors.

The regulation of the enzymes in the *de novo* biosynthetic route of PC synthesis by water-soluble phospholipid precursors is absolutely dependent upon inositol. The mechanism by which water-soluble precursors affect the enzymes is not yet clear. The fact that inositol is required for the coordinate regulation of these enzymes suggests that PC biosynthesis is coordinated with inositol synthesis.[41] Furthermore, mutants with lesions in genes whose wild-type products exert a positive (i.e., the *ino2* and *ino4* strains) or a negative (i.e., the *opi1* strain) effect upon IP synthase expression[20,50,59,60] exert pleiotropic effects upon the coordinately regulated enzymes CDP-DG synthase, PS synthase, and the PE methyltransferase pathway enzymes.

When wild-type cells are grown in inositol-containing medium with exogenous ethanolamine or choline, the enzyme activities in the *de novo* biosynthetic route for PC synthesis are repressed, presumably in response to the utilization of the CDP-ethanolamine- and CDP-choline-based auxiliary pathways for PE and PC synthesis. PA phosphatase catalyzes the formation of the DG needed for these auxiliary reactions. The addition of inositol to the growth medium of wild-type cells results in an elevation in PA phosphatase activity.[70] The addition of ethanolamine and choline to medium with or without inositol has no effect on PA phosphatase activity.[70] Like the enzymes in the *de novo* route of PC biosynthesis, PA phosphatase activity is affected in the *ino2* and *opi1* regulatory mutants.[70] Little is known about the regulation of the CDP-ethanolamine- and CDP-choline-based pathway enzymes in response to water-soluble phospholipid precursors. However, it is known that the CDP-choline-based pathway is regulated by the rate of choline transport in polyamine-stimulated cells.[71]

B. REGULATION OF ENZYME ACTIVITIES

There are examples of PC biosynthesis being regulated by mechanisms other than by enzyme repression. When PS synthase deficient *(cho1)* mutants are starved for ethanolamine and choline, the rates of PE and PC biosynthesis decline rapidly in coordination with a decline in PI biosynthesis.[33] The overall regulation of phospholipid biosynthesis in the ethanolamine/choline-starved cells occurs too rapidly to be attributed to a repression mechanism.[33] Furthermore, the overall regulation of phospholipid biosynthesis in the ethanolamine/choline-starved cells occurs under conditions (absence of water-soluble phospholipid precursors)[33] which produce the maximum derepression of the coordinately regulated enzymes CDP-DG synthase, PS synthase, and the PE methyltransferase pathway enzymes.[41] Furthermore, the levels of PI can fluctuate in cells grown in the presence of inositol;[42,56,62] however, PI synthase is not regulated on a genetic level.[18,56] The results of *in vitro* and *in vivo* studies have shown that phospholipid biosynthetic enzyme activities are regulated by substrate availability and the phospholipid environment of the membrane.

The competition of PS synthase and PI synthase for their common substrate CDP-DG is a major determinant of phospholipid composition *in vivo*.[14] The addition of micromolar concentrations of inositol to the growth medium of wild-type cells results in a rapid increase in PI synthesis at the expense of PS synthesis, and its derivatives PE and PC.[14] This occurs because the K_m for inositol for PI synthase is higher than the intracellular concentration of inositol.[14] The addition of millimolar concentrations of serine has an opposite, but much lesser effect because intracellular concentration of serine is much higher than the K_m for serine for PS synthase.[14] The results of kinetic experiments with purified PS synthase and PI synthase indicate that the partitioning of CDP-DG to PS and PI is not governed by the affinities both enzymes have for their common substrate CDP-DG.[14] In addition, kinetic experiments with purified PS synthase have shown that inositol is a noncompetitive inhibitor of PS synthase.[14] Thus, it appears that inositol, in addition to its role in the repression of PS synthase,[32,63] regulates the partitioning of CDP-DG to PI at the expense of PS, and ultimately PE and PC, by inhibiting PS synthase activity.[14]

The activities of purified PS synthase and PI synthase reconstituted in unilamellar phospholipid vesicles are modulated by the phospholipid composition of the membrane. PS synthase has been reconstituted into PC/PE/PI/PS vesicles in which the ratios of PI to PS approximate *in vivo* conditions of both wild-type and mutant cells. Increases in the ratios of PI to PS in vesicles result in a two- to threefold decrease in PS synthase activity.[17] Changes in the ratios of PC to PE do not significantly affect reconstituted PS synthase activity.[17] This modulation in PS synthase activity by changes in the PI to PS ratio is enough to account for some of the *in vivo* fluctuations in the PI and PS contents. For example, the addition of inositol to wild-type cells results in an approximate twofold increase in the cellular ratio of PI to PS.[42,56,62] The cellular ratio of PI to PS also increases about twofold when wild-type cells grown in the absence of inositol enter the stationary phase of growth.[64] Under both of these growth conditions, the cellular ratio of PC to PE is not significantly affected.[42,56,62,64]

The *ino2* and *ino4* mutants have reduced PS synthase activity[32,56] which is due in part to reduced levels of *CHO1* mRNA.[32] However, the reduced PS synthase activity in the *ino2* and *ino4* mutants cannot be fully attributed to the reduction of the *CHO1* message.[32] It is possible that some of the reduced PS synthase activity in the *ino2* and *ino4* mutants is due to the elevated ratios of PI to PS found in the membranes of these cells.[2]

Reconstituted PI synthase is also modulated by the PI to PS ratio in PC/PE/PI/PS vesicles under conditions which reflect *in vivo* conditions.[18] An increase in the PS content of phospholipid vesicles stimulates PI synthase activity about 1.4-fold.[18] When wild-type cells are grown in medium without inositol, they rely on endogenous inositol synthesis[20] and have an elevated PS content.[42,56,62] Under these conditions, PS synthase would not be repressed in growing cells[32,62,63] and could compete fully with PI synthase for their common substrate CDP-DG. Therefore, the stimulation of PI synthase activity by PS may be one of the cell's mechanisms to insure sufficient PI synthesis, since PI is an essential membrane phospholipid.[38,51,52] However, as indicated above, the presence of inositol in the growth medium results in the repression of PS synthase[32,62,63] which ultimately results in a reduced PS content.[42,56,62] This may result in the removal of stimulation of PI synthase activity by PS. On the other hand, the available pool of CDP-DG would increase due to the reduced level of PS synthase. Thus, PI synthesis may increase as a result of mass action given increased pools of CDP-DG and readily available inositol accounting for elevated PI content. In this manner, PI synthesis and PC synthesis are ultimately coordinated through the synthesis of PS.

The phospholipid biosynthetic pathway is also coordinately regulated with triacylglycerol synthesis. In stationary phase where triacylglycerol synthesis is favored over phospholipid biosynthesis,[21,72] the enzyme activities in the *de novo* route of PC biosynthesis are reduced. As cells enter the stationary phase of growth, CDP-DG synthase, PS synthase, and the PE methyltransferase pathway enzymes decrease 2.5- to 5-fold.[64] The decrease in CDP-DG synthase and PS synthase activities in the stationary phase is not due to a decrease in the subunit levels of CDP-DG synthase and PS synthase.[64,73] This suggests that the reduction in activity for at least CDP-DG synthase and PS synthase is not due to a reduction in enzyme synthesis or enzyme turnover. Therefore, another mechanism must be involved with the regulation of these enzymes during the growth phase. As indicated above, the ratio of PI to PS doubles as cells enter the stationary phase of growth.[64] Increases in the PI to PS ratio in phospholipid vesicles result in a reduction in PS synthase activity reconstituted in these vesicles.[17] It is possible that the alterations in the PI to PS ratio of stationary-phase cells compared with exponential-phase cells may be responsible in part for the reduction in PS synthase activity as well as the other enzyme activities in the primary route of PC biosynthesis. This may be an oversimplification of the regulation occurring in these cells.

PA phosphatase activity increases two- to threefold as wild-type cells enter the stationary phase of growth.[21] The increase in PA phosphatase activity in the stationary phase is presumably utilized for triacylglycerol biosynthesis.[21] If triacylglycerol synthesis is increased in the stationary phase of growth as a result of induced PA phosphatase activity, then the pool of CDP-DG would decrease. This available pool of CDP-DG must be partitioned between PI and PS. Because PI is an essential membrane phospholipid,[51,52] continued synthesis of PI from CDP-DG would be an important reaction for the cell to maintain throughout growth. Indeed, PI synthase activity and subunit levels are constitutive throughout growth in the absence or presence of water-soluble phospholipid precursors.[64]

VI. CONCLUSIONS

The regulation of the *de novo* pathway for PC biosynthesis in *S. cerevisiae* is complex. Long-term (i.e., genetic regulation) and short-term (i.e., modulation of enzyme activities) mechanisms appear to work together to control the phospholipid composition of *S. cerevisiae* in response to changes in growth conditions. Under all conditions studied thus far, the synthesis of PI is always favored.

Many questions still remain concerning the regulation of phospholipid biosynthesis in *S. cerevisiae*. Based on the predicted amino acid sequence, some of the phospholipid biosynthetic enzymes have potential phosphorylation and glycosylation sites. What roles do phosphorylation/dephosphorylation and glycosylation have on the regulation of phospholipid synthesis? A number of the phospholipid biosynthetic enzymes are localized in both the mitochondrial and microsomal fractions of the cell. What directs the localization of these enzymes, and is the localization of enzymes involved in the regulation of the biosynthetic pathways? What mechanisms control the partitioning of PA to DG and CDP-DG and the partitioning of CDP-DG to PS and PI? The answers to these questions should emerge in the not-too-distant future, given the state of our understanding of the genetics and biochemistry of phospholipid metabolism in *S. cerevisiae*.

ACKNOWLEDGMENTS

I thank my colleagues Myongsuk Bae-Lee, Charles Belunis, Anthony Fischl, Paulette Gaynor, Michael Homann, Michael Kelley, Anthony Kinney, Yi-Ping Lin, Kelley Morlock, and Cathy Welsch for their helpful discussions. I also thank Susan Henry, Satoshi Yamashita, and Dennis Vance for their constructive criticisms of the manuscript. I am grateful to Russell Hjelmstad and Robert Bell for providing me with their unpublished findings.

This work was supported by New Jersey state funds, U.S. Public Health Service grant GM-28140 from the National Institutes of Health, and the Charles and Johanna Busch Memorial Fund.

REFERENCES

1. **Esko, J. D. and Raetz, C. R. H.,** Synthesis of phospholipids in animal cells, in *The Enzymes,* Vol. 16, Boyer, P. D., Ed., Academic Press, New York, 1983, 207.
2. **Henry, S. A.,** Membrane lipids of yeast: biochemical and genetic studies, in *The Molecular Biology of the Yeast Saccharomyces: Metabolism and Gene Expression,* Strathern, J. N., Jones, E. W., and Broach, J. R., Eds., Cold Spring Harbor Laboratory, Cold Spring Harbor, NY, 1982, 101.
3. **Kennedy, E. P.,** The biosynthesis of phospholipids, in *Lipids and Membranes: Past, Present and Future,* Op den Kamp, J. A. F., Roelofsen, B., and Wirtz, K. W. A., Eds., Elsevier Science Publishers, Amsterdam, 1986, 171.
4. **Bae-Lee, M. and Carman, G. M.,** Phosphatidylserine synthesis in *Saccharomyces cerevisiae.* Purification and characterization of membrane-associated phosphatidylserine synthase, *J. Biol. Chem.,* 259, 10857, 1984.
5. **Kanfer, J. N.,** The base exchange enzymes and phospholipase D of mammalian tissue, *Can. J. Biochem.,* 58, 1370, 1980.
6. **Waechter, C. J. and Lester, R. L.,** Differential regulation of the *N*-methyl transferases responsible for phosphatidylcholine synthesis in *Saccharomyces cerevisiae, Arch. Biochem. Biophys.,* 158, 401, 1973.
7. **Steiner, M. R. and Lester, R. L.,** *In vitro* studies of phospholipid biosynthesis in *Saccharomyces cerevisiae, Biochim. Biophys. Acta,* 260, 222, 1972.
8. **Kuchler, K., Daum, G., and Paltauf, F.,** Subcellular and submitochondrial localization of phospholipid synthesizing enzymes in yeast, *J. Bacteriol.,* 165, 901, 1986.
9. **Belendiuk, G., Mangnall, D., Tung, B., Westley, J., and Getz, G. S.,** CTP-phosphatidic acid cytidyltransferase from *Saccharomyces cerevisiae:* partial purification, characterization, and kinetic behavior, *J. Biol. Chem.,* 253, 4555, 1978.
10. **Kelley, M. J. and Carman, G. M.,** Purification and characterization of CDP-diacylglycerol synthase from *Saccharomyces cerevisiae, J. Biol. Chem.,* 262, 14563, 1987.
11. **Poole, M. A., Fischl, A. S., and Carman, G. M.,** Enzymatic detection of phospholipid biosynthetic enzymes after electroblotting, *J. Bacteriol.,* 161, 772, 1985.
12. **Kiyono, K., Miura, K., Kushima, Y., Hikiji, T., Fukushima, M., Shibuya, Y., and Ohta, A.,** Primary structure and product characterization of the *Saccharomyces cerevisiae CHO1* gene that encodes phosphatidylserine synthase, *J. Biochem.,* 102, 1093, 1987.
13. **Nikawa, J., Tsukagoshi, Y., Kodaki, T., and Yamashita, S.,** Nucleotide sequence and characterization of the yeast PSS gene encoding phosphatidylserine synthase, *Eur. J. Biochem.,* 167, 7, 1987.
14. **Kelley, M. J., Bailis, A. M., Henry, S. A., and Carman, G. M.,** Regulation of phospholipid biosynthesis in *Saccharomyces cerevisiae* by inositol. Inositol is an inhibitor of phosphatidylserine synthase activity, *J. Biol. Chem.,* 263, 18078, 1988.
15. **Raetz, C. R. H., Carman, G. M., Dowhan, W., Jiang, R.-T., Waszkuc, W., Loffredo, W., and Tsai, M.-D.,** Phospholipid chiral at phosphorus. Steric course of the reactions catalyzed by phosphatidylserine synthase from *Escherichia coli* and yeast, *Biochemistry,* 26, 4022, 1987.

16. **Enoch, H. G. and Stritmatter, P.,** Formation and properties of 1000 A diameter, single-bilayer phospholipid vesicles, *Proc. Natl. Acad. Sci. U.S.A.,* 76, 145, 1979.
17. **Hromy, J. M. and Carman, G. M.,** Reconstitution of *Saccharomyces cerevisiae* phosphatidylserine synthase into phospholipid vesicles. Modulation of activity by phospholipids, *J. Biol. Chem.,* 261, 15572, 1986.
18. **Fischl, A. S., Homann, M. J., Poole, M. A., and Carman, G. M.,** Phosphatidylinositol synthase from *Saccharomyces cerevisiae.* Reconstitution, characterization, and regulation of activity, *J. Biol. Chem.,* 261, 3178, 1986.
19. **Fischl, A. S. and Carman, G. M.,** Phosphatidylinositol biosynthesis in *Saccharomyces cerevisiae:* purification and properties of microsome-associated phosphatidylinositol synthase, *J. Bacteriol.,* 154, 304, 1983.
20. **Donahue, T. F. and Henry, S. A.,** Myo-inositol-1-phosphate synthase: characteristics of the enzyme and identification of its structural gene in yeast, *J. Biol. Chem.,* 256, 7077, 1981.
21. **Hosaka, K. and Yamashita, S.,** Regulatory role of phosphatidate phosphatase in triacylglycerol synthesis of *Saccharomyces cerevisiae, Biochim. Biophys. Acta,* 796, 110, 1984.
22. **Hosaka, K. and Yamashita, S.,** Partial purification and properties of phosphatidate phosphatase in *Saccharomyces cerevisiae, Biochim. Biophys. Acta,* 796, 102, 1984.
23. **Tillman, T. S. and Bell, R. M.,** Mutants of *Saccharomyces cerevisiae* defective in *sn*-glycerol-3-phosphate acyltransferase, *J. Biol. Chem.,* 261, 9144, 1986.
24. **Raetz, C. R. H.,** Isolation of *Escherichia coli* mutants defective in enzymes of membrane lipid synthesis, *Proc. Natl. Acad. Sci. U.S.A.,* 72, 2274, 1975.
25. **Homann, M. J. and Carman, G. M.,** Detection of phospholipid biosynthetic enzyme activities in *Saccharomyces cerevisiae* by colony autoradiography, *Anal. Biochem.,* 135, 447, 1983.
26. **Klig, L. S., Homann, M. J., Kohlwein, S., Kelley, M. J., Henry, S. A., and Carman, G. M.,** *Saccharomyces cerevisiae* mutant with a partial defect in the synthesis of CDP-diacylglycerol and altered regulation of phospholipid biosynthesis, *J. Bacteriol.,* 170, 1878, 1988.
27. **Atkinson, K., Fogel, S., and Henry, S. A.,** Yeast mutant defective in phosphatidylserine synthesis, *J. Biol. Chem.,* 255, 6653, 1980.
28. **Atkinson, K. D., Jensen, B., Kolat, A. I., Storm, E. M., Henry, S. A., and Fogel, S.,** Yeast mutants auxotropic for choline or ethanolamine, *J. Bacteriol.,* 141, 558, 1980.
29. **Kovac, L., Gbelska, I., Poliachova, V., Subik, J., and Kovacova, V.,** Membrane mutants: a yeast mutant with a lesion in phosphatidylserine biosynthesis, *Eur. J. Biochem.,* 111, 491, 1980.
30. **Nikawa, J. and Yamashita, S.,** Characterization of phosphatidylserine synthase from *Saccharomyces cerevisiae* and a mutant defective in the enzyme, *Biochim. Biophys. Acta,* 665, 420, 1981.
31. **Kennedy, E. P. and Weiss, S. B.,** The function of cytidine coenzyme in the biosynthesis of phospholipids., *J. Biol. Chem.,* 222, 193, 1956.
32. **Bailis, A. M., Poole, M. A., Carman, G. M., and Henry, S. A.,** The membrane-associated enzyme phosphatidylserine synthase of yeast is regulated at the level of mRNA abundance, *Mol. Cell. Biol.,* 7, 167, 1987.
33. **Letts, V. A. and Henry, S. A.,** Regulation of phospholipid synthesis in phosphatidylserine synthase-deficient *(chol)* mutants of *Saccharomyces cerevisiae, J. Bacteriol.,* 163, 560, 1985.
34. **Letts, V. A., Klig, L. S., Bae-Lee, M., Carman, G. M., and Henry, S. A.,** Isolation of the yeast structural gene for the membrane-associated enzyme phosphatidylserine synthase, *Proc. Natl. Acad. Sci. U.S.A.,* 80, 7279, 1983.
35. **Nasmyth, K. A. and Tatchell, K.,** The structure of transposable yeast mating type loci, *Cell,* 19, 753, 1980.
36. **Nikawa, J. and Yamashita, S.,** Yeast mutant defective in synthesis of phosphatidylinositol. Isolation and characterization of a CDP-diacylglycerol-inositol 3-phosphatidyltransferase K_m mutant, *Eur. J. Biochem.,* 125, 445, 1982.
37. **Nikawa, J. and Yamashita, S.,** Molecular cloning of the gene encoding CDP-diacylglycerol-inositol 3-phosphatidyl transferase in *Saccharomyces cerevisiae, Eur. J. Biochem.,* 143, 251, 1984.
38. **Nikawa, J., Kodaki, T., and Yamashita, S.,** Primary structure and disruption of the phosphatidylinositol synthase gene of *Saccharomyces cerevisiae, J. Biol. Chem.,* 262, 4876, 1987.
39. **Yamashita, S. and Oshima, A.,** Regulation of phosphatidylethanolamine methyltransferase level by *myo*-inositol in *Saccharomyces cerevisiae, Eur. J. Biochem.,* 104, 611, 1980.
40. **Yamashita, S., Oshima, A., Nikawa, J., and Hosaka, K.,** Regulation of the phosphatidylethanolamine methylation pathway in *Saccharomyces cerevisiae, Eur. J. Biochem.,* 128, 589, 1982.
41. **Henry, S. A., Klig, L. S., and Loewy, B. S.,** Genetic regulation and coordination of biosynthetic pathways in yeast: amino acid and phospholipid synthesis, *Annu. Rev. Genet.,* 18, 207, 1984.
42. **Greenberg, M. L., Klig, L. S., Letts, V. A., Loewy, B. S., and Henry, S. A.,** Yeast mutant defective in phosphatidylcholine synthesis, *J. Bacteriol.,* 153, 791, 1983.

43. **Kodaki, T. and Yamashita, S.,** Yeast phosphatidylethanolamine methylation pathway: cloning and characterization of two distinct methyltransferase genes, *J. Biol. Chem.,* 262, 15428, 1987.
44. **Hosaka, H. and Yamashita, S.,** Choline transport in *Saccharomyces cerevisiae, J. Bacteriol.,* 143, 176, 1980.
45. **Hosaka, K. and Yamashita, S.,** Isolation and characterization of a yeast mutant defective in cholinephosphotransferase, *Eur. J. Biochem.,* 162, 7, 1987.
46. **Nikawa, J., Yonemura, K., and Yamashita, S.,** Yeast mutant with thermolabile CDP-choline synthesis: isolation and characterization of a cholinephosphate cytidyltransferase mutant, *Eur. J. Biochem.,* 131, 223, 1983.
47. **Tsukagoshi, Y., Nikawa, J., and Yamashita, S.,** Molecular cloning and characterization of the gene encoding cholinephosphate cytidylyltransferase in *Saccharomyces cerevisiae, Eur. J. Biochem.,* 169, 477, 1988.
48. **Hjelmstad, R. H. and Bell, R. M.,** Mutants of *Saccharomyces cerevisiae* defective in *sn*-1,2-diacylglycerol cholinephosphotransferase: isolation, characterization, and cloning of the *CPT1* gene, *J. Biol. Chem.,* 262, 3909, 1987.
49. **Culbertson, M. R. and Henry, S. A.,** Inositol requiring mutants of *Saccharomyces cerevisiae, Genetics,* 80, 23, 1975.
50. **Culbertson, M. R., Donahue, T. F., and Henry, S. A.,** Control of inositol biosynthesis in *Saccharomyces cerevisiae:* inositol-phosphate synthetase mutants, *J. Bacteriol.,* 126, 243, 1976.
51. **Becker, G. W. and Lester, R. L.,** Changes in phospholipids of *Saccharomyces cerevisiae* associated with inositol-less death, *J. Biol. Chem.,* 252, 8684, 1977.
52. **Henry, S. A., Atkinson, K. D., Kolat, A. J., and Culbertson, M. R.,** Growth and metabolism of inositol-starved *Saccharomyces cerevisiae, J. Bacteriol.,* 130, 472, 1977.
53. **Klig, L. S. and Henry, S. A.,** Isolation of the yeast *INO1* gene: located on an autonomously replicating plasmid, the gene is fully regulated, *Proc. Natl. Acad. Sci. U.S.A.,* 81, 3816, 1984.
54. **Lowey, B. S. and Henry, S. A.,** The *INO2* and *INO4* loci of *Saccharomyces cerevisiae* are pleiotropic regulatory genes, *Mol. Cell. Biol.,* 4, 2479, 1984.
55. **Donahue, T. F. and Henry, S. A.,** Inositol mutants of *Saccharomyces cerevisiae:* mapping the *ino1* locus and characterizing alleles of the *ino1, ino2,* and *ino4* loci, *Genetics,* 98, 491, 1981.
56. **Homann, M. J., Bailis, A. M., Henry, S. A., and Carman, G. M.,** Coordinate regulation of phospholipid biosynthesis by serine in *Saccharomyces cerevisiae, J. Bacteriol.,* 169, 3276, 1987.
57. **Klig, L. S., Hoshizaki, D. K., and Henry, S. A.,** Isolation of the yeast *INO4* gene, a positive regulator of phospholipid biosynthesis, *Curr. Genet.,* 13, 7, 1988.
58. **Hoshizaki, D. K. and Henry, S. A.,** The *Saccharomyces cerevisiae* gene *INO4* encodes a small highly basic protein necessary for the derepression of inositol-1-phosphate synthase and the phospholipid *N*-methyltransferases, *J. Mol. Biol.,* in press.
59. **Greenberg, M., Reiner, B., and Henry, S. A.,** Regulatory mutations of inositol biosynthesis in yeast: isolation of inositol excreting mutants, *Genetics,* 100, 19, 1982.
60. **Greenberg, M., Goldwasser, P., and Henry, S. A.,** Characterization of a yeast regulatory mutant constitutive for inositol-1-phosphate synthase, *Mol. Gen. Genet.,* 186, 157, 1982.
61. **Homann, M. J., Henry, S. A., and Carman, G. M.,** Regulation of CDP-diacylglycerol synthase activity in *Saccharomyces cerevisiae, J. Bacteriol.,* 163, 1265, 1985.
62. **Klig, L. S., Homann, M. J., Carman, G. M., and Henry, S. A.,** Coordinate regulation of phospholipid biosynthesis in *Saccharomyces cerevisiae:* pleiotropically constitutive *opi1* mutant, *J. Bacteriol.,* 162, 1135, 1985.
63. **Poole, M. A., Homann, M. J., Bae-Lee, M., and Carman, G. M.,** Regulation of phosphatidylserine synthase from *Saccharomyces cerevisiae* by phospholipid precursors, *J. Bacteriol.,* 168, 668, 1986.
64. **Homann, M. J., Poole, M. A., Gaynor, P. M., Ho, C. -T., and Carman, G. M.,** Effect of growth phase on phospholipid biosynthesis in *Saccharomyces cerevisiae, J. Bacteriol.,* 169, 533, 1987.
65. **Steiner, M. R. and Lester, R. L.,** *In vitro* studies of the methylation pathway of phosphatidylcholine synthesis and the regulation of this pathway in *Saccharomyces cerevisiae, Biochemistry,* 9, 63, 1969.
66. **Waechter, C. J. and Lester, R. L.,** Regulation of phosphatidylcholine biosynthesis in *Saccharomyces cerevisiae, J. Bacteriol.,* 105, 837, 1971.
67. **Carson, M. A., Atkinson, K. D., and Waechter, C. J.,** Properties of particulate and solubilized phosphatidylserine synthase activity from *Saccharomyces cerevisiae, J. Biol. Chem.,* 257, 8115, 1982.
68. **Carson, M. A., Emala, M., Hogsten, P., and Waechter, C. J.,** Coordinate regulation of phosphatidylserine decarboxylase activity and phospholipid *N*-methylation in yeast, *J. Biol. Chem.,* 259, 6267, 1984.
69. **Hirsch, J. P. and Henry, S. A.,** Expression of the *Saccharomyces cerevisiae* inositol-1-phosphate synthase *(INO1)* gene is regulated by factors that affect phospholipid synthesis, *Mol. Cell. Biol.,* 6, 3320, 1986.
70. **Morlock, K. R., Lin, Y. -P., and Carman, G. M.,** Regulation of phosphatidate phosphatase activity by inositol in *Saccharomyces cerevisiae, J. Bacteriol.,* 170, 3561, 1988.

71. **Hosaka, K. and Yamashita, S.,** Induction of choline transport and its role in the stimulation of the incorporation of choline into phosphatidylcholine by polyamine in a polyamine auxotroph of *Saccharomyces cerevisiae, Eur. J. Biochem.,* 116, 1, 1981.
72. **Taylor, F. R. and Parks, L. W.,** Triacylglycerol metabolism in *Saccharomyces cerevisiae* relation to phospholipid synthesis, *Biochim. Biophys. Acta,* 575, 204, 1979.
73. **Homann, M. J.,** Regulation of CDP-Diacylglycerol Synthesis and Utilization in *Saccharomyces cerevisiae,* Ph.D. thesis, Rutgers University, New Brunswick, NJ, 1987.
74. **Hjelmstad, R. H. and Bell, R. M.,** The *sn*-1,2-diacylglycerol ethanolaminephosphotransferase activity of *Saccharomyces cerevisiae*. Isolation of mutants and cloning of the EPT1 gene, *J. Biol. Chem.,* 263, 19748, 1988.

Chapter 11

SPHINGOMYELIN BIOSYNTHESIS AND CATABOLISM

Matthew W. Spence

TABLE OF CONTENTS

I. SPHINGOMYELIN

A. STRUCTURE

Sphingomyelin (SM) was initially described and partially characterized by Thudichum,[1] and the chemical structure was eventually elucidated in a series of studies culminating some 70 years later.[2-4] The SM molecule contains phosphocholine and a long-chain base, sphingenine (sphingosine; *trans*-4-sphingenine; sphing-4-enine; [2S,3R,4E]-2-amino-4-octadecene-1,3-diol) in amide linkage with a fatty acid.[2-4] Small amounts of sphinganine (dihydrosphingosine; [2S,3R]-2-amino-1,3-D-octadecanediol) with a saturated bond at the 4 position, and of 4-D-hydroxysphinganine (phytosphingosine; 4-D-hydroxysphinganine; [2S,3S,4R]-2-amino-1,3,4-octadecanetriol) are also found together with trace amounts of other sphingosine bases. (Throughout this chapter, the term sphingosine will be used to designate the entire group of long-chain bases. Where exact structure is known the more specific terms, sphingenine, sphinganine, etc. will be used). The fatty acid may vary in chain length from 12 to over 30 carbons;[5,6] the principal acyl-chain groups in most SM are 16:0, 18:0, 22:0, 24:0, and 24.1.[5-9] Hydroxy-fatty acids have not been reported[7] and, where specifically looked for, were not found.[9] At least half of the fatty acids in SM are greater than 20 carbons in length.[5-9] As sphingosine usually contributes a hydrocarbon tail of only 15 carbons, the N-acyl chain is generally longer, and this marked hydrocarbon chain-length asymmetry may lead to a mixed, interdigitated chain packing which is more ordered than the usual hydrated, noninterdigitated bilayers formed by complex lipids with symmetrical acyl chains.[10] The combination of sphingosine with fatty acid is called ceramide and is the parent molecule of both the glycosphingolipids and SM.

A phosphocholine (P-choline) moiety is attached to the primary hydroxyl of the sphingosine base and is identical in structure to the P-choline of phosphatidylcholine. However, the hydroxyl group of the sphingosine base and the amide bond confer important hydrogen bonding capabilities to SM that are not found in phosphatidylcholine, and the behavior of the two molecules in bilayer systems is different.[6]

B. TISSUE DISTRIBUTION

SM is found in serum lipoproteins and in membranes of all mammalian cells.[6,7,11] The amounts vary widely in cells, and in most, there is a concentration gradient, relative to other lipids and protein, between the SM-poor cell interior and the SM-rich plasma membrane.[6-8] A similar gradient has been shown for cholesterol, and there is preferential association of cholesterol and sphingomyelin in both natural membranes and artificial lipid mixtures.[6,7,10,11] In cells where membrane lipid asymmetry has been determined, almost all of the SM and the bulk of phosphatidylcholine are on the external surface.[6] In the human erythrocyte, the externally oriented sphingomyelin is enriched in long-chain fatty acids compared to the internally oriented pool.[12] The amount of sphingomyelin and cholesterol in membranes tends to increase with age, and this increase is greater in arterial intima with atherosclerosis.[6]

C. FUNCTION

In addition to a structural role in the membrane bilayer, SM has been implicated in several specific functions, based on its unique physical properties. Preferential association with cholesterol in natural membranes was noted earlier. *In vitro* evidence supporting an association comes from calorimetric studies of lipid mixtures,[6,11] as well as the observation that delipidated serum is more effective as an acceptor of cholesterol from cells when SM is added.[11] Further support for functional interrelationships between SM and cholesterol is provided by the recessively inherited group of SM lipidoses called Niemann-Pick disease.[13] In the sphingomyelinase-deficient forms of the disease, both SM and cholesterol accumulate in lysosomes.[13] In other forms where sphingomyelinase (Smase) activities are normal or moderately reduced, Pentchev et al.[14] have provided evidence for a defect in cellular cholesterol processing. In other studies of normal cultured fibroblasts, depletion of plasma membrane SM using exogenous Smase rapidly altered the distribution of cholesterol between plasma membranes and intracellular cholesterol pools.[15]

Tetradecanoylphorbol-13-acetate binds more specifically to SM vesicles than to those of other phospholipids, suggesting that SM may be involved at the binding sites for tumor promotors in cell membranes.[16] A related observation is that SM is particularly effective *in vitro* in inhibiting the diacylglycerol-stimulated activities of human platelet phosphatidylinositol phosphodiesterase, rat intestinal phospholipase A_2, and rat liver phospholipase A.[17] Preferential associations of SM with a variety of proteins have been reported also, including 5′-nucleotidase,[18] ATPase,[19] sea anemone toxin,[20] acetylcholinesterase,[21] apolipoprotein A-11,[22] and glucagon.[23] Such observations support an important role for SM in cell membranes and, as a consequence, the importance of the enzymes involved in the synthesis and catabolism of this lipid.

In addition to its role(s) as an intact molecule, SM may also be important as a source of other biologically active macromolecules, most notably sphingosine. Hannun et al. have shown that sphingosine bases inhibit the phospholipid-, Ca^{2+}-, and diacylglycerol-dependent protein kinase C.[24] In platelets, sphingosine inhibits phorbol ester binding and the phosphorylation of peptides at concentrations similar to those that inhibit protein kinase C *in vitro*.[25] In neutrophils, sphingenine and sphinganine block the oxidative burst in response to diacylglycerols, phorbol esters, and peptide agonists.[26] Sphinganine blocks cell adherence and growth inhibition characteristic of the differentiation response of HL-60 cells in culture.[27] Sphingosine bases also inhibit phorbol ester and insulin-induced changes in hexose uptake by 3T3-L1 cells.[28]

The source of the sphingosine that serves as a modulator is not known. It could be synthesized or mobilized in response to an agonist. If the latter, an attractive potential source is the sphingosine bases that are part of the membrane sphingolipids. Of the latter, SM is quantitatively the most important,[7] and membrane-bound enzyme systems that degrade sphingomyelin to sphingenine and fatty acid have been described.[29-32] Exogenous diacylglycerols, but not phorbol esters, stimulate hydrolysis of sphingomyelin by an acid sphingomyelinase in GH_3 cells,[33] and the addition of exogenous Smase or free sphingosine bases reverses phorbol-ester-induced translocation of protein kinase C activity.[34] Slife et al.[35] have reported that endogenous sphingosine levels rise in rat liver plasma membranes incubated at neutral pH in the presence of Mg^{2+}. A dramatic increase in sphingosine production was observed when the membranes were treated with sphingomyelinase or were incubated with exogenous ceramides. These studies suggest that Smase (described in later sections) may modulate the levels of membrane sphingosine bases and, thus, indirectly, the processes controlled by protein kinase C.

II. BIOSYNTHESIS OF SPHINGOMYELIN

A. SPHINGOSINE

The first step in sphingosine synthesis is the pyridoxal phosphate-dependent condensation of palmitoyl-CoA and serine to form 3-ketodihydrosphinganine.[36-37] The enzyme catalyzing this reaction (serine palmitoyltransferase) is relatively specific to palmitoyl-CoA, which may explain the limited size range of the sphingosine bases.[38] The 3-ketodihydrosphinganine is reduced to sphinganene by an NADPH-dependent reductase. The 4-*trans*-double bond is introduced in a flavoprotein-catalyzed dehydrogenation, either at the 3-keto stage[39] or following incorporation of the sphinganine into ceramide.[40,41]

B. CERAMIDE

At least three possible pathways of ceramide biosynthesis have been described. One is the acyl-CoA dependent acylation of the sphingosine amino group.[42-46] Studies of crude microsomal fractions from rodent brain suggest that at least four acyltransferases with varying acyl chain specificities are involved.[43] A second pathway is the reversal of the ceramidase reaction which normally cleaves ceramide, and involves the condensation of free fatty acid and sphingenine.[31] The reversibility of the ceramidase reaction has been shown *in vitro*.[31,45] Stoffel et al.[47] concluded that the product of the reverse reaction is *N*-acyl-ethanolamine, formed from the buffer mixture. However, others have demonstrated the reverse reaction in other buffer systems.[48,49] The reverse reaction is generally felt to be of limited physiological significance.[50,51] The third pathway is also CoA independent, but rather than involving ceramidase, requires a reduced pyridine nucleotide, various factors, and is associated with or is part of the system forming α-hydroxy fatty acids for the ceramides of glycolipids in brain.[46]

C. SPHINGOMYELIN

1. Synthesis via CDP-Choline (CDP-Choline:Ceramide Phosphocholinetransferase)

The original description of SM biosynthesis[52] was a reaction between CDP choline and the primary hydroxyl of ceramide in analogy to phosphatidylcholine biosynthesis from diacylglycerol.[53] CDP-choline is formed by the same series of reactions (i.e., choline kinase followed by cytidylyltransferase) as for phosphatidylcholine biosynthesis. The CDP-choline:ceramide phosphocholinetransferase reaction was more active with the threo (2S,3S) isomers than the natural erythro isomers (2S,3R) and, in addition, was more active with ceramides with very short acyl chains. Subsequent investigators demonstrated activity with erytho isomers.[54-56] Stoffel and Melzner[57] suggested that the phosphocholinetransferase enzyme is the same for both phosphatidylcholine and SM biosynthesis, based on (1) the similarities in properties with the phosphocholinetransferase catalyzing phosphatidylcholine biosynthesis and (2) the mutual inhibition of SM biosynthesis by diacylglycerol and of phosphatidylcholine biosynthesis by ceramide.

2. Synthesis via Sphingosylphosphocholine (Sphingosylphosphocholine:Acyl-CoA *N*-Acyltransferase)

A second pathway has been postulated involving acylation of sphingenylphosphocholine by acyl-CoA.[58,59] Evidence for this pathway includes the report of sphingenylphosphocholine formation from sphingenine and CDP-choline. However, others have shown that sphingosine bases do not act as acceptors for CDP-choline, and acylation of sphingenylphosphocholine could not be demonstrated with acyl-CoAs or free fatty acid, CoA, and ATP.[57] If this pathway is present, it makes a very modest contribution to SM biosynthesis.

3. Synthesis from Phosphatidylcholine (Phosphatidylcholine:Ceramide Phosphocholinetransferase)

The first clear evidence of SM biosynthesis by a phosphatidylcholine-ceramide exchange reaction came from studies by Diringer et al.[60] In ^{32}Pi and [^{3}H-methyl]choline pulse-chase experiments with SV40 transformed mouse fibroblasts, they demonstrated that the ratio of ^{3}H/^{32}Pi was similar in both phosphatidylcholine and SM, and that label continued to accumulate in SM after the level of label was dropping in phosphatidylcholine. They postulated a precursor-product relationship between the lipids involving some unspecified intervening intermediates. Ullman and Radin[61] reported formation of SM from phosphatidylcholine in mouse liver microsomes. CDP-choline did not act as a donor, and the reaction was not slowed or stopped by the trapping of the potential reaction intermediates CDP-choline, P-choline, and choline by dilution with carrier. The transfer reaction was found in kidney, spleen, and heart, but not in brain. Kanfer and Spielvogel[62] suggested that the transfer was due to a phospholipase C attack on phosphatidylcholine with ceramide as the phosphocholine acceptor rather than H_2O and provided evidence for this reaction with commercial preparations of *Clostridium perfringens* phospholipase C. Others have not confirmed this observation.[63] Subsequent studies[63-72] have added the following knowledge of the transfer reaction. The pH optimum is 7 to 8 in liver[61] and kidney membranes[68] and lung lamellar bodies,[71] and 6 in plasma membrane fractions from mouse fibroblast.[66] Divalent cations have been reported to activate,[63,66] inhibit,[71] or have no effect,[68] and EDTA and EGTA have been reported to stimulate,[63] to be without effect,[68] or to inhibit transfer activity.[66] Transfer can be demonstrated with exogenous phosphatidylcholine or ceramide coated on microsomal particles,[61,63] with phosphatidylcholine as a liposomal suspension[66,67] and with[68,69] or without[61,63,66,67] exogenous phospholipid exchange proteins. Activity is greater with unsaturated phosphatidylcholines than with saturated species.[72] Phosphatidylcholine formed by methylation with [^{3}H]methionine is also a substrate for the reaction.[68] Activity can be stimulated by addition of exogenous ceramide in some circumstances,[67,68,71] but not in others.[66] Where the activity has been examined in subcellular fractions, it is concentrated in the plasma membrane, with some activity in Golgi.[67,68] The plasma membrane activity may not be exposed on the external surface of the cell, since treatment of intact cells with trypsin causes little change in activity, whereas activity in the isolated plasma membranes could be reduced >70%.[67-69,72] In lung, the lamellar body fraction has higher activity than the plasma membrane.[71]

Activity has not been demonstrated in brain.[63] A reaction product formed from [^{14}C]phosphatidylethanolamine was tentatively identified as ceramide phosphoethanolamine, leading to the suggestion that a pathway of SM formation in brain might be via this intermediate, with subsequent methylation or base exchange with choline.[63]

In 3T3-L1 fibroblasts, transfer activity can be stimulated by treatment of cells with 0.1 μM dexamethasone;[73] in Novikoff rat hepatoma cells, activity is stimulated with 0.3 mM neophenoxine and inhibited by *p*-chlormercuribenzoate.[70] When examined, there has been no effect on the transfer reaction of exogenous choline, P-choline, CDP-choline, and glycerophosphocholine.[61,66]

This substantial body of evidence strongly supports a role for phosphatidylcholine in the biosynthesis of SM. However, the mechanism of the phosphatidylcholine:ceramide phosphocholinetransferase reaction remains unclear. Where both ^{32}Pi and [^{3}H]choline incorporation into phosphatidylcholine have been studied, the relative rates of labeling and transfer support the movement of the intact P-choline from phosphatidylcholine to ceramide.[64,68] One possibility might be a phospholipase-C catalyzed hydrolysis of phosphatidylcholine with release of phosphocholine, followed by its entry into the CTP pathway of choline lipid synthesis.[53] A second possibility is the well-documented capability for reversal of the phosphocholinetransferase reaction, with formation of CDP choline.[74,75] Since this enzyme may catalyze the transfer of P-choline to either ceramide or diacylglycerol,[57] the entire reaction could remain enzyme-linked, without the appearance of free CDP-choline. Arguments against these two possibilities include the rapid fall in cell P-choline and CDP-choline radioactivity during chase experiments when radioactivity is still rising,[64,68] and the failure to influence the activity by incubation *in vitro* with relatively large amounts of P-choline and CDP-choline.[66] Genetic evidence suggsting that P-choline is not an intermediate has come from studies by Esko et al.[76,77] of Chinese hamster ovary (CHO) cells with a mutation affecting the cytidylyltransferase reaction. When shifted to the nonpermissive temperature, this mutant shows an immediate fall in phosphatidylcholine and CDP-choline labeling, but SM synthesis continues for at least another 15 to 20 h. Taken together, the evidence suggests that free P-choline is not involved.

The evidence that CDP-choline is not a transfer intermediate is not as strong. In experiments with intact and disrupted mouse fibroblasts incubated with ^{32}Pi-phosphatidyl-^{3}H-choline, Marggraf and Anderer[65] showed similar ^{3}H/^{32}Pi ratios in phosphatidylcholine, CDP-choline, and SM. Simultaneous incubation with ^{14}C-choline suggested that part of the CDP-choline was synthesized from P-choline, but the possibility that part of the pool arose from phosphatidylcholine was not excluded. CDP-choline is at very low levels in cells,[78-81] and so, even low levels of radioactivity may reflect a high specific activity compatible with a precursor role in SM biosynthesis. In the absence of mass measurement of CDP-choline in these experiments, such a possibility cannot be excluded. Failure to reduce incorporation of phosphatidylcholine label into SM by the addition of CDP-choline or to label SM with labeled exogenous CDP-choline may be due to a limited mixing with an enzyme associated CDP-choline pool involved in the transfer. However, if CDP-choline is the transfer intermediate, the catalytic protein involved is probably not the same phosphocholinetransferase concerned with phosphatidylcholine biosynthesis. This enzyme is concentrated in the endoplasmic reticulum, and in contrast to the transfer reaction, is at very low levels in the plasma membrane.[82,83]

A third possibility is that transfer is carried out via an unknown, and hitherto uncharacterized, protein and mechanism. Marggraf and Kanfer[72] have studied the kinetics of the phosphatidylcholine:ceramide phosphocholinetransferase reaction in plasma membrane preparations from Erlich ascites tumor cells over a range of donor and acceptor concentrations, and in the presence of varied concentrations of products. Double-reciprocal plots of the results are compatible with a Ping-Pong reaction mechanism and the formation of an enzyme-bound intermediate. The data must be interpreted with caution, however, because of the uncertainties introduced by a tightly bound enzyme in a crude membrane preparation, a liposomal substrate in an aqueous milieu, and the complex nature of the interactions between them. The isolation and further characterization of this activity may prove difficult because of the apparent dependence on an intact membrane environment and the lack of a sensitive, facile assay. Activities reported with crude tissue preparations and subcellular fractions are low, even when the specific activity of the endogenous membrane phosphatidylcholine that seems to be the preferred substrate is artificially raised with a phospholipid transfer protein.[68] Such low activities are expected since the rates of sphingomyelin biosynthesis in most tissues are also low; perhaps an artificial acceptor might serve to increase activity.

It is still not clear whether the donor of the P-choline is exclusively phosphatidylcholine or whether one of its degradation products — lysophosphatidylcholine or glycerophosphocholine — may be a precursor also.[78] In pulse-chase experiments, the specific activity of these compounds closely parallels the parent phosphatidylcholine.[78] Where examined, lysophosphatidylcholine has inhibited transfer to ceramide;[60,71] this could be interpreted as diluting the precursor pools for SM biosynthesis, but might also be due to the well-documented detergent properties of the lyso lipid. Addition of exogenous glycerophosphocholine had no effect on SM formation from exogenous [^{14}C]phosphatidylcholine by mouse fibroblast homogenates *in vitro*.[66] These homogenates also failed to respond to exogenous ceramide, raising the possibility that substrates in the membranes participating in the transfer may not have mixed with the exogenously added substrates.

4. Synthesis of Phosphatidylcholine from Sphingomyelin

The reverse reaction, the transfer of P-choline from SM to diglyceride, has been described in experiments where human fibroblasts in culture have been incubated with exogenous labeled SM.[84-86] Experiments with [^{3}H-methyl]choline-^{32}Pi-SM indicate that the P-choline is transferred intact.[84] A decrease in transfer when the intracellular pools of P-choline are expanded, an intracellular P-choline specific activity compatible with a precursor role in phosphatidylcholine biosynthesis, and a decrease in the transfer in CHO cells with a temperature-sensitive mutation in cytidylyltransferase activity,[87] all strongly suggest that the transfer from SM proceeds via the classical Kennedy pathway:[53]

CTP → (consumed), PPi ← (released)

SM → P-choline → CDP-choline → phosphatidylcholine
(SM → ceramide)

Thus, the transfer of phosphocholine from SM to diacylglycerol is probably a different reaction sequence from that of the phosphatidylcholine/ceramide transfer.

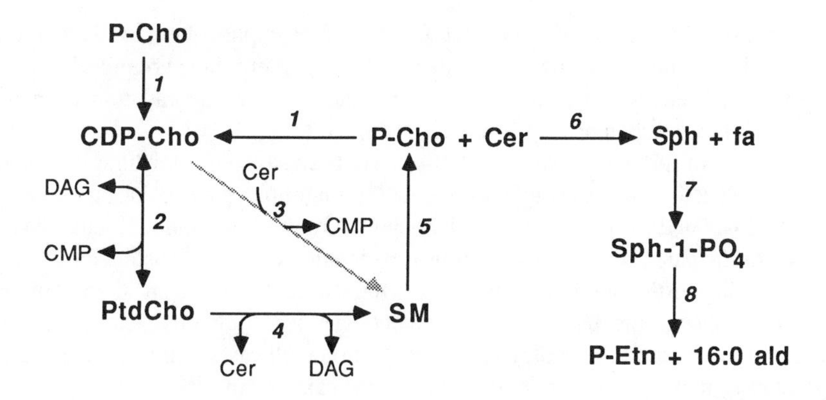

FIGURE 1. Pathways of sphingomyelin synthesis and catabolism. 1. CTP:phosphocholine cytidylyltransferase. 2. CDP-choline:1,2-diacylglycerol phosphocholinetransferase. 3. CDP-choline:ceramide phosphocholinetransferase. 4. Phosphatidylcholine:ceramide cholinephosphotransferase. 5. Sphingomyelinase. 6. Ceramidase. 7. Sphingosine kinase. 8. Sphingosine-1-PO_4 aldolase. Abbreviations: P-Cho, phosphocholine; CDP-Cho, cytidine diphosphate choline; PtdCho, phosphatidylcholine; SM, sphingomyelin; Cer, ceramide; Sph, sphingosine; fa, fatty acid; P-Etn, phosphoethanolamine; 16:0 ald, palmitaldehyde.

5. Relative Importance of the Possible Biosynthetic Pathways

As reviewed above, and presented as a diagram in Figure 1, present evidence indicates that the major pathway for SM biosynthesis is the transfer of P-choline from phosphatidylcholine to ceramide. The reaction mechanism(s) and the nature of the intermediates are not known. The reaction is particularly concentrated in the plasma membrane and the Golgi.[67,68] In many cells, SM is concentrated on the external surface of the plasma membane.[88-91] In such circumstances, ceramide, synthesized in the microsomes together with other lipids, may remain in the apolar portion of the membrane bilayer[92] until required as a substrate for transfer from PC. The latter reaction takes place near the external surface of the plasma membrane, essentially anchoring the completed SM molecule on the external surface of the cell.

The plasma membrane is not the sole site of SM synthesis, however; studies of the intracellular translocation of fluorescent sphingolipids and their precursors suggest that both SM and glucocerebroside are synthesized internally and are translocated through the Golgi apparatus to the cell surface.[93] Whether this internally synthesized sphingomyelin is made by direct transfer from phosphatidylcholine, by a reversal of the cholinephosphotransferase that is concentrated in these internal organelles,[82] or from choline via the cytidylyltransferase and cholinephosphotransferase is not known. The weight of evidence suggests that transfer from phosphatidylcholine is the predominant route and that, where both pathways have been examined simultaneously, synthesis from choline via CDP-choline accounts for <5% of overall synthesis.[65] However, it is possible that cells in which SM is also oriented internally[94] or in cells where SM is destined for secretion, as in liver,[4] in some tissues with a particularly high SM content,[5] or in tissues, such as brain, where transfer from phosphatidylcholine cannot be demonstrated,[63] the synthesis of SM from CDP-choline via the cholinephosphotransferase reaction may play a larger role.

III. CATABOLISM OF SPHINGOMYELIN

A. POSSIBLE PATHWAYS OF SPHINGOMYELIN DEGRADATION

There are at least three possible routes by which SM degradation might be initiated: (1) a phospholipase D reaction with formation of ceramide-PO_4 and choline; (2) removal of amide-linked fatty acid to form sphingosylphosphocholine and free fatty acid; and (3) a phospholipase C reaction (sphingomyelinase; Smase; sphingomyelin phosphodiesterase), yielding phosphocholine and ceramide.

SM is a substrate for some of the bacterial[95] and cabbage phospholipase D activities.[96] In plants, the activity with SM is lower than that with the glycerophospholipids.[96] To date, this pathway has not been shown to contribute significantly to SM degradation in mammalian tissues, nor has base-exchange with SM been described.[97]

Small amounts of sphingenylphosphocholine have been isolated from the spleen of a patient with Niemann-Pick disease.[98] A similar deacylated sphingolipid, psychosine (sphingenylgalactose) has been isolated from tissues of patients with Krabbe's globoid cell leukodystrophy.[99] Deacylation of the parent lipid, galactocerebroside, could not be demonstrated by two laboratories,[99,100] leading to the suggestion that it is synthesized from galactose and sphingenine. Sphingenylphosphocholine may also be a product of synthesis rather than catabolism. Thus, in the absence of evidence to the contrary, the deacylation pathway should be regarded as either nonexistent or as making a minor contribution to SM catabolism.

All evidence to date is consistent with a phospholipase C type reaction or Smase as the first step in SM catabolism. These enzymes hydrolyze SM to ceramide and P-choline. Schneider and Kennedy[101] injected labeled dihydrosphingomyelin intravenously into rats. The isolated metabolic products were compatible with the reaction sequence: dihydrosphingomyelin → dihydroceramide → sphinganine. Patients with inherited defects of Smases accumulate SM in tissues.[13] In drug-induced lipidoses in cells in culture, SM accumulates as Smase activity is reduced.[102] It seems clear, therefore, that the major catabolic route for SM is via this route, and the number and nature of the various Smases is discussed in the next few sections.

B. MAMMALIAN SPHINGOMYELINASES

1. Acid (Lysosomal) Sphingomyelinase

This Smase is ubiquitously distributed in all mammalian tissues that have been examined.[29,103,104] It has a pH optimum of 4.6 to 5.0 and no requirement for metal ions.[29,103] Latency and subcellular distribution indicate a lysosomal origin.[105,106] Activity can be demonstrated with liposomal substrates, but is markedly stimulated by the addition of neutral or anionic detergents[29,103,107] used to suspend the substrate. D-*erythro* sphingomyelin is a preferred substrate over L-*erythro* or D,L-*threo* sphingomyelin;[29] the enzyme will hydrolyze phosphatidylcholine,[108] phosphatidylglycerol,[108] 2 hexadecanoylamino-4-nitrophenyl-phosphocholine,[109] *N*-trinitrophenyl aminolauryl-sphingosylphosphocholine,[110] *N*-(10[1-pyrene]-decanoyl) sphingomyelin,[111] 11-(9-anthroyloxy)undecanoyl sphingomyelin,[110] and bis(4-methylumbelliferyl) phosphate.[112] Activator proteins for Smase have been demonstrated in assays with liposomal SM as a substrate.[113,114] The enzyme is strongly inhibited by 5′AMP.[30,115] The enzyme has been purified from human placenta,[116-119] brain,[120] and human urine.[121] It is a glycoprotein with a molecular mass of 180 to 200 kDa[116,119,122] and is probably a dimer of a single polypeptide chain. Size estimates of the polypeptide have varied between 70 and 120 kDa, depending on enzyme source, method, and the reporting laboratory.[116,118,121,123] The reasons for this heterogeneity are not known. Kinetic studies of the purified enzyme suggest that a carboxyl group and a protonated histidine are involved at the active site, and that the hydrophobic binding site requires linear aliphatic molecules at least eight carbons long.[115,121]

Deficiencies in the activity of this enzyme are part of the biochemical phenotype of the Type I forms of the recessively inherited SM lipidosis, Niemann-Pick disease.[13] These patients present with a wide-ranging spectrum of disease, varying from an acute infantile form with death in the first decade of life to more chronic forms which involve chiefly visceral organs. This phenotypic heterogeneity argues for more than one mutation and more than one functional form of the enzyme in various tissues. Such observations are supported by the lack of complementation between fibroblasts of patients with acute and chronic forms of the disease.[124] At least some of the mutations may be single base substitutions in the structural gene, since inactive enzyme protein in apparently normal amounts and of normal size has been demonstrated in cultured fibroblasts from some patients.[125,126]

2. Neutral Sphingomyelinase

Mammalian tissues have a second Smase with a neutral pH optimum (pH 7.4). The enzyme is particularly enriched in brain,[30,104] but has been described in a wide variety of tissues[104] including liver,[127] adrenal tissues,[128] and kidney.[129] The enzyme is inhibited by EDTA and stimulated by Mg^{2+}.[30] It hydrolyzes SM and dihydrosphingomyelin, as well as *N*-(10[1-pyrene]-decanoyl sphingomyelin but not 2*N*-hexadecanoylamino-4-nitrophenyl-phosphocholine.[30,130,131] In liver[127] and in cultured neuroblastoma cells,[132] the enzyme is concentrated in plasma membrane and microsomes, and in the neuroblastoma cells, the plasma membrane activity is oriented externally.[133] Activity levels are particularly high in brain, especially in gray matter[30,104] and activity levels rise in brain during development in parallel with neuronal maturation.[134,135] Even within gray matter, activity is not uniformly distributed and is particularly high in the basal ganglia.[104] The activity levels in the basal ganglia are not changed by acute pharmacological manipulation of the dopaminergic system.[136] Levels do rise in the rat neurohypophysis during dehydration, suggesting a possible role in neurosecretion.[137] In calf-brain synaptosomes and cultured neuroblastoma cells, activity is stimulated by volatile anesthetics.[138,139] In neuroblastoma cells the stimulation was positively correlated with anesthetic dose and potency. In 3T3-L1 cells, plasma membrane neutral Smase activity is increased threefold by incubation for 4 h with $10^{-7}M$ dexamethasone.[140] Activity levels are not affected by the mutation in Niemann-Pick disease.[107]

The neutral enzyme is a lipophilic, integral membrane protein.[30,31] Normally particulate, it can be solubilized by mild neutral detergents.[30,107] Activity is generally measured in the presence of Triton X-100, and with varying Triton concentrations, the enzyme activity can display complex kinetics.[107] Gatt has suggested that the Triton used to suspend the substrate may also displace lipid from the active site of the enzyme;[107] both phosphatidylcholine[141] and phosphatidylserine[142] have been implicated as lipid activators. The molecular weight by radiation inactivation was 165 kDa in samples from three human brains[143] and 740 kDa in a fourth brain.[144] Other studies have shown a size equivalent to a globular protein >600 kDa by gel filtration and <150 kDa by sucrose density gradient centrifugation.[135] The latter observations suggest relatively extensive hydrophobic areas in the enzyme with lipid or detergent binding. Aggregation of hydrophobic units might also explain the large variations in molecular weight observed with radiation inactivation.[143,144] An abstract reporting tenfold purification of the enzyme indicates a molecular weight of 38.5 kDa on polyacrylamide gel electrophoresis and a tendency for aggregation.[145]

3. Intestinal Sphingomyelinase

Nilssen[48] described a Smase with an alkaline pH optimum (8.9 to 9.2) that is enriched in brush-border preparations of rat intestinal mucosa and is present in the crude protein fraction of human duodenal contents and also in meconium. Activity was not found in rat pancreatic juice or pancreatic homogenates. Hydrolysis was stimulated by taurocholate; the effects of metal ions were not examined. The activity may be concerned with the degradation of SM in ingested foods, and, thus, the intestinal brush border may be the main site of hydrolysis of dietary SM.

4. Fetal Bovine Serum Sphingomyelinase

This enzyme is in the lipoprotein-free infranatant of serum, has an acid pH optimum (pH 5.5), and is inhibited by EDTA.[146] Of the metal ions examined, Zn^{2+} was most effective in stimulating activity. The enzyme was specific for SM; phosphatidylcholine, lysophosphatidylcholine, and glycerophosphocholine were not hydrolyzed. Activity levels were highest in fetal bovine serum which was greater than newborn bovine > horse > human serum. The properties are clearly distinct from those of other mammalian Smases described to date. This Smase may play a role at the surface of cells exposed to serum or its products, or in the tissue of origin.

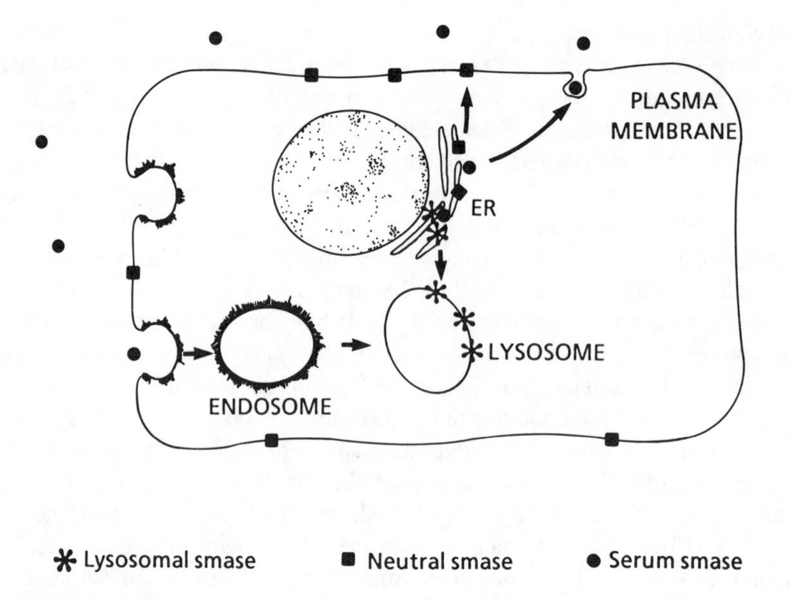

FIGURE 2. Sphingomyelinases (Smases) of mammalian tissues. All Smases are synthesized in the endoplasmic reticulum (ER). The acid sphingomyelinase is incorporated into the membranes of lysosomes and degrades sphingomyelin accumulating during endocytosis and autophagy. The neutral Smase is incorporated into plasma membrane where it modifies the SM content and the availability of ceramide for sphingosine biosynthesis. A similar plasma membrane enzyme is in intestinal cells and may function to degrade dietary sphingomyelin. The serum Smase is exported from the cell, and may subserve some special developmentally regulated function in the circulation.

5. Relationship between Mammalian Sphingomyelinases

A possible relationship between the mammalian Smases is depicted in Figure 2. All of the Smases are separate proteins, and are synthesized in the endoplasmic reticulum. The lysosomal enzyme is packaged in part of the lysosome, where it is concerned with the degradation of SM entering the lysosomal system through endocytosis and/or autophagy. The neutral, Mg^{2+}-stimulated enzyme is incorporated into the plasma membrane, where it serves to modulate local concentrations of SM, and to provide ceramide and, hence, indirectly, sphingosine for other synthetic processes and regulatory roles. The Zn^{2+}-stimulated enzyme is exported from the cells of origin, and subserves some unknown function in the serum compartment, possibly related to growth. The intestinal enzyme is tissue-specific and acts at the surface of the intestinal cell to hydrolyze dietary SM.

C. AVIAN SPHINGOMYELINASES (CHICKEN ERYTHROCYTE)

Hirshfield and Loyter[147] described a Smase in chicken erythrocyte plasma membrane that hydrolyzes either endogenous SM, or exogenous SM-Triton X-100 or sphingomyelin-cholate micelles. The pH optimum was 7 to 9, and activity was stimulated by Mn^{2+} and Mg^{2+} and inhibited by EDTA. Normally latent, this enzyme is activated by incubation in hypotonic medium. Once activated, activity continues even when the medium is made iso- or hypertonic and can degrade >70% of the membrane sphingomyelin. Record et al.[148] have demonstrated that the enzyme in chicken erythrocyte membranes will hydrolyze sphingomyelin in other exogenous membranes, such as human erythrocyte ghosts. No endogenous activity was found in human erythrocyte ghosts.[147] The physiological role(s) of the enzyme is not known.

D. BACTERIAL SPHINGOMYELINASES

A phospholipase-C type enzyme that hydrolyzes both micellar SM and membrane SM, and which has no demonstrable activity with phosphatidylcholine, has been purified from *Staphylococcus aureus*.[149-150] The purified enzyme has a pH optimum of 7.4, and activity is enhanced by addition of Mg^{2+}. The extracellular fluid from *C. perfringens* contains two phospholipase-C activities resolvable by gel chromatography.[151] One activity has a Ca^{2+} requirement and hydrolyzes phosphatidylcholine ten times faster than SM. A second activity, specific for SM, requires Mg^{2+} for activity and is inhibited by Ca^{2+}. The Zn^{2+}-containing phospholipase C from *Bacillus cereus* hydrolyzes glycerophospholipids, but has little activity with SM except in aged erythrocytes.[152] Substitution of Zn^{2+} by Co^{2+} led to SM hydrolysis at 30% of the rate of phosphatidylcholine hydrolysis.

E. CERAMIDASES

Ceramidase (*N*-acylsphingosine deacylase) catalyzes the hydrolysis of the amide linkage of ceramide to form sphingosine and fatty acid. At least three different ceramidases have been described in mammalian tissues.[31,32,48-50] There is a lysosomal form with an acid pH optimum that is widely distributed in mammalian tissues.[32,51] A deficiency of this enzyme is responsible for the inherited lysosomal storage disease Farbers lipogranulomatosis.[51] A second activity with a neutral optimum (pH 7.0 to 8.0) has been described in pig small intestine[48] and rat liver microsome.[57] Alkaline ceramidase activity with a pH optimum at 8 to 9 has been described in several mammalian tissues and often exceeds the activity of the acid lysosomal enzyme.[32]

The acid lysosomal activity has been purified 200-fold from rat brain.[31] The molecular weight was >150,000, and the enzyme activity withstands prolonged treatment with trypsin and chymotrypsin. Both the ceramidase and ceramide synthetase activities of the brain extracts copurified together.

Ceramide is a precursor of both sphingomyelin and glycosphingolipids, and so a major route of removal could be through conversion to other sphingolipids. Studies of intracardiac[153] or intracerebral[154] injection of ceramides labeled in both the fatty-acid and sphingenine portion of the molecule indicate that although there was significant incorporation into SM and cerebroside, extensive fatty-acid replacement occurred, suggesting recycling at the level of sphingenine rather than ceramide. Interestingly, no labeled ceramide was incorporated into gangliosides, and the free ceramide pool in brain contains little of the C_{20}-sphingenine characteristic of brain gangliosides.[44] In cultured fibroblasts incubated with fluorescent derivatives of ceramide, incorporation of ceramide into ceramide monohexoside and SM was observed.[155] It seems likely that there is selective recycling of degradation products at all steps in degradation.

F. FURTHER METABOLISM OF SPHINGOSINE

Sphingosine can be either incorporated into ceramide, as described earlier, or further degraded. The first step in further degradation is phosphorylation to sphingosine-1-phosphate. Sphingosine kinase has been studied in rat liver,[156] human and rabbit erythrocytes,[157] and human and pig platelets.[158] The enzyme is cytosolic, ATP-dependent, has a pH optimum around 7.0, and is most active with long-chain bases of 14 to 16 carbons in length. The platelet enzyme exhibited some sterospecificity, and affinity decreased in the order D(+)erythro-4t-sphingenine > D(+)erythro-and L(-)threo sphinganine.[158]

Sphingosine-1-phosphate is hydrolyzed to ethanolamine-1-phosphate and palmitaldehyde by sphingosine-1-phosphate-alkanal lyase.[159-161] The enzyme is membrane bound and is widely distributed in mammalian tissues and has been described in the yeast *Hansenula cifferrii*. The enzyme is pyridoxal-phosphate dependent and is inhibited by sulfhydryl reagents.

IV. FUTURE DIRECTIONS

A rapid perusal of the indices of scientific articles in the past decade will show that SM has always been of interest, but has occupied less attention than the phosphoglycerides and glycolipids. That situation is rapidly changing, and one can anticipate greatly increased research activity in the future along several related lines.

The first will concern the role of the intact SM molecule in the normal physiological functions of the cell. An intimate and important role in cholesterol metabolism has long been suspected, but has received new impetus with the discovery that some variants of the SM lipidosis, Niemann-Pick disease, show defects in cholesterol processing, and the observations that some of the regulatory roles of lipoproteins can be mimicked by perturbing SM metabolism. The intriguing reports of relatively specific interactions between SM and a variety of enzyme and nonenzyme proteins suggest that this area may also assume increased importance in the future. Important new insights in these areas will be aided by the increasing sophistication of microscopic and physicochemical techniques for the study of lipid-lipid and lipid-protein interactions.

In the immediate future, another area to watch is the role of sphingolipids in general, and of SM in particular, as precursors of sphingosine. The situation is analogous to the surge of interest in phospholipases and arachidonic acid initiated by the discovery of eicosanoids and the stimulation of research on phosphoinositides as a result of the discovery of the regulatory roles of their degradation products. Sphingosine bases are regulators of protein kinase C, and their levels in tissues may relate to the availability of SM and SM-degrading enzymes. The latter will be of particular interest, and the number and properties of Smases, in particular the nonlysosomal enzymes, will occupy increasing attention.

The major unresolved question in the biosynthesis of SM remains the reaction mechanism for the transfer of P-choline from phosphatidylcholine to ceramide. A fundamental knowledge of this reaction is critical to our understanding of how nature maintains the balance between the choline-containing lipids in the membranes of mammalian cells.

With the exception of the lysosomal Smases, none of the enzymes of mammalian SM biosynthesis and catabolism have been purified and characterized, and little is known of their regulation. Even for the lysosomal Smase, there have been no reports of successful cloning of the cDNA. Thus, detailed knowledge of environmental and genetic control of SM biosynthesis and catabolism remains an elusive goal, and one to which increasing attention will be paid in the next while.

Thudichum is purported to have assigned the name "sphingosin" to this class of lipids because the enigmas that they present are like those of the Sphinx of Thebes. The mysteries of the latter remain veiled in antiquity; the mysteries of the former are slowly being revealed. The pace of revelation will undoubtedly increase in the next few years, and SM may rival the phosphoinositides as the "leading lady" of the "lipid stage".

ACKNOWLEDGMENTS

The author's research is supported by a Program Grant (PG-16) and a Career Investigatorship from the Medical Research Council of Canada. The helpful advice and criticism of Drs. David M. Byers, Harold W. Cook, and Dennis E. Vance, and the excellent assistance of Ms. Diann Nicholson in preparing this manuscript is gratefully acknowledged.

REFERENCES

1. **Thudicum, J. L. W.,** *A Treatise on the Chemical Constitution of the Brain,* (a facsimile edition of the original published by Balliere, Tindall and Cox, London, with a new historical introduction: "Reflections upon a Classic", by D. L. Drabkin), 1962.
2. **Pick, L. and Bielschowsky, M.,** Uber lipoidzellige splenomegalie (Typus Niemann-Pick) und amourotische idiotie, *Klin. Wochenschr.,* 6, 1631, 1927.

3. **Rouser, G., Berry, J. F., Marinetti, G. V., and Stolz, E.,** Studies on the structure of sphingomyelin. I. Oxidation of products of partial hydrolysis, *J. Am. Chem. Soc.*, 75, 310, 1953.
4. **Shapiro, D. and Flowers, H. M.,** Studies on sphingolipids. VII. Synthesis and configuration of natural sphingomyelins, *J. Am. Chem. Soc.*, 84, 1047, 1962.
5. **Rouser, G., Kitchevsky, G., and Yamamoto, A.,** Lipids in the nervous system of different species as a function of age: brain, spinal cord, peripheral nerve, purified whole cell preparations, and subcellular particulates: regulatory mechanisms and membrane structure, *Adv. Lipid Res.*, 10, 261, 1972.
6. **Barenholz, Y. and Thompson, T. E.,** Sphingomyelins in bilayers and biological membranes, *Biochim. Biophys. Acta*, 604, 129, 1980.
7. **White, D. A.,** The phospholipid composition of mammalian tissues, in *Form and Function of Phospholipids*, Ansell, G. B., Hawthorne, J. N., and Dawson, R. M. C., Eds., Elsevier Science, Amsterdam, 1974, 441.
8. **Rouser, G., Nelson, G. J., Fleischer, S., and Simon, G.,** Lipid composition of animal cell membranes, organelles and organs, in *Biological Membranes*, Chapman, D., Ed., Academic Press, London, 1968, 5.
9. **O'Brien, J. S. and Sampson, E. L.,** Fatty acid and fatty aldehyde composition of the major brain lipids in normal human gray matter, white matter, and myelin, *J. Lipid Res.*, 6, 545, 1965.
10. **Levin, I. W., Thompson, T. E., Barenholz, Y., and Huang, C.,** Two types of hydrocarbon chain interdigitation in sphingomyelin bilayers, *Biochemistry*, 24, 6282, 1985.
11. **Barenholz, Y. and Gatt, S.,** Sphingomyelin: metabolism, chemical synthesis, chemical and physical properties, in *Phospholipids*, Hawthorne, J. N. and Ansell, G. B., Eds., Elsevier Biomedical Press, New York, 1982, 129.
12. **Boegheim, J. P. J., Jr., Van Linde, M., Op Den Kamp, J. A. F., and Roelofsen, B.,** The sphingomyelin pools in the outer and inner layer of the human erythrocyte membrane are composed of different molecular species, *Biochim. Biophys. Acta*, 735, 438, 1983.
13. **Spence, M. W. and Callahan, J. W.,** Sphingomyelin/cholesterol lipidoses: the Niemann-Pick group of diseases, in *The Metabolic Basis of Inherited Diseases*, Scriver, C. R., Beaudet, A. L., Sly, W. S., and Valle, D., Eds., 6th ed., McGraw-Hill, New York, in press.
14. **Pentchev, P. G., Kruth, H. S., Comly, M. E., Butler, J. D., Vanier, M. R., Wenger, D. A., and Patel, S.,** Type C Niemann-Pick disease: a parallel loss of regulatory responses in both the uptake and esterification of low density lipoprotein-derived cholesterol in cultured fibroblasts, *J. Biol. Chem.*, 261, 16775, 1986.
15. **Slotte, J. P. and Bierman, E. L.,** Depletion of plasma-membrane sphingomyelin rapidly alters the distribution of cholesterol between plasma membranes and intracellular cholesterol pools in cultured fibroblasts, *Biochem. J.*, 250, 653, 1988.
16. **Esumi, M. and Fujiki, H.,** Sphingomyelin is a possible constituent of binding sites for the tumor promoters phorbol ester, indole alkaloids and polyacetates, *Biochem. Biophys. Res. Commun.*, 112, 709, 1983.
17. **Dawson, R. M. C., Hemington, N., and Irvine, R. F.,** The inhibition of diacylglycerol-stimulated intracellular phospholipases by phospholipids with a phosphocholine-containing polar group, *Biochem. J.*, 230, 61, 1985.
18. **Widnell, C. C. and Unkeless, J. C.,** Partial purification of a lipoprotein with 5′nucleotidase activity from membranes of rat liver cells, *Proc. Natl. Acad. Sci. U.S.A.*, 61, 1050, 1968.
19. **Sood, C. K., Sweet, C., and Zull, J. E.,** Interaction of kidney (Na^+-K^+)-ATPase with phospholipid model membrane systems, *Biochim. Biophys. Acta*, 282, 429, 1972.
20. **Linder, R., Bernheimer, A. W., and Kim, K.,** Interaction between sphingomyelin and a cytolysin from the sea anemone Stoichactis Hellianthus, *Biochim. Biophys. Acta*, 467, 290, 1977.
21. **Cohen, R. and Barenholz, Y.,** Characterization of the association of electrophorus electricus acetylcholinesterase with sphingomyelin liposomes, *Biochim. Biophys. Acta*, 778, 94, 1984.
22. **Stoffel, W., Zierenberg, O., Tunggal, B., and Schreiber, E.,** ^{13}C nuclear magnetic resonance spectroscopic evidence for hydrophobic lipid-protein interactions in human high density lipoproteins, *Proc. Natl. Acad. Sci. U.S.A.*, 71, 3696, 1974.
23. **Epand, R. M., Boni, L. T., and Hui, S. W.,** Interaction of glucagon with sphingomyelins, *Biochim. Biophys. Acta*, 692, 330, 1982.
24. **Hannun, Y. A., Loomis, C. R., Merrill, A. H., Jr., and Bell, R. M.,** Sphingosine inhibition of protein kinase C activity and of phorbol dibutyrate binding *in vitro* and in human platelets, *J. Biol. Chem.*, 261, 12604, 1986.
25. **Gabellec, M. M., Steffan, A. M., Dodeur, M., Durand, G., and Rebel, G.,** Membrane lipids of hepatocytes, kupffer cells and endothelial cells, *Biochem. Biophys. Res. Commun.*, 113, 845, 1983.
26. **Wilson, E., Rice, W. G., Kinkade, J. M., Jr., Merrill, A. H., Jr., Arnold, R. R., and Lambeth, J. D.,** Protein kinase C inhibition by sphingoid long-chain bases: effects on secretion in human neutrophils, *Arch. Biochem. Biophys.*, 259, 204, 1987.
27. **Merill, A. H., Jr., Sereni, A. M., Stevens, V. L., Hannun, Y. A., Bell, R. M., and Kinkade, J. M., Jr.,** Inhibition of phorbol ester-dependent differentiation of human promyelocytic leukemic (HL-60) cells by sphinganine and other long-chain bases, *J. Biol. Chem.*, 261, 12610, 1986.

28. **Nelson, D. H. and Murray, D. K.,** Sphingolipids inhibit insulin and phorbol ester stimulated uptake of 2-deoxyglucose, *Biochem. Biophys. Res. Commun.,* 138, 463, 1986.
29. **Barnholz, Y., Roitman, A., and Gatt, S.,** Enzymatic hydrolysis of sphingolipids, *J. Biol. Chem.,* 241, 3731, 1966.
30. **Rao, B. G. and Spence, M. W.,** Sphingomyelinase activity at pH 7.4 in human brain and a comparison to activity at pH 5.0, *J. Lipid Res.,* 17, 506, 1976.
31. **Yavin, E. and Gatt, S.,** Enzymatic hydrolysis of sphingolipids. VII. Further purification and properties of rat brain ceramidase, *Biochemistry,* 6, 1692, 1969.
32. **Spence, M. W., Beed, S., and Cook, H. W.,** Acid and alkaline ceramidases of rat tissues, *Biochem. Cell Biol.,* 64, 400, 1986.
33. **Kolesnick, R. N.,** 1,2-Diacylglycerols but not phorbol esters stimulate sphingomyelin hydrolysis in GH3 pituitary cells, *J. Biol. Chem.,* 262, 16759, 1987.
34. **Kolesnick, R. N. and Clegg, S.,** 1,2-Diacylglycerols, but not phorbol esters, activate a potential inhibitory pathway for protein kinase C in GH3 pituitary cells. Evidence for involvement of a sphingomyelinase, *J. Biol. Chem.,* 263, 6534, 1988.
35. **Slife, C. W., Wang, E., Want, S., and Merrill, A.,** Sphingosine formation from endogenous substrates in a plasma membrane enriched liver fraction, *FASEB J.,* 2, A1416, 1988.
36. **Brady, R. O., Formica, J. V., and Koval, G. J.,** The enzymatic synthesis of sphingosine. II. Further studies on the mechanism of the reaction, *J. Biol. Chem.,* 233, 1072, 1958.
37. **Braun, P. E. and Snell, E. E.,** Biosynthesis of sphingolipid bases. II. Keto intermediates in synthesis of sphingosine and dihydrosphingosine by cell-free extracts of Hansenula ciferri, *J. Biol. Chem.,* 243, 3775, 1968.
38. **Willians, R. D., Wang, E., and Merrill, A. H.,** Enzymology of long-chain base synthesis by liver: characterization of serine palmitoyltransferase in rat liver microsomes, *Arch. Biochem. Biophys.,* 228, 282, 1984.
39. **Stoffel, W.,** Studies on the biosynthesis and degradation of sphingosine bases, *Chem. Phys. Lipids,* 5, 139, 1970.
40. **Ong, D. E. and Brady, R. N.,** *In vivo* studies on the introduction of the 4-t-double bond of the sphingenine moiety of rat brain ceramides, *J. Biol. Chem.,* 248, 3884, 1973.
41. **Merrill, A. H., Jr. and Wang, E.,** Biosynthesis of long-chain (sphingoid)bases from serine by LM cells. Evidence for introduction of the 4-trans-double bond after de novo biosynthesis of n-acylsphinganine(s), *J. Biol. Chem.,* 261, 3764, 1986.
42. **Sribney, M.,** Enzymatic synthesis of ceramide, *Biochim. Biophys. Acta,* 125, 542, 1966.
43. **Morell, P. and Radin, N. S.,** Specificity in ceramide biosynthesis from long chain bases and various fatty acyl coenzyme A's by brain microsomes, *J. Biol. Chem.,* 245, 342, 1970.
44. **Ulman, M. D. and Radin, N. S.,** Enzymatic formation of hydroxy ceramides and comparison with enzymes forming nonhydroxy ceramides, *Arch. Biochem. Biophys.,* 152, 767, 1972.
45. **Narimatsu, S., Soeda, S., Tanaka, T., and Kishimoto, Y.,** Solubilization and partial characterization of fatty acyl-CoA: sphingosine acyltransferase (ceramide synthetase) from rat liver and brain, *Biochim. Biophys. Acta,* 877, 334, 1986.
46. **Kishimoto, Y. and Kawamura, N.,** Ceramide metabolism in brain, *Mol. Cell Biochem.,* 23, 17, 1979.
47. **Stoffel, W., Kruger, E., and Melzner, I.,** Studies on the biosynthesis of ceramide. Does the reversed ceramidase reaction yield ceramides?, *Hoppe-Seyler's Z. Physiol. Chem.,* 361, 773, 1980.
48. **Nilsson, A.,** The presence of sphingomyelin and ceramide-cleaving enzymes in the small intestinal tract, *Biochim. Biophys. Acta,* 176, 339, 1969.
49. **Sugita, M., Williams, M., Dulaney, J. T., and Moser, H. W.,** Ceramidase and ceramide synthesis in human kidney and cerebellum: description of a new alkaline ceramidase, *Biochim. Biophys. Acta,* 398, 125, 1975.
50. **Momoi, T., Ben-Yoseph, Y., and Nadler, H. L.,** Substrate-specificities of acid and alkaline ceramidases in fibroblasts from patients with Farber disease and controls, *Biochem. J.,* 205, 419, 1982.
51. **Moser, H. and Chen, W. W.,** Ceramidase deficiency: Farber's lipogranulomatosis, in *The Metabolic Basis of Inherited Disease,* Stanbury, J. B., Wyngaarden, J. B., Fredrickson, D. S., Goldstein, J. L., and Brown, M. S., Eds., 5th ed., McGraw-Hill, New York, 1982, 820.
52. **Sribney, M. and Kennedy, E. P.,** The enzymic synthesis of sphingomyelin, *J. Biol. Chem.,* 233, 1315, 1958.
53. **Pelech, S. and Vance, D. E.,** Regulation of phosphatidylcholine biosynthesis, *Biochim. Biophys. Acta,* 779, 217, 1984.
54. **Sribney, M.,** Stimulation and inhibition of sphingomyelin synthetase, *Arch. Biochem. Biophys.,* 126, 954, 1968.
55. **Fujino, Y. and Negishi, T.,** Investigation of the enzymatic synthesis of sphingomyelin, *Biochim. Biophys. Acta,* 152, 428, 1968.
56. **Fujino, Y., Negishi, T., and Ito, S.,** Enzymatic synthesis of sphingosylphosphorylcholine, *Biochem. J.,* 109, 310, 1968.

57. **Stoffel, W. and Melzner, I.,** Studies in vitro on the biosynthesis of ceramide and sphingomyelin. A reevaluation of proposed pathways, *Hoppe-Seyler's Z. Physiol. Chem.,* 361, 755, 1980.
58. **Brady, R. O., Bradley, R. M., Young, O. M., and Kaller, H.,** An alternative pathway for the enzymatic synthesis of sphingomyelin, *J. Biol. Chem.,* 240, 3693, 1965.
59. **Fujino, Y., Nakano, M., Negishi, T., and Ito, S.,** Substrate specificity for ceramide in the enzymatic formation of sphingomyelin, *J. Biol. Chem.,* 243, 4650, 1968.
60. **Diringer, H., Marggraf, W. D., Koch, M. A., and Anderer, F. A.,** Evidence for a new biosynthetic pathway of sphingomyelin in SV 40 transformed mouse cells, *Biochem. Biophys. Res. Commun.,* 47, 1345, 1972.
61. **Ullman, D. M. and Radin, N. S.,** The enzymatic formation of sphingomyelin from ceramide and lecithin in mouse liver, *J. Biol. Chem.,* 249, 1506, 1974.
62. **Kanfer, J. N. and Spielvogel, C. H.,** Phospholipase C catalyzed formation of sphingomyelin-14C from lecithin and N-(14C)-oleoyl-sphingosine, *Lipids,* 10, 391, 1975.
63. **Bernert, J. T., Jr. and Ullman, D. M.,** Biosynthesis of sphingomyelin from erythro-ceramides and phosphatidylcholine by a microsomal cholinephosphotransferase, *Biochim. Biophys. Acta,* 666, 99, 1981.
64. **Marggraf, H. D., Koch, M. A., and Anderer, F. A.,** Evidence for a new biosynthetic pathway of sphingomyelin in SV 40 transformed mouse cells, *Biochem. Biophys. Res. Commun.,* 47, 1345, 1972.
65. **Marggraf, W.-D. and Anderer, F. A.,** Alternative pathways in the biosynthesis of sphingomyelin and the role of phosphatidylcholine, CDPcholine and phosphorylcholine as precursors, *Hoppe-Seyler's Z. Physiol. Chem.,* 355, 803, 1974.
66. **Marggraf, W. D., Anderer, F. A., and Kanfer, J. N.,** The formation of sphingomyelin from phosphatidylcholine in plasma membrane preparations from mouse fibroblasts, *Biochim. Biophys. Acta,* 664, 61, 1981.
67. **Marggraf, W. D., Zertani, R., Anderer, F. A., and Kanfer, J. N.,** The role of endogenous phosphatidylcholine and ceramide in the biosynthesis of sphingomyelin in mouse fibroblasts, *Biochim. Biophys. Acta.,* 710, 314, 1982.
68. **Voelker, D. R. and Kennedy, E. P.,** Cellular and enzymic synthesis of sphingomyelin, *Biochemistry,* 21, 2753, 1982.
69. **van den Hill, A., van Heusden, P. H., and Wirtz, K. W. A.,** The synthesis of sphingomyelin in the Morris hepatomas 777 and 5123D is restricted to the plasma membrane, *Biochim. Biophys. Acta,* 833, 354, 1985.
70. **Eppler, C. M., Malewicz, B., Jenkin, H. M., and Baumann, W. J.,** Phosphatidylcholine as the choline donor in sphingomyelin synthesis, *Lipids,* 22, 351, 1987.
71. **Lecerf, J., Fouilland, L., and Gagniarre, J.,** Evidence for a high activity of sphingomyelin biosynthesis by phosphocholine transfer from phosphatidylcholine to ceramides in lung lamellar bodies, *Biochim. Biophys. Acta,* 918, 48, 1987.
72. **Marggraf, W. D. and Kanfer, J. N.,** Kinetic and topographical studies of the phosphatidylcholine: ceramide cholinephosphotransferase in plasma membrane particles from mouse ascites cells, *Biochim. Biophys. Acta,* 897, 57, 1987.
73. **Nelson, D. H. and Murray, D. K.,** Dexamethasone increases the synthesis of sphingomyelin in 3T3-L1 cell membranes, *Proc. Natl. Acad. Sci. U.S.A.,* 79, 6690, 1982.
74. **Kanoh, H. and Ohno, K.,** Utilization of endogenous phospholipids by the back-reaction of CDP-choline (ethanolamine): diglyceride choline (ethanolamine)-phosphotransferase in rat liver microsomes, *Biochim. Biophys. Acta,* 306, 203, 1973.
75. **Goracci, G., Francescangeli, E., Horrocks, L. A., and Porcellati, G.,** The reverse reaction of cholinephosphotranferase in rat brain microsomes. A new pathway for degradation of phosphatidylcholine, *Biochim. Biophys. Acta,* 664, 373, 1981.
76. **Esko, J. D. and Raetz, C. R. H.,** Autoradiographic detection of animal cell membrane mutants altered in phosphatidylcholine biosynthesis, *Proc. Natl. Acad. Sci. U.S.A.,* 77, 5192, 1980.
77. **Esko, J. D., Wermuth, M. M., and Raetz, C. R. H.,** Thermolabile CDP-choline synthetase in an animal cell mutant defective in lecithin formation, *J. Biol. Chem.,* 256, 7388, 1981.
78. **Morash, S. M., Cook, H. W., and Spence, M. W.,** Phosphatidylcholine metabolism in cultured cells. Catabolism via glycerophosphocholine, *Biochim. Biophys. Acta,* 961, 194, 1988.
79. **Warden, C. H. and Friedkin, M.,** Regulation of phosphatidylcholine biosynthesis by mitogenic growth factors, *Biochim. Biophys. Acta,* 792, 270, 1984.
80. **Zelinski, T. A., Savard, J. D., Man, R. Y. K., and Choy, P. S.,** Phosphatidylcholine biosynthesis in isolated hamster heart, *J. Biol. Chem.,* 255, 11423, 1980.
81. **Vance, D. E., Trip, E. M., and Paddon, H.,** Poliovirus increases phosphatidylcholine biosynthesis in HeLa cells by stimulation of the rate-limiting reaction catalyzed by CTP: phosphocholine cytidylyltransferase, *J. Biol. Chem.,* 255, 1064, 1980.
82. **Cornell, R. B., Ishidate, K., Ridgway, N. D., Sanghera, J. S., and Vance, D.,** The enzymes of phosphatidylcholine biosynthesis, in *Enzymes of Lipid Metabolism II,* Freysz, L., Dreyfus, H., Massarelli, R., and Gatt, S., Eds., Plenum Press, New York, 1986, 47.

83. **Vance, D. E., Choy, P., Fanen, S. B., Lim, P. H., and Schneider, W. J.,** Asymmetry of phospholipid biosynthesis, *Nature,* 270, 268, 1977.
84. **Spence, M. W., Clarke, J. T. R., and Cook, H. W.,** Pathways of sphingomyelin metabolism in cultured fibroblasts from normal and sphingomyelin lipidosis subjects, *J. Biol. Chem.,* 258, 8595, 1983.
85. **Maziere, J. C., Maziere, C., Mora, L., Routier, J. D., and Polonovski, J.,** *In situ* degradation of sphingomyelin by cultured normal fibroblasts and fibroblasts from patients with Niemann-Pick disease type A and C, *Biochem. Biophys. Res. Commun.,* 108, 1101, 1982.
86. **Beaudet, A. L. and Manschreck, A. A.,** Metabolism of sphingomyelin by intact cultured fibroblasts: differentiation of Niemann-Pick disease types A and B, *Biochem. Biophys. Res. Commun.,* 105, 14, 1982.
87. **Spence, M. W. and Cook, H. W.,** unpublished observations.
88. **Zwaal, R. F. A., Roelofsen, B., Confurius, P., and Van Deenen, L. L. M.,** Organization of phospholipids in human red cell membranes as detected by the action of various purified phospholipases, *Biochim. Biophys. Acta,* 406, 83, 1975.
89. **Renooij, W., Van Golde, L. M. G., Zwaal, R. F. A., and Van Deenen, L. L. M.,** Topological asymmetry of phospholipid metabolism in rat erythrocyte membranes, *Eur. J. Biochem.,* 61, 53, 1976.
90. **Sandra, A. and Pagano, R. E.,** Phospholipid asymmetry in LM cell plasma membrane derivatives: polar head group and acyl chain distributions, *Biochemistry,* 17, 332, 1978.
91. **Patzer, E. J., Moore, N. F., Barenholz, Y., Shaw, J. M., and Wagner, R. R.,** Lipid organization of the membrane of vesicular stomatitis virus, *J. Biol. Chem.,* 253, 4544, 1978.
92. **Verkleij, A. J., Zwaal, R. F. A., Roelofsen, B., Comfurius, P., Kastelijn, D., and Van Deenen, L. L. M.,** The asymmetric distribution of phospholipids in the human red cell membrane. A combined study using phospholipases and freeze-etch electron microscopy, *Biochim. Biophys. Acta,* 323, 178, 1973.
93. **Lipsky, N. G. and Pagano, R. E.,** Intracellular translocation of fluorescent sphingolipids in cultured fibroblasts: endogenously synthesized sphingomyelin and glucocerebroside analogues pass through the Golgi apparatus en route to the plasma membrane, *J. Cell Biol.,* 100, 27, 1985.
94. **Rothman, J. E., Tsai, D. K., Davidowicz, E. A., and Lenard, J.,** Transbilayer phospholipid asymmetry and its maintenance in the membrane of influenza virus, *Biochemistry,* 15, 2361, 1976.
95. **Berheimer, A. W., Linder, R., and Avigad, L. S.,** Stepwise degradation of membrane sphingomyelin by corynebacterial phospholipases, *Infect. Immunol.,* 29, 123, 1980.
96. **Heller, M.,** Phospholipase D, *Adv. Lipid Res.,* 16, 267, 1978.
97. **Kanfer, J. N.,** The base exchange enzymes and phospholipase D of mammalian tissue, *Can. J. Biochem.,* 58, 1370, 1980.
98. **Strasberg, P. M. S. and Callahan, J. W.,** Psychosine and sphingosylphosphorylcholine bind to mitochondrial membranes and disrupt their function, in *Lipid Storage Disorders (Biological and Medical Aspects),* Salvayre, R. L., Douste-Blazy, L., and Gatt, S., Eds., Plenum Press, New York, in press.
99. **Suzuki, K. and Suzuki, Y.,** Calactosylceramide lipidosis: globoid cell leukodystrophy (Krabbe's Disease), in *The Metabolic Basis of Inherited Disease,* Stanbury, J. B., Wyngaarden, J. B., Fredrickson, D. S., Golstein, J. L., and Brown, M. S., Eds., McGraw-Hill, New York, 1982, 857.
100. **Lin, Y. N. and Radin, N. S.,** Alternate pathways of cerebroside catabolism, *Lipids,* 8, 732, 1973.
101. **Schneider, P. B. and Kennedy, E. P.,** Metabolism of labelled dihydrosphingomyelin *in vivo, J. Lipid Res.,* 9, 58, 1968.
102. **Aubert-Tulkens, G., Van Hoof, F., and Tulkens, P.,** Gentamicin-induced lysosomal phospholipidosis in cultured rat fibroblasts: quantitative ultrastructural and biochemical study, *Lab. Invest.,* 40, 481, 1979.
103. **Kanfer, J. N., Young, O. M., Shapiro, D., and Brady, R. O.,** The metabolism of sphingomyelin. I. Purification and properties of a sphingomyelin-cleaving enzyme from rat liver tissue, *J. Biol. Chem.,* 240, 1081, 1966.
104. **Spence, M. W., Burgess, J. K., and Sperker, E. R.,** Neutral and acid sphingomyelinases: somatotopographical distribution in human brain and distribution in rat organs: a possible relationship with the dopamine system, *Brain Res.,* 168, 543, 1979.
105. **Weinreb, N. J., Brady, R. O., and Tappel, A. L.,** The lysosomal localization of sphingolipid hydrolases, *Biochim. Biophys. Acta,* 159, 141, 1968.
106. **Fowler, S.,** Lysosomal localization of sphingomyelinase in rat liver, *Biochim. Biophys. Acta,* 191, 481, 1969.
107. **Gatt, S.,** Studies on sphingomyelinase, in *Phospholipids in the Nervous System,* Vol. 1, Horrocks, L. A., Ansell, G. B., and Porcellati, G., Eds., Raven Press, New York, 1982, 181.
108. **Huterer, S., Wherrett, J. T., Poulos, A., and Callahan, J. W.,** Deficiency of phospholipase C acting on phosphatidylglycerol in Niemann-Pick disease, *Neurology,* 33, 67, 1983.
109. **Gal, A. E., Brady, R. O., Hibbert, S. R., and Pentchev, P. G.,** A practical chromogenic procedure for the detection of homozygotes and heterozygous carriers of Niemann-Pick disease, *N. Engl. J. Med.,* 293, 632, 1975.

110. **Gatt, S., Dinur, T., and Barenholz, Y.,** A fluorometric determination of sphingomyelinase by use of fluorescent derivatives of sphingomyelin, and its application to diagnosis of Niemann-Pick disease, *Clin. Chem.,* 26, 93, 1980.
111. **Levade, T., Salvayre, R., Bes, J.-C., Nezri, M., and Douste-Blazy, L.,** New tools for the study of Niemann-Pick disease: analogues of natural substrate and Epstein-Barr virus-transformed lymphoid cell lines, *Pediatr. Res.,* 19, 153, 1985.
112. **Jones, C. S., Davidson, D. J., and Callahan, J. W.,** Complex kinetics of bis(4-methylumbelliferyl)phosphate and hexadecanoyl(nitrophenyl)phosphorylcholine hydrolysis by purified sphingomyelinase in the presence of Triton X-100, *Biochim. Biophys. Acta,* 701, 261, 1982.
113. **Christomanou, H. and Kleinschmidt, T.,** Isolation of two forms of an activator protein for the enzymic sphingomyelin degradation from human Gaucher spleen, *Biol. Chem. Hoppe Seyler,* 366, 245, 1985.
114. **Poulos, A., Ranieri, E., Shankaran, P., and Callahan, J. W.,** Studies on the activation of sphingomyelinase activity in Niemann-Pick type A, B, and C fibroblasts: enzymological differentiation of types A and B, *Pediatr. Res.,* 18, 1088, 1984.
115. **Callahan, J. W., Jones, C. S., Davidson, D. J., and Shankaran, P.,** The active site of lysosomal sphingomyelinase: evidence for the involvement of hydrophobic and ionic groups, *J. Neurosci. Res.,* 10, 151, 1983.
116. **Jones, C. S., Shankaran, P., and Callahan, J. W.,** Purification of sphingomyelinase to apparent homogeneity by using hydrophobic chromatography, *Biochem. J.,* 195, 373, 1981.
117. **Rousson, R., Vanier, M.-T., and Louisot, P.,** Chromatofocusing of purified placental sphingomyelinase, *Biochimie,* 65, 115, 1983.
118. **Sakuragawa, N.,** Acid sphingomyelinase of human placenta: purification, properties, and 125iodine labeling, *J. Biochem. (Tokyo),* 92, 637, 1982.
119. **Pentchev, P. G., Brady, R. O., Gal, A. E., and Hibbert, S. R.,** The isolation and characterization of sphingomyelinase from human placental tissue, *Biochim. Biophys. Acta,* 488, 312, 1977.
120. **Yamanaka, T. and Suzuki, K.,** Acid sphingomyelinase of human brain: purification to homogeneity, *J. Neurochem.,* 38, 1753, 1982.
121. **Weitz, G., Quintern, L. E., Schram, A. W., Barranger, J. A., Tager, J. M., and Sandhoff, K.,** Purification and properties of acid sphingomyelinase from human urine, in *Enzymes of Lipid Metabolism II,* Freysz, L., Dreyfus, H., Massarelli, R., and Gatt, S., Eds., Plenum Press, New York, 1986, 261.
122. **Gatt, S. and Gottesdiner, T.,** Solubilization of sphingomyelinase by isotonic extraction of rat brain lysosomes, *J. Neurochem.,* 26, 421, 1976.
123. **Weitz, G., Lindl, T., Hinrichs, U., and Sandhoff, K.,** Release of sphingomyelin phosphodiesterase (acid sphingomyelinase) by ammonium chloride from CL 1D mouse L-cells and human fibroblasts and partial purification and characterization of the exported enzymes, *Hoppe-Seyler's Z. Physiol. Chem.,* 364, 863, 1983.
124. **Besley, G. T. N., Hoogeboom, A. J. M., Hoogeveen, A., Kleimjer, W. J., and Galjaard, H.,** Somatic cell hybridisation studies showing different gene mutations in Niemann-Pick variants, *Hum. Genet.,* 54, 409, 1980.
125. **Jobb, E. and Callahan, J. W.,** The subunit of human sphingomyelinase is not the same size in all tissues: studies with a polyclonal rabbit serum, *J. Inherited Metab. Dis.,* 10 (Suppl. 2), 326, 1987.
126. **Rousson, R., Vanier, M. T., and Louisot, P.,** Immunological studies on acidic sphingomyelinase, in *Enzymes of Lipid Metabolism II,* Freysz, L., Dreyfus, H., Massarelli, R., and Gatt, S., Eds., Plenum Press, New York, 1986, 273.
127. **Hostetler, K. Y. and Yazaki, P. J.,** The subcellular localization of neutral sphingomyelinase in rat liver, *J. Lipid Res.,* 20, 456, 1979.
128. **Bartolf, M. and Franson, R. C.,** Characterization and localization of neutral sphingomyelinase in bovine adrenal medulla, *J. Lipid Res.,* 26, 57, 1986.
129. **Ghosh, P. and Chatterjee, S.,** Effects of gentamicin on sphingomyelinase activity in cultured human renal proximal tubular cells, *J. Biol. Chem.,* 262, 12550, 1987.
130. **Gatt, S.,** Magnesium-dependent sphingomyelinase, *Biochem. Biophys. Res. Commun.,* 68, 235, 1976.
131. **Levade, T., Salvayre, R., and Douste-Blazy, L.,** Sphingomyelinases and Niemann-Pick Disease, *J. Clin. Chem. Clin. Biochem.,* 24, 205, 1986.
132. **Spence, M. W., Wakkary, J., Clarke, J. T. R., and Cook, H. W.,** Localization of neutral magnesium-stimulated sphingomyelinase in plasma membrane of cultured neuroblastoma cells, *Biochim. Biophys. Acta,* 719, 162, 1982.
133. **Mohan Das, D. V., Cook, H. W., and Spence, M. W.,** Evidence that neutral sphingomyelinase of cultured murine neuroblastoma cells is oriented externally on the plasma membrane, *Biochim. Biophys. Acta,* 777, 339, 1984.

134. **Spence, M. W. and Burgess, J. K.,** Acid and neutral sphingomyelinases of rat brain: activity in developing brain and regional distribution in adult brain, *J. Neurochem.*, 30, 917, 1978.
135. **Spence, M. W., Burgess, J. K., Sperker, E. R., Hamed, L., and Murphy, M. G.,** Neutral sphingomyelinases of brain, in *Lysosomes and Lysosomal Storage Diseases*, Callahan, J. W. and Lowden, J. A., Eds., Raven Press, New York, 1981, 219.
136. **Sperker, E. R. and Spence, M. W.,** Neutral and acid sphingomyelinases of rat brain: somatotopographical distribution and activity following experimental manipulation of the dopaminergic system *in vivo, J. Neurochem.*, 40, 1182, 1983.
137. **Guy, N. C., Clarke, J. T. R, Spence, M. W., and Cook, H. W.,** Stimulation of neutral, magnesium-stimulated sphingomyelinase activity in the neurohypophysis of the rat by hypertonic saline ingestion, *Brain Res. Bull.*, 10, 603, 1983.
138. **Pellkofer, R. and Sandhoff, K.,** Halothane increases membrane fluidity and stimulates sphingomyelin degradation by membrane-bound neutral sphingomyelinase of synaptosomal plasma membranes from calf brain already at clinical concentrations, *J. Neurochem.*, 34, 988, 1980.
139. **Mooibroek, M. J., Cook, H. W., Clarke, J. T., and Spence, M. W.,** Catabolism of exogenous and endogenous sphingomyelin and phosphatidylcholine by homogenates and subcellular fractions of cultured neuroblastoma cells. Effects of anesthetics., *J. Neurochem.*, 44, 1551, 1985.
140. **Ramachandran, C. K., Murray, D. K., and Nelson, D. H.,** Increased activity of neutral sphingomyelinase in dexamethasone treated 3T3-L1 cells, *FASEB J.*, 2, A581, 1988.
141. **Petkova, D. H., Momchilova, A. B., and Koumanov, K. S.,** Phospholipid dependence of the neutral sphingomyelinase in rat liver plasma membranes, *Biochimie*, 68, 1195, 1986.
142. **Tamiya-Koizumi, K. and Koima, K.,** Activation of magnesium-dependent, neutral sphingomyelinase by phosphatidylserine, *J. Biochem. (Tokyo)*, 99, 1803, 1986.
143. **Levade, T., Potier, M., Salvayre, R., and Douste-Blazy, L.,** Molecular weight of human brain neutral sphingomyelinase determined *in situ* by the radiation inactivation method, *J. Neurochem.*, 45, 630, 1985.
144. **Levade, T., Salvayre, R., Potier, M., and Douste-Blazy, L.,** Interindividual heterogeneity of molecular weight of human brain neutral sphingomyelinase determined by radiation inactivation method, *Neurochem. Res.*, 11, 1131, 1986.
145. **Noguchi, S., Sakuragawa, N., and Arima, M.,** Partial purification and characterization of rat brain microsomal neutral sphingomyelinase, *J. Neurochem.*, 41, S14, 1983.
146. **Spence, M. W. and Cook, H. W.,** A new Zn^{++}-stimulated sphingomyelinase in fetal bovine serum, *FASEB J.*, 2, A1416, 1988.
147. **Hirshfield, D. and Loyter, A.,** Sphingomyelinase of chicken erythrocyte membranes, *Arch. Biochem. Biophys.*, 167, 186, 1975.
148. **Record, M., Loyter, A., and Gatt, S.,** Utilization of membranous lipid substrates by membranous enzymes. Hydrolysis of sphingomyelin in erythrocyte 'ghosts' and liposomes by the membranous sphingomyelinase of chicken erythrocyte 'ghosts', *Biochem. J.*, 187, 115, 1980.
149. **Wadstrom, T. and Mollby, R.,** Studies on extracellular proteins from Staphylococcus aureus. VI. Production and purification of beta-haemolysin in large scale, *Biochim. Biophys. Acta*, 242, 288, 1971.
150. **Wadstron, T. and Mollby, R.,** Studies on extracellular proteins from *Staphylococcus aureus*. VII. Studies on beta-haemolysin, *Biochim. Biophys. Acta*, 242, 308, 1971.
151. **Pastan, I., Macchia, V., and Katzen, R.,** A phospholipase specific for sphingomyelin from *Clostridium perfringens, J. Biol. Chem.*, 243, 3750, 1968.
152. **Otnaess, A.-B.,** The hydrolysis of sphingomyelin by phospholipase C from *Bacillus cereus, FEBS Lett.*, 114, 202, 1980.
153. **Shoyama, Y. and Kishimoto, Y.,** *In vivo* conversion of 3-ketoceramide to ceramide in rat liver, *Biochem. Biophys. Res. Commun.*, 70, 1035, 1976.
154. **Okabe, H. and Kishimoto, Y.,** *In vivo* metabolism of ceramides in rat brain: fatty acid replacement and esterification of ceramide, *J. Biol. Chem.*, 252, 7068, 1977.
155. **Sutrina, S. L. and Chen, W. W.,** Lysosomal involvement in cellular turnover of plasma membrane sphingomyelin, *Biochim. Biophys. Acta*, 793, 169, 1984.
156. **Hirschberg, C. B., Kisic, A., and Schroepfer, G. J., Jr.,** Enzymatic formation of dihydrosphingosine 1-phosphate, *J. Biol. Chem.*, 245, 3084, 1970.
157. **Stoffel, W., Assmann, G., and Binczek, E.,** Enzymatic synthesis of 1-phosphate esters of 4t-sphingenine (sphingosine), sphinganine (dihydrosphingosine), 4-hydroxysphinganine (phytosphingosine) and 3-dehydrosphinganine by erythrocytes, *Hoppe-Seyler's Z. Physiol. Chem.*, 351, 635, 1970.
158. **Stoffel, W., Hellenbroich, B., and Heimann, G.,** Properties and specificities of sphingosine kinase from blood platelets, *Hoppe-Seyler's Z. Physiol. Chem.*, 354, 1311, 1973.

159. **Stoffel, W., LeKim, D., and Sticht, G.,** Metabolism of sphingosine bases. XI. Distribution and properties of dihydrosphingosine-1-phosphate adolase (sphinganine-1-phosphate alkanal-lyase), *Hoppe-Seyler's Z. Physiol. Chem.*, 350, 1233, 1969.
160. **Stoffel, W.,** Sphingosine metabolism and its link to phospholipid biosynthesis, *Mol. Cell. Biochem.*, 1, 147, 1973.
161. **Shimojo, T. and Schroepfer, G. J., Jr.,** Sphingolipid base metabolism. Sphinganine-1-phosphate lyase: identification of ethanolamine-1-phosphate as product, *Biochim. Biophys. Acta*, 431, 433, 1976.

Chapter 12

METABOLISM OF PHOSPHATIDYLCHOLINE IN LUNG

Fred Possmayer

TABLE OF CONTENTS

I. INTRODUCTION

As indicated in the first chapter, initial studies on the biosynthesis of phosphatidylcholine (PC) and other phospholipids were conducted with liver. It was only after the major pathways had been defined that younger scientists, a number of whom had trained under scientists working predominantly with liver and brain, chose to examine the possibility that novel concepts might arise from other mammalian systems. Tissues such as lung were usually examined as part of metabolic "surveys". Still, as early as 1950, Popjack and Beekmans[1] observed that lung possessed an unusual capacity for incorporating labeled precursors into acyl groups. In 1958, Lands[2] noted that neutral triacylglycerols and PC are labeled in lung to a similar degree as in liver with radioactive glycerol, but not with acetate. Even earlier, Thannhauser and his associates[3] had observed the presence of dipalmitoylphosphatidylcholine (DPPC) in beef lung, probably the first isolation of a distinct molecular species of PC from a biological source. The discovery of the surfactant system by Pattle[4] and Clements[5] in the mid-1950s and the observation by Avery and Mead,[6] that the lungs of infants dying from the neonatal respiratory distress syndrome were surfactant-deficient, stimulated interest in the pulmonary system. This led to the recognition of PC in surfactant and DPPC as a critical component in 1961.[7] The first studies on enzymes involved in PC synthesis in developing lung were reported by Artom in 1968.[8] Since that time considerable literature has accumulated on the formation and, to a lesser extent, the degradation of pulmonary phospholipids. This review will focus on special aspects of phospholipid metabolism in lung and identify critical questions which must still be addressed. Due to space considerations, previous reviews will be cited whenever possible.

II. PULMONARY SURFACTANT

A. COMPOSITION

Pulmonary surfactant stabilizes the lung by reducing the surface tension at the air-liquid interface of the alveolar surface. This stabilization involves the adsorption and spreading of surfactant lipids at the surface with the formation of a phospholipid monolayer enriched in DPPC. A number of excellent reviews are available which discuss the biophysical[9-12] and biological[10,12-14] properties of pulmonary surfactant.

Surfactant can be isolated by centrifuging the material obtained by lavaging lungs with saline via the trachea. The composition of bovine surfactant, depicted in Figure 1, is similar to that from human lung and from experimental species.[15,16] The major phospholipid is PC, accounting for approximately 80% of the total phospholipid. Half or slightly more than half of the PC is present as the 1,2-*sn*-dipalmitoyl species. Monoenoic PC accounts for most of the unsaturated species. Surfactant also contains the acidic phospholipids, phosphatidylglycerol (PG) and phosphatidylinositol (PI). PI is the major anionic component in surfactant from fetal lung, but is largely replaced by PG in the adult. The remaining phospholipids are phosphatidylethanolamine (PE), phosphatidylserine (PS), lyso-*bis*-phosphatidic acid (LBPA), sphingomyelin (SM), and lysophosphatidylcholine (LPC).

The presence in surfactant of a major glycoprotein species of approximately 35 kDa[15] was observed in early studies. Surfactant also contains two low molecular weight, hydrophobic proteins which remain associated with the lipids extracted by organic solvents. These three protein families will be referred to as surfactant-associated protein-A (SP-A), SP-B, and SP-C.[17] Several recent reviews have described these proteins in detail.[12,17,18] Only a brief summary of the more salient findings will be included here.

SP-A refers to the protein family which includes the major 35-kDa surfactant glycoprotein. Examination of SP-A from various species reveals two or more electrophoretic bands apparently

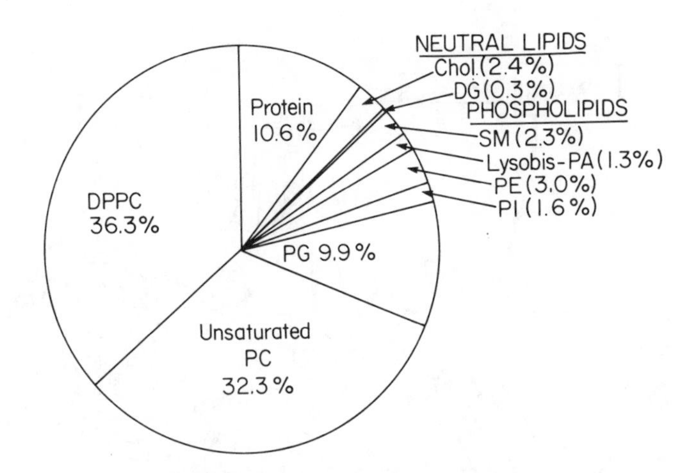

FIGURE 1. Composition of bovine pulmonary surfactant. Chol., cholesterol; DG, diacylglycerol; PA, phosphatidic acid; DPPC, dipalmitoylphosphatidylcholine; PC, phosphatidylcholine; PE, phosphatidylethanolamine; PG, phosphatidylglycero; PF, phosphatidylinositol; SM, sphingomyelin. (From Possmayer, F., in *Phosphatidate Phosphohydrolase,* Vol. 2, Brindley, D. N., Ed., CRC Press, Boca Raton, FL, 1988, 39. With permission.)

arising from differences in the extent of glycosylation.[18] In addition to N-glycosylation, this protein possesses a number of other secondary modifications including sialylation, proline hydroxylation, and glutamate carboxylation.[17,18] SP-A is composed of a hydrophobic signal sequence, a short amino-terminal section followed by two collagen-like sections separated by a single glycine, and a long globular section at the carboxy terminus of the protein.[19] The carboxy portion of SP-A shows extensive sequence homology with a number of animal lectins and can bind mannose. SP-A also binds to lipid.

SP-B is produced as a primary translation product of approximately 41 kDa, glycosylated to approximately 43 kDa, and then proteolytically processed at both the amino and carboxy termini to yield the alveolar forms of approximately 18 kDa nonreduced and 8 kDa reduced.[20,21] Due to the posttranslational clipping, the precise molecular mass cannot be predicted from the cDNAs. Alveolar forms which are 79 amino acids long have been reported in bovine and porcine surfactant.[17]

The cDNAs for human SP-C predict a relatively large primary translation product of approximately 20 kDa.[22,23] The alveolar forms of SP-B migrate with nominal molecular masses of 5 kDa and 10 kDa. The latter band appears to represent a dimer of the smaller protein. Alveolar forms of SP-C which are 35 amino acids long have been reported.[24] This protein is characterized by a long hydrophobic segment which contains six consecutive valine residues.

Initial studies indicated that SP-A promoted the adsorption and spreading of DPPC at the air-liquid interface.[11,15,17,20] However, the resulting decrease in surface tension was small compared to that observed with natural surfactant or its lipid extract. Reconstitution studies have revealed that the essential biophysical properties inherent in surfactant are dependent upon the presence of the low molecular weight hydrophobic proteins SP-B and SP-C. Further studies have concluded that SP-A synergistically promotes the adsorption and spreading of surfactant lipids observed in the presence of SP-B and/or SP-C.[20]

The critical role of the hydrophobic proteins in the biophysical activity of surfactant has been supported by a number of physiological studies revealing that lipid extracts of pulmonary surfactant and reconstituted preparations containing SP-B and SP-C promote lung expansion and survival of prematurely delivered animals.[10-14] The biological effectiveness of lipid extract surfactant has also been documented in clinical trials (see References 13 and 14).

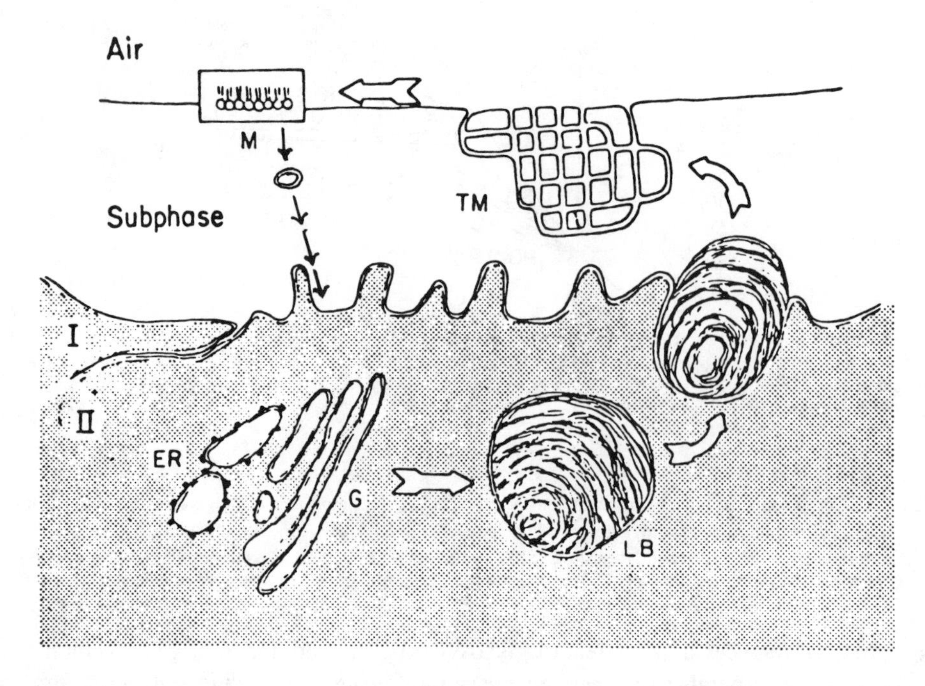

FIGURE 2. Diagrammatic representation of the formation and excretion of pulmonary surfactant. After exocytosis, the lamellar bodies form tubular myelin which is thought to be the source of the monolayer. Only the monolayer can act in the reduction of the surface tension. ER, endoplasmic reticulum; G, Golgi apparatus; LB, lamellar bodies; M, monolayer; TM, tubular myelin. (Modified from Goerke, J., *Biochim. Biophys. Acta,* 344, 241, 1974. With permission.)

B. METABOLISM AND TURNOVER OF PULMONARY SURFACTANT

It is now accepted that pulmonary surfactant is produced in the type II cells of the alveolus which account for 10 to 15% of the total cellular content of the lung.[25] Histochemical and autoradiographic studies indicate that surfactant components are synthesized in the rough endoplasmic reticulum of these cells and assembled into lamellar bodies (see Figure 2).[26] Specialized structures known as multivesicular bodies appear to be involved in the transport of nonlipid components. The lamellar bodies are composed of alternating lipid bilayers separated by aqueous spaces. In some species, including the human, the lamellar bodies resemble multilamellar liposomes. The lamellar bodies are extruded from the type II cells by exocytosis. The limiting membrane of the lamellar bodies fuses with the plasma membrane of the type II cells during this process. The expelled lamellar bodies subsequently undergo a transition to form a unique structure known as tubular myelin. Tubular myelin consists of a regular lattice-like structure composed of a continuous series of elongated tubes apparently formed through the fusion of alternating phospholipid bilayers of the lamellar bodies. It is presently thought that tubular myelin represents an intermediate between the lamellar bodies and the surface monolayer (Figure 2).[11,12,18,27]

It has long been thought that the surface-lining layer loses its more unsaturated components during repeated compressions and becomes enriched in DPPC. This ultimately leads to the loss of DPPC from the surface. A number of recent studies have revealed that DPPC and some of the other phospholipid components of surfactant are taken up and reutilized for surfactant production (Figure 2). (See References 12, 13, and 18 for review.) Type II cells in culture can take up liposomes from the media, and this process is stimulated by both SP-A and the low-molecular-weight hydrophobic proteins. SP-A also inhibits the secretion of surfactant from type II cells. The control of surfactant secretion and our present appreciation of the uptake processes have been extensively reviewed.[12,13,18,28-31]

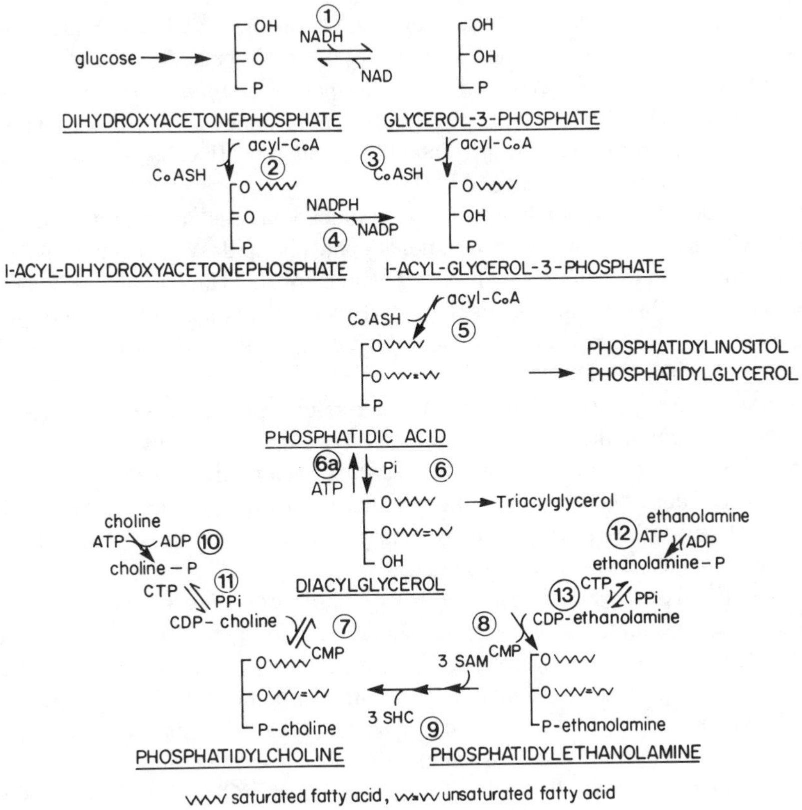

SCHEME I

III. BIOSYNTHESIS OF PHOSPHATIDYLCHOLINE IN LUNG

A. THE *DE NOVO* PATHWAY FOR THE BIOSYNTHESIS OF PHOSPHATIDYLCHOLINE IN LUNG

1. Formation of Diacylglycerol

The initial step in the formation of PC and, in fact, all glycerolipids involves the acylation of either the glycolytic intermediate, DHAP, or its reduced counterpart, *sn*-glycerol-3-phosphate (GP)(Reactions 2 or 3, Scheme I). The primary acylation of GP occurs at the 1-position. The resulting 1-acyl-*sn*-glycerol-3-phosphate can then be acylated at the 2-position (Reaction 5, Scheme I) to yield phosphatidic acid (PA). The DHAP pathway requires reduction of the 1-acyl-DHAP intermediate with NADPH prior to the second acylation (Reaction 5). Studies conducted with microsomes from whole lung and from type II cells have revealed little preference for palmitate over oleate at the 1-position or oleate over palmitate at the 2-position.[32-35]

While the observed lack of specificity for unsaturated fatty acids at the 2-position is consistent with the ability of lung and type II cells to produce DPPC, the surprising lack of acyl specificity at the 1-position indicates that these results must be interpreted cautiously. Recent studies on the nature of the acyl-CoAs in type II cells from rat lung have revealed a high proportion of the palmitoyl species.[29]

Although the microsomal acyltransferase can utilize either substrate, the mitochondrial activity exhibits a preference for GP, and the peroxisomal activity is highly selective for DHAP.[36,37] It has generally been concluded that the endoplasmic reticulum is responsible for the production of PA for surfactant synthesis.[12,16,28,29,38] However, the possibility that in some species, such as the guinea pig, extramicrosomal acyltransferase activity could function in the production of surfactant lipids during development must still be addressed.[39,40]

The peroxisomal activity is required for the formation of alkyl ethers and plasmalogens. Although significant amounts of these glycerolipids have not been reported for whole lung or isolated surfactant, preparations from rabbit lung and type II cells possess the ability to synthesize platelet-activating factor (PAF).[41]

Studies conducted with rabbit lung[42] and type II adenoma cells[32] reveal that both radioactive GP and DHAP can be incorporated into pulmonary phospholipids. Whether the DHAP pathway provides a significant contribution to glycerolipid metabolism in any tissue is still controversial (see References 16, 28, 29, and 38 for review). However, studies with isolated type II cells incubated with [2-^{3}H]glycerol and [^{14}C]glycerol indicate that the glycerol moiety of PC and PG can be derived from GP or DHAP.[43]

In addition to the acylation of the primary water-soluble precursors GP and DHAP, PA can also be produced through the phosphorylation of diacyl-*sn*-glycerols by diacylglycerol kinase (Reaction 6a, Scheme 1). This reaction, which represents a reutilization rather than a *de novo* pathway, has been documented in cytosolic and microsomal fractions from rat lung.[44] In the rat, the specific activity of this enzyme displays a sharp, but transient increase prior to birth, but the significance of this observation remains unclear.

PA is hydrolyzed by phosphatidic acid phosphohydrolase (PAPase) to provide the 1,2 diacyl-*sn*-glycerol moiety required for the *de novo* generation of PC.[16,28,38,45] Early studies focused on the Mg^{2+}-independent phosphohydrolase activities primarily associated with the microsomal fraction, but present in all fractions including the lamellar bodies.[45,47] The Mg^{2+}-dependent, but not the Mg^{2+}-independent, PAPase activity present in lung microsomes can be removed by washing with salt solution. Salt-washed microsomes lost their ability to hydrolyze endogenously generated PA to diacylglycerol or to produce PC from radioactive GP. These properties could be restored by adding either the microsomal washings or partially purified cytosolic Mg^{2+}-dependent PAPase.[48] These observations are consistent with the view that the Mg^{2+}-dependent PAPase activities present in lung microsomes and cytosolic fractions are specifically involved in glycerolipid metabolism. The Mg^{2+}-independent PAPase activities appear to represent a number of nonspecific phosphohydrolase activities whose function(s) remain(s) unknown.[38]

The observation that the Mg^{2+}-dependent PAPase activities observed in liver and lung cytosolic and microsomal preparations shared a number of properties suggested that this enzyme might be controlled through translocation of cytosolic enzyme to the endoplasmic reticulum. Exposure of A549 cells, a human type II cell line derived from a benign tumor, to free fatty acids results in the translocation of Mg^{2+}-dependent PAPase from the cytosol to the endoplasmic reticulum.[49]

2. The CDP-Choline Pathway

Diacylglycerols can be converted to PC through the transfer of a cholinephosphate moiety from CDP-choline (Reaction 7, Scheme 1) or to PE via transfer of ethanolaminephosphate from CDP-ethanolamine (Reaction 8). PC can also be generated via the stepwise methylation of PE (Reaction 9, Scheme 1), but this pathway appears to have a low activity in lung.

Cholinephosphotransferase (CPT), the enzyme responsible for the terminal reaction in the formation of PC in lung, has been extensively examined in lung.[16,28,50] Although this Mg^{2+}-requiring enzyme has been considered a marker for the endoplasmic reticulum, recent studies[51,52] indicate that in guinea pig lung, the outer mitochondrial membrane may possess significant activity. Early studies suggested that pulmonary CPT discriminated against disaturated diacylglycerols.[16,28,50] However, little preference for unsaturated species can be detected with endogenous substrates.[53-56] In addition, exogenous dipalmitin dispersed with PG can be readily utilized by CPT.[57]

The CDP-choline utilized for the formation of PC arises from choline kinase (CK) and cholinephosphate cytidylyltransferase (CPCT) reactions (Reactions 10 and 11, Scheme 1). A similar ethanolamine pathway provides CDP-ethanolamine for the formation of PE (Reactions 12 and 13). Choline can be transported from the extracellular fluid, but this has not been extensively studied in lung. CK, a soluble enzyme (Chapter 2),[58] appears to be enriched in type II cells.[59] As discussed in Chapter 3, CPCT is present in both the microsomal and cytosolic fractions. In adult lung the cytosolic activity is primarily present in a high molecular weight (H-M_r) form.[60] In contrast, cytosols isolated from fetal lungs contain CPCT in a low molecular weight (L-M_r) form. The L-M_r form can be activated by the addition of acidic phospholipids or lysoPE. Only a small increment in activity is observed with the H-M_r form of cytidylyltransferase, indicating that this form is already activated by its association with lipid. CPCT from fetal lung can also be activated by free fatty acids.[61] The increase in the incorporation of radioactive choline into PC which occurs in rat lung after delivery of premature fetuses is associated with the translocation of CPCT activity from the cytosol to the endoplasmic reticulum.[62] However, as discussed subsequently, studies with term fetuses and with cells in culture indicate that, in lung, microsomal CPCT may be activated directly.[63]

B. THE FORMATION OF DPPC IN LUNG

The disaturated PC in lung can be synthesized partly via the *de novo* pathway, but also through a progressive conversion of unsaturated PCs generated by the *de novo* pathway into their disaturated counterparts.[56,64-66] Calculations based on studies with type II cells isolated from rat lung suggest approximately three quarters of the DPPC produced *in vitro* arises through remodeling.[65] Since palmitate is readily incorporated into the 1-position of PA during *de novo* synthesis, the remodeling pathways need to introduce a second palmitate at the 2-position. Three auxiliary pathways have been proposed:

1. A deacylation-reacylation cycle in which the unsaturated fatty acid is removed from the 2 position of PC by phospholipase A_2 and the resulting 1-palmitoyl-lysoPC is reacylated with palmitoyl-CoA
2. A deacylation-transacylation cycle in which, following phospholipase A_2, the palmitate at the 1-position of one molecule of 1-palmitoyl-lysoPC is transferred to the 2-position of another lysoPC
3. A novel acyl exchange reaction in which either free or esterified palmitate is enzymatically exchanged for the unsaturated fatty acid at the 2-position of PC

Although supported by early studies, it has become evident that the transacylation or "Marinetti pathway" does not function *in vivo* (see References 16, 28, 29, and 31). The most striking argument against this mechanism arose from studies from van den Bosch's group[67] which demonstrated the lack of stereospecificity of the transacylation reaction *in vitro* and the absence of any significant incorporation of the nonphysiological 3-palmitoyl-*sn*-lysoPC isomer into pulmonary PC *in vivo*. These experiments support the view that the transacylase reaction represents a minor enzymatic activity of lysoPC acylhydrolase.[68]

The deacylation-reacylation, or "Land's cycle", can account for many of the observations made with fractions from whole lung or type II cells. The relative increase in lysoPC acyltransferase observed with type II cells compared to whole lung[69,70] and the higher levels of saturated than unsaturated acyl-CoAs in whole lung[71] and type II cells[29] support the potential importance of the acyltransferase activity in the production of DPPC. Incubation of microsomes isolated from whole lung or type II cells with mixtures of acyl-CoAs at the relatively low levels anticipated *in vivo* results in preferential acylation by palmitoyl-CoA.[33,70,71]

Recent evidence suggests that DPPC may be formed through an exchange reaction involving the unsaturated fatty acyl groups at the 2-position of PC and free CoASH through a reversal of

1-acyl-lysoPC acyltransferase.[64,72,73] Since, as indicated above, this enzyme selectively utilizes palmitoyl-CoA in the forward reaction, this exchange reaction would favor the elimination of molecular species of PC with unsaturated fatty acids at the 2-position and the accumulation of DPPC. LysoPE, and to a lesser extent lysoPG and lysoPS, acted as acyl acceptors for the unsaturated acyl-CoAs generated by the backward reaction.

The acyl exchange mechanism differs from the transacylation and reacylation mechanisms in not requiring a PC-specific phospholipase A_2. However, an A_2 activity is required for generating acceptors for the unsaturated acyl-CoAs arising from the 2-position of PC. Phospholipase A_2 activities which could function in remodeling have been documented in microsomal and cytosolic preparations from whole lung.[74-77] A preference for PE over PC has been observed in the absence of detergents.[75,77]

Infante[78] has recently reported that the mitochondrial fractions from a number of tissues, including lung, contain an enzymatic activity which transfers the choline moiety from CDP-choline to GP to yield glycerophosphorylcholine (GPC). Infante[79,80] has speculated that the GPC moiety resulting from this reaction can be acylated to produce DPPC for surfactant. No evidence has been presented for this latter reaction and attempts to demonstrate the acylation of GPC with subcellular fractions from rat lung or type II cells have not been successful.[135]

IV. CONTROL OF PC SYNTHESIS IN LUNG

The synthesis of PC is primarily affected by the concentrations of diacylglycerol and CDP-choline available to CPT and the activity of this enzyme. However, the supply of these substrates will be modified by those factors affecting the individual steps of the Kennedy *de novo* and CDP-choline pathways. The insoluble nature of diacylglycerol implies that topographical features must be incorporated into the supply mechanism.[81,82] Both CPCT and CPT catalyze reversible reactions. In fact, given the presence of diacylglycerol kinase, the flow of diacylglycerol from PA can also be considered a reversible process. These complications will influence the kinetic analysis of the events occurring in whole lung *in vivo* or in isolated type II cells *in vitro*. Due to the absolute requirement for pulmonary surfactant to stabilize the terminal airways at birth, the metabolic adaptations accounting for the large increase in the levels of PC and DPPC in lung during the perinatal period have received considerable attention. This developmental transformation involves two related, but metabolically distinct processes: (1) the conversion from the "immature" nonsurfactant-producing to the "mature" surfactant-producing and -storing fetal lung during the so-called "transition" period and (2) the metabolic adaption to air-breathing which occurs with delivery. The major approaches applied to this problem have been to examine the alterations in precursor incorporation, enzymatic activity *in vitro,* and the pool sizes of intermediates and products which occur either during fetal development or after the induction of pulmonary maturation with one or more of the increasing number of hormones and agents[12,29,31,83,84] reported to affect lung. The present review will attempt to focus on those observations which have contributed to our present perceptions of the underlying mechanisms or which have added to our insight by challenging apparent principles.

A. PC SYNTHESIS IN LUNG DURING FETAL DEVELOPMENT

In the rabbit, the transition period extends from days 25 to 29 of gestation (term 31).[16] The number of prematurely delivered rabbit fetuses that can establish regular breathing increases from 10% on day 27 to 50% on day 28 and essentially 100% on day 29. Fetal pulmonary maturation is accompanied by a marked increase in the level of PC and, to an even greater extent, DPPC. This increase, which continues during the remaining fetal period and in the neonate,[85,86] is accompanied by a striking four- to sixfold increase in the rate of incorporation of choline into PC with slices of fetal rabbit lung.[87]

Surprisingly, the marked increase in choline incorporation into PC in fetal rabbit lung is not

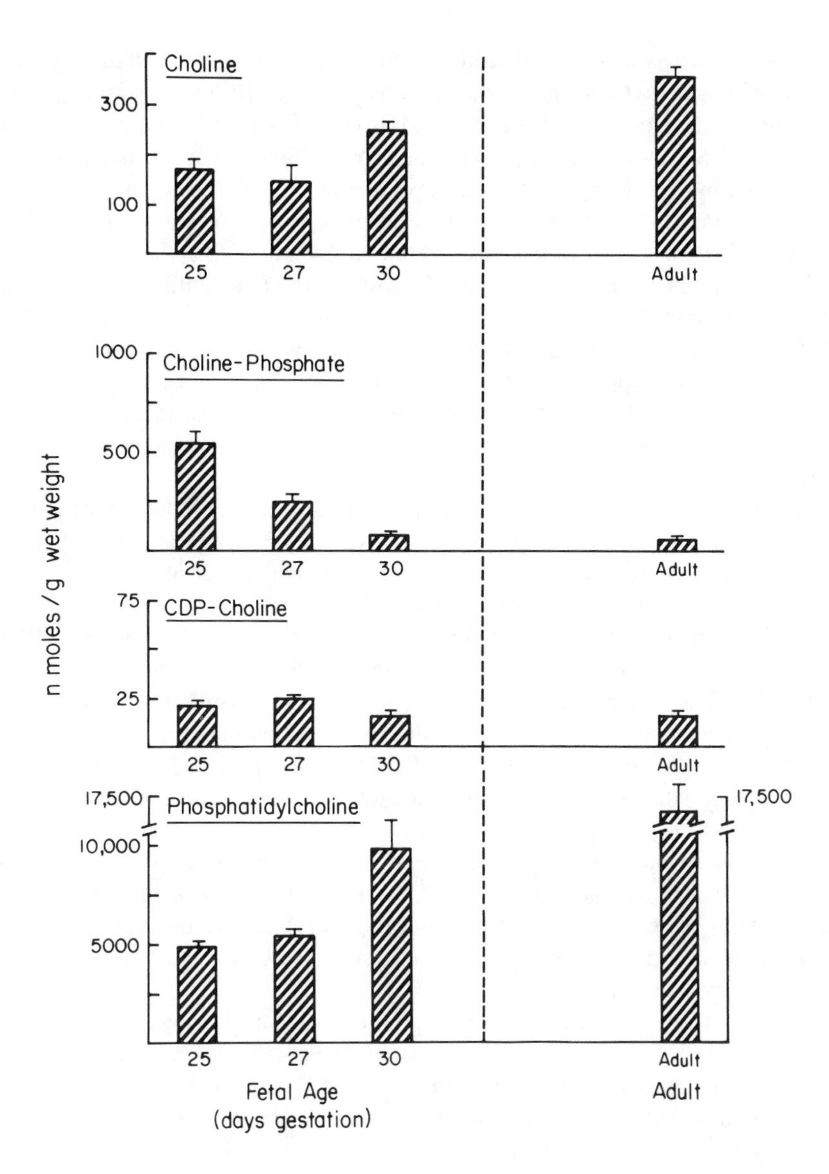

FIGURE 3. Pool sizes of choline and its derivatives in fetal and adult rabbit lung. (Adapted from Tokmajian, S., Haines, D. S. M., and Possmayer, F., *Biochim. Biophys. Acta,* 663, 557, 1981. With permission.)

paralleled by increases in the specific or the total activity of CK, CPCT, or CPT when these enzymes are assayed under optimal conditions *in vitro*.[88-91] A small increase in cytosolic CPCT activity in the absence of lipid activators is sometimes observed at term, but in the author's experience, this increased activity can be variable. A moderate increase has been reported in the specific activity of the Mg^{2+}-dependent PAPase activity associated with the microsomal fraction.[92] No change was observed in the larger cytosolic activity.

The lack of correlation between the increase in the incorporation of choline into PC and the activities of the enzymes involved in PC production prompted examination of the pool sizes of choline and its intermediates in rabbit lung (Figure 3). The pool size of choline remained relatively constant, but the level of cholinephosphate fell dramatically during the transition period and less so after birth, resulting in adult values which are only 10% of those observed on day 25 of gestation.[93] CDP-choline levels increased during the initial part of the transition period,[27] but fell toward term and in the adult.[93,94] The level of CTP fell to half its original value during the transition period and then remained constant.[94] CMP levels exhibit a moderate increase during fetal development. These alterations in the pool sizes of cholinephosphate, CTP, CDP-choline, and CMP indicate a shift toward the forward direction occurs in the equilibrium reaction catalyzed by CPCT during the transition period of fetal development. The marked increase in the labeling of PC, mentioned earlier, reflects the decrease in cholinephosphate levels and represents only a modest increase in the overall production of PC.

The decrease in the concentration of a pathway substrate for a nonequilibrium reaction during an increase in flux through the pathway is considered diagnostic for a regulatory pathway. Calculations based on the equilibrium constant reported for CPCT from rat liver[95] suggest that nonequilibrium conditions apply prior to and, perhaps, after the transition period. The combined decrease in cholinephosphate and CTP provides strong evidence for an enhanced synthesis of CDP-choline due to an increase in CPCT activity, despite the lack of an increase in the activity of this enzyme.

An important role for CPCT is also supported by studies on the induction of pulmonary maturation by estradiol. Treatment of pregnant rabbits on day 25 with estradiol-17β leads to a significant increase in the incorporation of choline into PC with slices of fetal lung on day 26 or 27.[96,97] Extensive examination of the enzymes related to PC synthesis revealed the only alteration was a 50% increase in CPCT activity in the cytosolic fraction. Microsomal CPCT was not affected. Addition of PG or other lipid activators to the cytosols resulted in less of a stimulation of CPCT with treated than control fetuses, so the resulting specific activities were identical. Thus, the increased cytosolic activity due to steroid treatment can be attributed to an enhanced activation rather than an increase in CPCT protein. An overall increase in cytosolic lipids has been observed, but extensive studies failed to identify a change in any specific lipid component.[98,99]

Pool size measurements revealed a small, nonsignificant increase in choline, a marked significant decrease in cholinephosphate, a small but significant elevation in CDP-choline, and a significant increase in PC.[97] The concurrent fall in the pool size of cholinephosphate while the levels of CDP-choline and PC rise suggested the elevation in the cytosolic activity observed *in vitro* reflects activity changes *in vivo*.

The crucial role of CPCT in the increase in PC production is further suggested by studies with fetal rat lung. In this species, the transition from immature to mature lung occurs between the 19th and 20th day of gestation. Pulmonary maturation is accompanied by a 6- to 8-fold increase in the incorporation of choline and a 2 1/2- to 3-fold increase in the incorporation of glycerol into PC between day 19 of gestation and the first day of life before falling to adult levels.[100,101]

Choline levels remain constant between days 19 to 22 of gestation and then rise slightly after birth.[102] Cholinephosphate levels fall during the transition period and more slowly after birth. The levels of CDP-choline also fall. PC levels increase significantly between day 19 and term with a further increase during the neonatal period.

Studies on the activities of the enzymes associated with PC production reveal a small increase in CPT in the whole homogenate and a more marked change in the microsomal activity during the transition period.[100,101] Little alteration was observed in CK.[58,100] Although cytosolic CPCT activity did not increase, elevations in the activity measured with lipid activators and in the microsomal activity have been observed before birth.[100,101] A slight increase has been reported in the specific activity of Mg^{2+}-dependent PAPase during the transition period followed by a fall at term.[103] The levels of diacylglycerol present in fetal rat lung exhibit an even greater relative increase than that of PC between 19 to 22 days gestation.[55,102]

A regulatory role for CPCT is further indicated by studies with isolated type II cells. The pool sizes of choline and CDP-choline in type II cells isolated from fetal rat lung of 19 days gestation are approximately 10-fold higher than in whole lung.[104,105] Cholinephosphate levels were 20-fold greater, but PC was only slightly elevated. Comparison between type II cells from fetal and adult lung revealed that adult cells contained a 10-fold increase in CDP-choline, a slight increase in PC, but cholinephosphate levels were similar. It was suggested that the high levels of CDP-choline could reflect a reversal of CPT. Kinetic analysis of the pool sizes indicated that both CK and CPCT are displaced from equilibrium and hence may catalyze regulatory steps.[105]

It is interesting to note that adult type II cells are approximately 25-fold enriched over whole lung with respect to CTP, but only 2-fold with respect to ATP.[105] This suggests a highly specialized adaptation of these cells with respect to the production of PC and other phospholipids. Surprisingly, even higher levels of CMP were noted. However, because pyrophosphate levels are less than 10 μmol/g lung, the CPCT step is held at a marked disequilibrium, indicating CDP-choline is rapidly utilized.

Administration of glucocorticoids to pregnant rats at day 18 or 19 of gestation leads to a marked increase in the incorporation of choline into PC and disaturated PC in lung slices. A specific increase is observed in the activity of CPCT *in vitro,* while CK and CPT activities remain unaltered.[106-108] Although the investigations with whole lung suggested an increase in cytosolic CPCT activity, studies with fetal type II cells isolated after maternal treatment with dexamethasone or exposure of organotypic cultures (containing fibroblasts as well as type II cells) to cortisol found the elevation was limited to the microsomal fraction.[108] Since the cytosolic activity was unaltered, translocation was not indicated. Studies by Smith and colleagues (reviewed in Reference 83) have demonstrated that this glucocorticoid effect on PC formation does not arise from a direct effect of the steroid on type II cells, but rather through the stimulation of the production by fibroblasts of a fibroblast-pneumonocyte factor (FPF) which subsequently acts on the fetal type II cell. Exposure of fetal type II cells to FPF resulted in a doubling of CPCT activity and a 1.5-fold increase in the labeling of disaturated PC from choline.[107] Pulse-chase experiments indicated the enhanced labeling of PC arose from an increased utilization of cholinephosphate. The effect of FPF on choline incorporation was evident within 60 min after treatment, suggesting an activation process rather than increased CPCT synthesis.

B. CONTROL OF PC SYNTHESIS IN LUNG DURING THE NEONATAL PERIOD

Neonatal survival is dependent upon the presence of sufficient surfactant in the terminal airways to prevent alveolar collapse. Surfactant secretion is initiated prior to birth, but is accentuated by labor[89] and continues after delivery, resulting in the presence of approximately two thirds of the endogenous surfactant pool in the extracellular spaces.[13,86] Birth is also accompanied by an overall increase in total lung PC and DPPC.[13,86,87]

Although CPT activity remains relatively constant during fetal life, some authors have reported postnatal increases in the rabbit[88,89,91] and rat.[101] Increases in the specific activity of this enzyme in rabbit lung are observed with labor[89] and after fetal injection with saline.[109] The activity of the Mg^{2+}-independent PAPase activity in rat microsomes, but not cytosol, is elevated after birth. However, little effect is noted in either fraction with the Mg^{2+}-dependent activities thought to be involved in glycerolipid metabolism.[103] A number of workers have reported a marked increase in CPCT activity in cytosols from rabbit[88,91] and rat[62,63] lung during the postnatal period.

Weinhold and associates[62,63] observed that there was an increase in the incorporation of choline into PC and DPPC during the first 3 h after delivery of premature, term, and postmature rat fetuses. Only minor alterations were observed in the specific and total activities of CK or CPT. The specific activity of CPCT in the microsomal fraction prepared by centrifugation of the postmitochondrial supernatant at 250,000 *g* for 60 min was increased. This fraction includes the H-M_r but not the L-M_r form of CPCT. With prematurely delivered rats at 20 days gestation, there was a corresponding decrease in CPCT activity in the cytosol, indicating translocation of enzyme to the microsomes. However, no decrease was observed in the cytosolic activity when delivery occurred closer to term or with postmature fetuses. A correlation was observed between the levels of free fatty acid in the microsomal fraction and CPCT activity during the neonatal period.[63] The authors concluded that the free fatty acids arose from the elevation in the levels of free fatty acids and triacylglycerols in serum which occurs in the neonatal period.[110]

Short-term exposure of type II cells from fetal[111] or adult lung[112] to free fatty acids increased the incorporation of choline into PC and disaturated PC by accelerating the utilization of cholinephosphate. The increased labeling of PC was accompanied by an increase in CPCT activity in the whole homogenates. With fetal type II cells this increase was only associated with the particulate fraction, but with adult cells a slight increase in CPCT activity was also observed in the cytosol. The absence of any decrease in the cytosolic activity argued against translocation. Interestingly, although palmitate is rather ineffective in enhancing microsomal CPCT activity or promoting translocation of this enzyme *in vitro*,[63] both saturated and unsaturated fatty acids affected CPCT with the intact type II cells.[111,112] When taken together with the effects of air-breathing, these experiments indicate that, with lung, free fatty acids preferentially stimulate the cytidylyltransferase associated with the microsomes rather than promoting translocation of cytosolic enzyme. It may be that with systems such as hepatocytes and HeLa cells where most of the CPCT is cytosolic, increases in translocation are more apparent than stimulation of the microsomal activity, whereas with systems such as adult lung and fetal or adult type II cells, where most of the CPCT is membrane-associated, the stimulatory effect on the microsomal enzyme is more easily detected. Alternatively, the fatty acid effects could be dose dependent with activation of microsomal CPCT being susceptible at lower cellular concentrations of free fatty acid.

C. OTHER FACTORS INFLUENCING PC SYNTHESIS IN LUNG

It has been proposed (Chapter 3) that phosphorylation of CPCT results in a decreased affinity for the endoplasmic reticulum, but free fatty acids can override the inhibitory effect of phosphorylation. Exposure of microsomal and cytosolic preparations from fetal and adult lung to phosphorylating conditions produced a sharp reduction in CPCT activity.[113] However, cAMP and phosphodiesterase inhibitors promote rather than inhibit the incorporation of choline into PC with explants of fetal lung[114] and with type II cells in culture.[115,116] In addition, stimulation of β-adrenergic receptors and purinoreceptors, which elevate intracellular cAMP, promotes surfactant secretion in the neonatal and adult periods without any indication of a suppression of surfactant synthesis.[30,31,84] Hence, it appears unlikely that phosphorylation of CPCT by a cAMP-dependent mechanism restricts PC production in lung.

Although initially thought to be of relatively minor significance, it has become apparent that fetal lung possesses the capacity to synthesize considerable amounts of fatty acids, particularly palmitate. Lactate, which is relatively abundant in fetal serum, is utilized preferentially to glucose *in utero*.[117] Degradation of glycogen in type II cells during the transition period may also provide substrate.[12,29,31,84] Fatty acid synthetase activity and fatty acid synthesis monitored with ^3H_2O show parallel increases during the transition period and fall in the neonate.[118] The ability of fetal rat lung explants to incorporate ^3H_2O parallels the increase in the incorporation of choline into disaturated PC.[119] Fatty acid synthesis is promoted by glucocorticoids, but inhibited by thyroxine.[119,120] The possibility should be examined that the increased PC synthesis and the elevated CPCT observed in fetal type II cells after hormonal induction[107,108] could be related to enhanced fatty acid synthesis.

In contrast to liver, fatty acid synthesis by fetal rat lung explants was resistant to short-term inhibition by cAMP and was stimulated by long-term exposure to this cyclic nucleotide.[119] Thus, although acetyl-CoA carboxylase and fatty acid synthetase are inhibited under phosphorylating conditions *in vitro*,[121] little effect is apparent *in situ*.[119,122] High levels of palmitate can inhibit fatty acid synthesis by type II cells in culture,[123,124] but levels of exogenous palmitate similar to those anticipated *in vivo* had no effect on fetal rat explants, suggesting pulmonary PC synthesis may be highly dependent upon endogenous production of palmitate.[119,122]

Little is known about the effect of the uptake of surfactant lipids from the alveolus for

Little is known about the effect of the uptake of surfactant lipids from the alveolus for reutilization on the control of surfactant synthesis. Treatment of type II cells from adult rats with surfactant resulted in an inhibition of the incorporation of choline into PC and of acetate and palmitate into all phospholipids.[125-127] The activity of DHAP acyltransferase was reduced in treated type II cells, but glycerophosphate acyltransferase (GPAT) was not affected. The low-molecular-weight, hydrophobic apoproteins associated with pulmonary surfactant, rather than the lipids, appear to be responsible for this effect. These results suggest a continued production of PA is required to support the synthesis of PC and other lipids in type II cells. Treatment of isolated type II cells with ozone leads to a decrease in the incorporation of acetate, palmitate, glucose, and glycerol into PC, but choline incorporation is not affected.[59] GPAT activity was specifically inhibited. Interestingly, pulmonary GPAT activity is enhanced during the increased phospholipid production which can follow exposure of rat to hyperoxia[128] or nitrogen dioxide.[129]

Incubation of adult rat type II cells with PG or other acidic phospholipids resulted in a marked increase in the rate of incorporation of choline and glycerol into PC.[130] This effect, which appeared specific for PC, was largely limited to unsaturated species. CPCT activity in type II cell sonicates was specifically elevated.[131] Since this elevation was noted in both the presence and absence of lipid activators, the increase cannot be explained by an alteration in the proportion of enzyme activated by lipids. The effect was detected within 60 min, indicating a different activation process, such as dephosphorylation, could be involved. In addition to PG, a number of 1,2-*sn*-diacylglycerols and 1-oleoyl-2-acetyl, glycerol also promoted the incorporation of choline into PC and an elevation in CPCT activity.[132] The effects of PG and diacylglycerols or 1-oleoyl-2-acetyl, glycerol were not additive, implying a common mechanism.

Bleasdale and co-workers observed that CPCT activity can be increased reversibly by depleting the choline levels in type II cells through treatment with choline oxidase and catalase.[126,133] The increase in CPCT was observed in both the presence and absence of lipid activators, and the proportion of enzyme in the cytosol was elevated. Under these conditions, choline supplementation increased the incorporation of glycerol into PC and PG.[133] Batenburg et al.[134] observed that the addition of choline to type II cells promoted an increase in the labeling of PC, PG, and PI from glycerol. These results are consistent with the view that the biosynthesis of lipids for pulmonary surfactant may be coordinated.

V. CONCLUSION

This chapter has summarized the present status of our understanding of the control of the synthesis of PC and DPPC in lung. While differences exist between the mechanisms regulating pulmonary PC metabolism and those reported for other systems, the author has the impression that the number of parallels is increasing more rapidly than the number of contrasts. Both are contributing to our appreciation. The availability of methods for isolating type II cells from fetal and adult lung has provided the means for a more focused examination of lipid metabolism. This approach has clarified a number of phenomena previously observed in whole lung and led to new concepts such as FPF. However, so far, few investigators have attempted to establish whether the lipids being synthesized are intended for surfactant. Future studies with type II cells will need to address the nature of the processes involved in the assembly of surfactant lipid and proteins into lamellar bodies.

ACKNOWLEDGMENTS

The author would like to thank Mr. Mark Quirie and Ms. Amanda Cockshutt for helpful comments. Ms. Barbara McDougall and Mrs. Nancy Wilson provided editorial services. This work was supported by grants from the Medical Research Council of Canada and the Ontario Thoracic Society.

REFERENCES

1. **Popjak, G. and Beeckmans, M. L.,** Extrahepatic lipid synthesis, *Chem. Phys. Lipids,* 9, 51, 1950.
2. **Lands, W. E. M.,** Metabolism of glycerides: a comparison of lecithin and triglyceride synthesis, *J. Biol. Chem.,* 231, 883, 1958.
3. **Thannhauser, S. J., Benotti, J., and Boncoddo, N. F.,** Isolation and properties of hydrolecithin (dipalmityl lecithin) from lung, its occurrence in the sphingomyelin fraction of animal tissues, *J. Biol. Chem.,* 166, 669, 1946.
4. **Pattle, R. E.,** Properties, function and origin of the alveolar lining layer, *Nature (London),* 175, 1125, 1955.
5. **Clements, J. A.,** Surface tension of lung extracts, *Proc. Soc. Exp. Biol. Med.,* 95, 170, 1957.
6. **Avery, M. E. and Mead, J.,** Surface properties in relation to atelectasis and hyaline membrane disease, *Am. J. Dis. Child.,* 97, 517, 1959.
7. **Klaus, M. H., Clements, J. A., and Havel, R. J.,** Composition of surface active material isolated from beef lung, *Proc. Natl. Acad. Sci. U.S.A.,* 47, 1858, 1961.
8. **Artom, C.,** Enzymes for the synthesis of lecithin from choline in tissues of developing rats, *Fed. Proc. Fed. Am. Soc. Exp. Biol.,* 27, 457, 1968.
9. **Notter, R. H.,** Surface chemistry of pulmonary surfactant: the role of individual components, in *Pulmonary Surfactant,* Robertson, B., Van Golde, L. M. G., and Batenburg, J. J., Eds., Elsevier, Amsterdam, 1984, 17.
10. **Possmayer, F., Yu, S.-H., Weber, J. M., and Harding, P. G. R.,** Pulmonary surfactant, *Can. J. Biochem. Cell Biol.,* 62, 1121, 1984.
11. **Goerke, J. and Clements, J. A.,** Alveolar surface tension and lung surfactant, in *Handbook of Physiology: The Respiratory System: Mechanics of Breathing,* Macklem, P. T. and Mead, J., Eds., American Physiological Society, Bethesda, MD, 1986, 247.
12. **Van Golde, L. M. G., Batenburg, J. J., and Robertson, B.,** The pulmonary surfactant system: biochemical aspects and functional significance, *Physiol. Rev.,* 68, 374, 1988.
13. **Jobe, A. and Ikegami, M.,** Surfactant for the treatment of respiratory distress syndrome, *Am. Rev. Respir. Dis.,* 136, 1256, 1987.
14. **Robertson, B. and Lachmann, B.,** Experimental evaluation of surfactants for replacement therapy, *Exp. Lung Res.,* 14, 279, 1988.
15. **King, R. J.,** Isolation and chemical composition of pulmonary surfactant, in *Pulmonary Surfactant,* Robertson, B., Van Golde, L. M. G., and Batenburg, J. J., Eds., Elsevier, Amsterdam, 1984, 1.
16. **Possmayer, F.,** Biochemistry of pulmonary surfactant during fetal development and in the perinatal period, in *Pulmonary Surfactant,* Robertson, B., Van Golde, L. M. G., and Batenburg, J. J., Eds., Elsevier, Amsterdam, 1984, 295.
17. **Possmayer, F.,** A proposed nomenclature for pulmonary surfactant-associated proteins, *Am. Rev. Respir. Dis.,* 138, 990, 1988.
18. **Wright, J. R. and Clements, J. A.,** Metabolism and turnover of lung surfactant, *Am. Rev. Respir. Dis.,* 135, 426, 1987.
19. **Benson, B., Hawgood, S., Schilling, J., Clements, J., Damm, D., Cordell, B., and Tyler, R. T.,** Structure of canine pulmonary surfactant apoprotein: cDNA and complete amino acid sequence, *Proc. Natl. Acad. Sci. U.S.A.,* 82, 6379, 1985.
20. **Hawgood, S., Benson, B. J., Schilling, J., Damm, D., Clements, J. A., and White, R. T.,** Nucleotide and amino acid sequences of pulmonary surfactant SP18 and evidence for cooperation between SP-18 and SP28-36 in surfactant lipid adsorption, *Proc. Natl. Acad. Sci. U. S. A.,* 84, 66, 1987.
21. **Whitsett, J. A., Weaver, T. E., Clark, J. C., Sawtell, N., Glasser, S. W., Korfhagen, T. R., and Hull, M. W.,** Glucocorticoid enhances surfactant proteolipid Phe and pVal synthesis and RNA in fetal lung, *J. Biol. Chem.,* 262, 15618, 1987.
22. **Warr, R. G., Hawgood, S., Buckley, D. I., Crisp, T. M., Schilling, J., Benson, B. J., Ballard, P. L., Clements, J. A., and White, R. T.,** Low molecular weight human surfactant protein (SP 5): isolation, characterization, and cDNA and amino acid sequences, *Proc. Natl. Acad. Sci. U. S. A.,* 84, 7915, 1987.
23. **Glasser, S. W., Korfhagen, T. R., Weaver, T. E., Clark, J., Pilot-Matias, T., Meuth, J., Fox, J. L., and Whitsett, J. A.,** cDNA, deduced polypeptide structure and chromosomal assignment of human pulmonary surfactant proteolipid: SPL p val, *J. Biol. Chem.,* 263, 9, 1988.
24. **Johannson, J., Jornvall, H., Eklund, A., Christensen, N., Robertson, B., and Curstedt, T.,** Hydrophobic 3.7 kDa surfactant polypeptide: structural characterization of the human and bovine forms, *Fed. Eur. Biochem. Soc.,* 232, 61, 1988.
25. **Crapo, J. D., Young, S. L., Fram, E. K., Pinkerton, K. E., Barry, B. E., and Crapo, R. D.,** Morphometric characteristics of cells in the alveolar region of mammalian lungs, *Am. Rev. Respir. Dis.,* 128 (Suppl. 2), S42, 1983.

26. **Chevalier, G. and Collet, A. J.,** *In vivo* incorporation of choline-^{3}H, leucine ^{3}H and galactose-^{3}H in alveolar type II pneumocytes in relation to surfactant synthesis. A quantitative radioautographic study in mouse by electron microscopy, *Anat. Rec.,* 174, 289, 1972.
27. **Goerke, J.,** Lung surfactant, *Biochim. Biophys. Acta,* 344, 241, 1974.
28. **Batenburg, J. J.,** Biosynthesis and secretion of pulmonary surfactant, in *Pulmonary Surfactant,* Robertson, B., Van Golde, L. M. G., and Batenburg, J. J., Eds., Elsevier, Amsterdam, 1984, 237.
29. **Post, M. and van Golde, L. M. G.,** Metabolic and developmental aspects of the pulmonary surfactant system, *Biochim. Biophys. Acta,* 947, 249, 1988.
30. **Hollingsworth, M. and Gilfillan, A. M.,** The pharmacology of lung surfactant secretion, *Pharmacol. Rev.,* 36, 69, 1984.
31. **Rooney, S. A.,** The surfactant system and lung phospholipid biochemistry, *Am. Rev. Respir. Dis.,* 131, 439, 1985.
32. **Snyder, F. and Malone, B.,** Acyltransferases and the biosynthesis of pulmonary surfactant lipid in adenoma alveolar type II cells, *Biochem. Biophys. Res. Commun.,* 66, 914, 1975.
33. **Yamada, K. and Okuyama, H.,** Possible involvement of acyltransferase systems in the formation of pulmonary surfactant in the lung, *Arch. Biochem. Biophys.,* 196, 209, 1979.
34. **Finkelstein, J. N., Maniscalco, W. M., and Shapiro, D. L.,** Properties of freshly isolated type II alveolar epithelial cells, *Biochim. Biophys. Acta,* 762, 398, 1983.
35. **Batenburg, J. J., Den Breejen, J. N., Yost, R. W., Haagsman, H. P., and Van Golde, L. M. G.,** Glycerol 3-phosphate acylation in microsomes of type II cells isolated from adult rat lung, *Biochim. Biophys. Acta,* 878, 301, 1986.
36. **Schlossman, D. M. and Bell, R. M.,** Microsomal *sn*-glycerol 3-phosphate and dihydroxyacetone phosphate acyltransferase activities from liver and other tissues, *Arch. Biochem. Biophys.,* 182, 732, 1977.
37. **DeClerq, P. E., Haagsman, H. P., Van Veldhoven, P., Debeer, L. J., Van Golde, L. M. G., and Mannaerts, G. P.,** Rat liver dihydroxyacetone-phosphate acyltransferases and their contribution to glycerolipid synthesis, *J. Biol. Chem.,* 259, 9064, 1984.
38. **Possmayer, F.,** Pulmonary phosphatidate phosphohydrolase and its relation to the surfactant system of the lung, in *Phosphatidate Phosphohydrolase,* Vol. 2, Brindley, D. N., Ed., CRC Press, Boca Raton, FL, 1988, 39.
39. **Das, S. K., McCullough, M. S., and Haldar, D.,** Acyl-CoA:*sn*-glycerol 3-phosphate acyltransferase in mitochondria and microsomes of adult and fetal guinea pig lung, *Biochem. Biophys. Res. Commun.,* 101, 237, 1981.
40. **Das, S. K., Kakked, P. B., and McCullough, M. S.,** Development of glycerophosphate acyltransferase in guinea pig lung mitochondria and microsomes, *Biochim. Biophys. Acta,* 802, 423, 1984.
41. **Hoffman, D. R., Truong, C. T., and Johnston, J. M.,** Metabolism, regulation and function of ether-linked glycerolipids, *Biochim. Biophys. Acta,* 879, 88, 1986.
42. **Fisher, A. B., Huber, G. A., Furia, L., Bassett, D., and Rabinowitz, J. L.,** Evidence for lipid synthesis by the dihydroxyacetone phosphate pathway in rabbit lung cellular fractions, *J. Lab. Clin. Med.,* 87, 1033, 1976.
43. **Mason, R. J.,** Importance of the acyl dihydroxyacetone phosphate pathway in the synthesis of phosphatidylglycerol and phosphatidylcholine in alveolar type II cells, *J. Biol. Chem.,* 253, 3367, 1978.
44. **Ide, H. and Weinhold, P. A.,** Properties of diacylglycerol kinase in adult and fetal rat lung, *Biochim. Biophys. Acta,* 713, 547, 1982.
45. **Bleasdale, J. E. and Johnston, J. M.,** Phosphatidic acid production and utilization, in *Lung Development: Biological and Clinical Perspectives,* Vol. 1, Farrell, P. M., Ed., Academic Press, New York, 1982, 259.
46. **Johnston, J. M., Reynolds, G., Wylie, M. B., and MacDonald, P. C.,** The phosphohydrolase activity in lamellar bodies and its relationship to phosphatidylglycerol and lung surfactant formation, *Biochim. Biophys. Acta,* 531, 65, 1978.
47. **Okazaki, T. and Johnston, J. M.,** Distribution of the phosphatidate phosphohydrolase activity in the lamellar body and lysosomal fractions, *Lipids,* 15, 447, 1980.
48. **Walton, P. and Possmayer, F.,** The role of Mg^{2+}-dependent phosphatidate phosphohydrolase in pulmonary glycerolipid biosynthesis, *Biochim. Biophys. Acta,* 796, 364, 1984.
49. **Walton, P.A. and Possmayer, F.,** Translocation of Mg^{2+}-dependent phosphatidate phosphohydrolase between cytosol and endoplasmic reticulum in a permanent cell line from human lung, *Biochem. Cell Biol.,* 64, 1135, 1986.
50. **Van Golde, L. M. G.,** The CDPcholine pathway: cholinephosphotransferase, in *Lung Development: Biological and Clinical Perspectives,* Vol. 1, Farrell, P.M., Ed., Academic Press, New York, 1982, 337.
51. **Stith, I. E. and Das, S. K.,** Development of cholinephosphotransferase in guinea pig lung mitochondria and microsomes, *Biochim. Biophys. Acta,* 714, 250, 1982.

52. **Sikpi, M. O. and Das, S. K.,** The localization of cholinephosphotransferase in the outer membrane of guinea-pig lung mitochondria, *Biochim. Biophys. Acta,* 899, 35, 1987.
53. **Van Heusden, G. P. H. and Van den Bosch, H.,** Utilization of disaturated and unsaturated phosphatidylcholine and diacylglycerols by cholinephosphotransferase in rat lung microsomes, *Biochim. Biophys. Acta,* 711, 361, 1982.
54. **Ide, H. and Weinhold, P. A.,** Cholinephosphotransferase in rat lung: *in vitro* formation of dipalmitoylphosphatidylcholine and general lack of selectivity using endogenously generated diacylglycerol, *J. Biol. Chem.,* 257, 14926, 1982.
55. **Ishidate, K. and Weinhold, P. A.,** The content of diacylglycerol, triacylglycerol and monoacylglycerol and a comparison of the structural and metabolic heterogeneity of diacylglycerols and phosphatidylcholine during rat lung development, *Biochim. Biophys. Acta,* 664, 133, 1981.
56. **Post, M., Schuurmans, E. A. J. M., Batenburg, J. J., and Van Golde, L. M. G.,** Mechanisms involved in the synthesis of disaturated phosphatidylcholine by alveolar type II cells isolated from adult rat lung, *Biochim. Biophys. Acta,* 750, 68, 1983.
57. **Miller, J. C. and Weinhold, P. A.,** Cholinephosphotransferase in rat lung — the *in vitro* synthesis of dipalmitoylphosphatidylcholine from dipalmitoylglycerol, *J. Biol. Chem.,* 256, 12662, 1981.
58. **Ulane, R. E.,** The CDPcholine pathway: choline kinase, in *Lung Development: Biological and Clinical Perspectives,* Vol. 1, Farrell, P. M., Ed., Academic Press, New York, 1982, 295.
59. **Haagsman, H. P., Schuurmans, E. A. J. M., Batenburg, J. J., and van Golde, L. M. G.,** Synthesis of phosphatidylcholines in ozone-exposed alveolar type II cells isolated from adult rat lung: is glycerolphosphate acyltransferase a rate-limiting enzyme?, *Exp. Lung Res.,* 14, 1, 1988.
60. **Feldman, D. A., Kovac, C. R., Dranginis, P. L., and Weinhold, P. A.,** The role of phosphatidylglycerol in the activation of CTP:phosphocholine cytidylyltransferase from rat lung, *J. Biol. Chem.,* 253, 4980, 1978.
61. **Feldman, D. A., Brubaker, P. G., and Weinhold, P. A.,** Activation of CTP:phosphocholine cytidylyltransferase in rat lung by fatty acids, *Biochim. Biophys. Acta,* 665, 53, 1981.
62. **Weinhold, P. A., Feldman, D., Quade, M. M., Miller, J. C., and Brooks, R. L.,** Evidence for a regulatory role of CTP:cholinephosphate cytidylyltransferase in the synthesis of phosphatidylcholine in fetal lung following premature birth, *Biochim. Biophys. Acta,* 665, 134, 1981.
63. **Weinhold, P. A., Rounsifer, M. E., Williams, S. E., Brubaker, P. G., and Feldman, D. A.,** CTP:phosphorylcholine cytidylyltransferase in rat lung, *J. Biol. Chem.,* 259, 10315, 1984.
64. **Nijssen, J. G. and Van den Bosch, H.,** Cytosol-stimulated remodeling of phosphatidylcholine in rat lung microsomes, *Biochim. Biophys. Acta,* 875, 450, 1986.
65. **Mason, R. J. and Nellenbogen, J.,** Synthesis of saturated phosphatidylcholine and phosphatidylglycerol by freshly isolated rat alveolar type II cells, *Biochim. Biophys. Acta,* 794, 392, 1984.
66. **Voelker, D. R. and Snyder, F.,** Subcellular site and mechanism of synthesis of disaturated phosphatidylcholine in alveolar type II cell adenomas, *J. Biol. Chem.,* 254, 8628, 1979.
67. **Van Heusden, G. P. H., Vianen, G. M., and van den Bosch, H.,** Differentiation between acyl CoA:lysophosphatidylcholine acyltransferase and lysophosphatidylcholine:lysophosphatidylcholine transacylase in the synthesis of dipalmitoylphosphatidylcholine in rat lung, *J. Biol. Chem.,* 255, 9312, 1980.
68. **Vianen, G. M. and van den Bosch, H.,** Lysophospholipase and lysophosphatidylcholine:lysophosphatidylcholine transacylase from rat lung: evidence for a single enzyme and some aspects of its specificity, *Arch. Biochem. Biophys.,* 190, 373, 1978.
69. **Batenburg, J. J., Longmore, W. J., Klazinga, W., and Van Golde, L. M. G.,** Lysolecithin acyltransferase and lysolecithin:lysolecithin acyltransferase in adult rat lung alveolar type II epithelial cells, *Biochim. Biophys. Acta,* 573, 136, 1979.
70. **Crecelius, C. A. and Longmore, W. J.,** Acyltransferases in adult rat type II pneumonocytes-derived subcellular fractions, *Biochim. Biophys. Acta,* 795, 247, 1984.
71. **Okuyama, H., Yamada, K., Miyagawa, T., Suzuki, M., Prasad, R., and Lands, W. E. M.,** Enzymatic basis for the formation of pulmonary surfactant lipids by acyltransferase systems, *Arch. Biochem. Biophys.,* 221, 99, 1983.
72. **Stymne, S. and Stobart, A. K.,** Involvement of acyl exchange between acyl-CoA and phosphatidylcholine in the remodelling of phosphatidylcholine in microsomal preparations of rat lung, *Biochim. Biophys. Acta,* 837, 239, 1985.
73. **Nijssen, J. G. and van den Bosch, H.,** Coenzyme A-mediated transacylation of *sn*-2 fatty acids from phosphatidylcholine in rat lung microsomes, *Biochim. Biophys. Acta,* 875, 458, 1986.
74. **Ohta, M., Hasegawa, H., and Ohno, K.,** Calcium independent phospholipase A_2 activity in rat lung supernatant, *Biochim. Biophys. Acta,* 280, 552, 1972.
75. **Franson, R. C. and Weir, D. L.,** Isolation and characterization of a membrane-associated calcium-dependent phospholipase A_2 from rabbit lung, *Lung,* 160, 275, 1982.
76. **Longmore, W. J., Oldenburg, U., and van Golde, L. M. G.,** Phospholipase A_2 in rat lung microsomes: substrate specificity towards endogenous phosphatidylcholines, *Biochim. Biophys. Acta,* 572, 452, 1979.

77. **Filgueiras, O. M. O. and Possmayer, F.,** Characterization of phospholipase A_2 from rabbit lung microsomes, *Lipids,* 22, 731, 1987.
78. **Infante, J. P.,** *De novo* sn-glycerol-3-phosphorylcholine synthetase activity in lung and muscle and its subcellular location, *Mol. Cell. Biochem.,* 71, 135, 1987.
79. **Infante, J. P.,** Biosynthesis of acyl-specific glycerophospholipids in mammalian tissues: Postulation of new pathways, *FEBS Lett.,* 170, 1, 1984.
80. **Infante, J. P. and Huszagh, V. A.,** Is there a new biosynthetic pathway for lung surfactant phosphatidylcholine?, *Trends Biochem. Sci.* 23, 131, 1987.
81. **Rustow, B. and Kunze, D.,** Diacylglycerol synthesized *in vitro* from *sn*-glycerol 3-phosphate and the endogenous diacylglycerol are different substrate pools for the biosynthesis of phosphatidylcholine in rat lung microsomes, *Biochim. Biophys. Acta,* 835, 273, 1985.
82. **Rustow, B. and Kunze, D.,** Further evidence for the existence of different diacylglycerol pools of the phosphatidylcholine synthesis in microsomes, *Biochim. Biophys. Acta,* 921, 552, 1987.
83. **Smith, B. T.,** Pulmonary surfactant during fetal development and neonatal adaptation: hormonal control, in *Pulmonary Surfactant,* Robertson, B., Van Golde, L. M. G., and Batenburg, J. J., Eds., Elsevier, Amsterdam, 1984, 357.
84. **Kresch, M. J. and Gross, I.,** The biochemistry of fetal lung development, in *Clinics in Perinatology: The Respiratory System in the Newborn,* Vol. 14, Stern, L., Guest Ed., W. B. Saunders, Philadelphia, 1987, 481.
85. **Gross, I., Wilson, C. M., and Rooney, S. A.,** Phosphatidylcholine synthesis in newborn rabbit lung developmental pattern and the influence of nutrition, *Biochim. Biophys. Acta,* 528, 190, 1978.
86. **Oulton, M. and Dolphin, M.,** Subcellular distribution of disaturated phosphatidylcholine in developing rabbit lung, *Lipids,* 23, 55, 1988.
87. **Rooney, S. A., Gobran, L. I., Marino, P. A., Maniscalco, W. M., and Gross, I.,** Effects of betamethasone on phospholipid content, composition and biosynthesis in the fetal rabbit lung, *Biochim. Biophys. Acta,* 572, 64, 1979.
88. **Rooney, S. A., Wai-Lee, T. S., Gobran, L., and Motoyama, E. K.,** Phospholipid content, composition and biosynthesis during fetal lung development in the rabbit, *Biochim. Biophys. Acta,* 431, 447, 1976.
89. **Rooney, S. A., Gobran, L. I., and Wai-Lee, T. S.,** Stimulation of surfactant production by oxytocin-induced labor in the rabbit, *J. Clin. Invest.,* 60, 754, 1977.
90. **Tsao, F. H. and Zachman, R. D.,** Phosphatidylcholine-lysophosphatidylcholine cycle pathway enzymes in rabbit lung. II. Marked differences in the effect of gestational age on activity compared to the CDP cholinepathway, *Pediatr. Res.,* 11, 858, 1977.
91. **Heath, M. F. and Jacobson, W.,** Development changes in enzyme activities in fetal and neonatal rabbit lung: cytidylyltransferase, cholinephosphotransferase, phospholipases A_1 and A_2, β-galactosidase, and β-glucuronidase, *Pediatr. Res.,* 18, 395, 1984.
92. **Casola, P. G. and Possmayer, F.,** Pulmonary phosphatidic acid phosphohydrolase: developmental patterns in rabbit lung, *Biochim. Biophys. Acta,* 665, 186, 1981.
93. **Tokmakjian, S., Haines, D. S. M., and Possmayer, F.,** Pulmonary phosphatidylcholine biosynthesis — alterations in the pool sizes of choline and choline derivatives in rabbit fetal lung during development, *Biochim. Biophys. Acta,* 663, 557, 1981.
94. **Quirk, J. G., Bleasdale, J. E., MacDonald, P. C., and Johnston, J. M.,** A role for cytidine monophosphate in the regulation of the glycerophospholipid composition of surfactant in developing lung, *Biochem. Biophys. Res. Commun.,* 95, 985, 1980.
95. **Choy, P. C., Lim, P. H., and Vance, D. E.,** Purification and characterization of CTP:cholinephosphate cytidylyltransferase from rat liver cytosol, *J. Biol. Chem.,* 252, 7673, 1977.
96. **Khosla, S., Gobran, L. I., and Rooney, S. A.,** Stimulation of phosphatidylcholine synthesis by 17-β-estradiol in fetal rabbit lung, *Biochim. Biophys. Acta,* 617, 282, 1980.
97. **Possmayer, F., Casola, P. G., Chan, F., MacDonald, P., Ormseth, M. A., Wong, T., Harding, P. G. R., and Tokmakjian, S.,** Hormonal induction of pulmonary maturation in the rabbit fetus: effects of treatment with estradiol-17beta on the endogenous levels of cholinephosphate, CDP-choline and phosphatidylcholine, *Biochim. Biophys. Acta,* 664, 10, 1981.
98. **Chu, A. J. and Rooney, S. A.,** Stimulation of cholinephosphate cytidylyltransferase activity by estrogen in fetal rabbit lung is mediated by phospholipids, *Biochim. Biophys. Acta,* 834, 346, 1985.
99. **Chu, A. J. and Rooney, S. A.,** Developmental differences in activation of cholinephosphate cytidylyltransferase by lipids in rabbit lung cytosol, *Biochim. Biophys. Acta,* 835, 132, 1985.
100. **Maniscalco, W. M., Wilson, C. M., Gross, I., Gobran, L., Rooney, S. A., and Warshaw, J. B.,** Development of glycogen and phospholipid metabolism in fetal and newborn rat lung, *Biochim. Biophys. Acta,* 530, 333, 1978.
101. **Chan, F., Harding, P. G. R., Wong, T., Fellows, G. F., and Possmayer, F.,** Cellular distribution of enzymes involved in phosphatidylcholine synthesis in developing rat lung, *Can. J. Biochem. Cell Biol.,* 61, 107, 1983.

102. **Tokmakjian, S. and Possmayer, F.,** Pool sizes of the precursors for phosphatidylcholine synthesis in developing rat lung, *Biochim. Biophys. Acta*, 666, 176, 1981.
103. **Casola, P. G. and Possmayer, F.,** Separation and characterization of the membrane-bound and aqueously-dispersed phosphatidate-dependent phosphatidic acid phosphohydrolase activities in rat lung cytosol and microsomal fractions, *Biochim. Biophys. Acta*, 664, 298, 1981.
104. **Post, M., Batenburg, J. J., Van Golde, L. M. G., and Smith, B. T.,** The rate-limiting reaction in phosphatidylcholine synthesis by alveolar type II cells isolated from fetal rat lung, *Biochim. Biophys. Acta*, 795, 558, 1984.
105. **Post, M., Batenburg, J. J., Smith, B. T., and Van Golde, L. M. G.,** Pool sizes of precursors for phosphatidylcholine formation in adult rat lung type II cells, *Biochim. Biophys. Acta*, 795, 552, 1984.
106. **Rooney, S. A., Dynia, D. W., Smart, D. A., Chu, A. J., Ingleson, L. D., Wilson, C. M., and Gross, I.,** Glucocorticoid stimulation of cholinephosphate cytidylyltransferase activity in fetal rat lung: receptor-response relationships, *Biochim. Biophys. Acta*, 888, 208, 1986.
107. **Post, M., Barsoumian, A., and Smith, B. T.,** The cellular mechanism of glucocorticoid acceleration of fetal lung maturation: fibroblast-pneumonocyte factor stimulates choline-phosphate cytidylyltransferase activity, *J. Biol. Chem.*, 261, 2179, 1986.
108. **Post, M.,** Maternal administration of dexamethasone stimulates cholinephosphate cytidylyltransferase in fetal type II cells, *Biochem. J.*, 241, 291, 1987.
109. **Rooney, S. A., Gobran, L., Gross, I., Wai-Lee, T. S., Nardone, L. L., and Motoyama, E. K.,** Effects of cortisol administration to fetal rabbits on lung phospholipid content, composition and biosynthesis, *Biochim. Biophys. Acta*, 450, 121, 1976.
110. **Hietanen, E. and Hartiala, J.,** Developmental pattern of pulmonary lipoprotein lipase in growing rats, *Biol. Neonate*, 36, 85, 1979.
111. **Aeberhard, E. E., Barrett, C. T., Kaplan, S. A., and Scott, M. L.,** Stimulation of phosphatidylcholine synthesis by fatty acids in fetal rabbit type II pneumocytes, *Biochim. Biophys. Acta*, 875, 6, 1986.
112. **Chander, A. and Fisher, A. B.,** Cholinephosphate cytidyltransferase activity and phosphatidylcholine synthesis in rat granular pneumocytes are increased with exogenous fatty acids, *Biochim. Biophys. Acta*, 958, 343, 1988.
113. **Radika, K. and Possmayer, F.,** Inhibition of foetal pulmonary cholinephosphate cytidylyltransferase under conditions favouring protein phosphorylation, *Biochem. J.*, 232, 833, 1985.
114. **Gross, I. and Wilson, C. M.,** Fetal lung maturation: initiation and modulation, *J. Appl. Physiol.*, 55, 1725, 1983.
115. **Aeberhard, E. E., Scott, M. L., Barrett, C. T., and Kaplan, S. A.,** Effects of cyclic AMP analogues and phosphodiesterase inhibitors on phospholipid biosynthesis in fetal type II pneumocytes, *Biochim. Biophys. Acta*, 803, 29, 1984.
116. **Niles, R. M. and Makarski, J. S.,** Regulation of phosphatidylcholine metabolism by cyclic AMP in a model alveolar type II cell line, *J. Biol. Chem.*, 254, 4324, 1979.
117. **Patterson, C. E., Konicki, M. V., Selig, W. M., Owens, C. M. and Rhoades, R. A.,** Integrated substrate utilization by perinatal lung, *Exp. Lung Res.*, 10, 71, 1986.
118. **Pope, T. S., Smart, D. A., and Rooney, S. A.,** Hormonal effects of fatty-acid synthase in cultured fetal rat lung; induction by dexamethasone and inhibition of activity by triiodothyronine, *Biochim. Biophys. Acta*, 959, 169, 1988.
119. **Patterson, C. E., Davis, K. S., Beckman, D. E., and Rhoades, R. A.,** Fatty acid synthesis in the fetal lung: relationship to surfactant lipids, *Biochim. Biophys. Acta*, 878, 110, 1986.
120. **Pope, T. S. and Rooney, S. A.,** Effects of glucocorticoid and thyroid hormones on regulatory enzymes of fatty acid synthesis and glycogen metabolism in developing fetal rat lung, *Biochim. Biophys. Acta*, 918, 141, 1987.
121. **Maniscalco, W. M., Finkelstein, J. N., and Parkhurst, A. B.,** *De novo* fatty acid synthesis in developing rat lung, *Biochim. Biophys. Acta*, 711, 49, 1982.
122. **Patterson, C. E., Davis, K. S., and Rhoades, R. A.,** Regulation of fetal lung disaturated phosphatidylcholine synthesis by *de novo* palmitate supply, *Biochim. Biophys. Acta*, 958, 60, 1988.
123. **Maniscalco, W. M., Finkelstein, J. N., and Parkhurst, A. B.,** *De novo* fatty acid synthesis by freshly isolated alveolar type II epithelial cells, *Biochim. Biophys. Acta*, 751, 462, 1983.
124. **Geppert, E. F. and Elstein, K. H.,** Short-term regulation of fatty acid synthesis in isolated alveolar type II cells from adult rat lung: effects of free fatty acids and hormones, *Exp. Lung Res.*, 4, 281, 1983.
125. **Miles, P. R., Wright, J. R., Bowman, L., and Castranova, V.,** Incorporation of [^{3}H]palmitate into disaturated phosphatidylcholines in alveolar type II cells isolated by centrifugal elutriation, *Biochim. Biophys. Acta*, 753, 107, 1983.
126. **Tesan, M., Anceschi, M. M., and Bleasdale, J. E.,** Regulation of CTP:phosphocholine cytidylyltransferase activity in type II pneumonocytes, *Biochem. J.*, 232, 705, 1985.
127. **Thakur, N. R., Tesan, M., Tyler, N. E., and Bleasdale, J. E.,** Altered lipid synthesis in type II pneumonocytes exposed to lung surfactant, *Biochem. J.*, 240, 679, 1986.

128. **Merrill, A. H., Wang, E., Stevens, J., and Brumley, G. W.,** Activities of the initial enzymes of glycerolipid and sphingolipid synthesis in lung microsomes from rats exposed to air or 85% oxygen, *Biochem. Biophys. Res. Commun.,* 119, 995, 1984.
129. **Wright, E. S., Vang, M. J., Finkelstein, J. N., and Mavis, R. D.,** Changes in phospholipid biosynthetic enzymes in type II cells and alveolar macrophages isolated from rat lungs after NO_2 exposure, *Toxicol. Appl. Pharmacol.,* 66, 305, 1982.
130. **Gilfillan, A. M., Chu, A. J., and Rooney, S. A.,** Stimulation of phosphatidylcholine synthesis by exogenous phosphatidylglycerol in primary cultures of type II pneumocytes, *Biochim. Biophys. Acta,* 794, 269, 1984.
131. **Gilfillan, A. M., Smart, D. A., and Rooney, S. A.,** Phosphatidylglycerol stimulates cholinephosphate cytidylyltransferase activity and phosphatidylcholine synthesis in type II pneumocytes, *Biochim. Biophys. Acta,* 835, 141, 1985.
132. **Rosenberg, I. L., Smart, D. A., Gilfillan, A. M., and Rooney, S. A.,** Effect of 1-oleoyl-2-acetylglycerol and other lipids on phosphatidylcholine synthesis and cholinephosphate cytidylyltransferase activity in cultured type II pneumocytes, *Biochim. Biophys. Acta,* 921, 473, 1987.
133. **Anceschi, M. M., DiRenzo, G. C., Venincasa, M. D., and Bleasdale, J. E.,** The choline-depleted type II pneumonocyte: a model for investigating the synthesis of surfactant lipids, *Biochem. J.,* 224, 253, 1984.
134. **Batenburg, J. J., Klazinga, W., and van Golde, L. M. G.,** Regulation of phosphatidylglycerol and phosphatidylinositol synthesis in alveolar type II cells isolated from adult rat lung, *Fed. Eur. Biochem. Soc. Lett.,* 147, 171, 1982.
135. **Veldhuizen, R. and Possmayer, F.,** Examination of the potential role of the glycerophosphorylcholine pathway in the biosynthesis of phosphatidylcholine by liver and lung, *Biochim. Biophys. Acta,* submitted.

Chapter 13

REGULATORY AND FUNCTIONAL ASPECTS OF PHOSPHATIDYLCHOLINE METABOLISM

Dennis E. Vance

TABLE OF CONTENTS

I. INTRODUCTION

The previous chapters of this book present authoritative summaries of most aspects of phosphatidylcholine (PC) metabolism. Thus, the various enzymes involved in PC biosynthesis have been discussed with emphasis on the properties of these enzymes and where applicable, regulatory features. The enzymes involved in PC catabolism have been introduced and the complexities of this frontier in research have been identified. Finally, special aspects of PC metabolism have been reviewed with respect to the ether lipids, sphingomyelin metabolism, and PC metabolism in yeast. The purpose of this chapter is to provide a summary and synthesis of the preceding chapters with emphasis on regulation of PC metabolism and the function of PC.

II. REGULATION OF PHOSPHATIDYLCHOLINE BIOSYNTHESIS

A. CONTROL AT THE LEVEL OF GENE EXPRESSION — IS THIS EVER AN IMPORTANT REGULATORY MECHANISM?

Regulation of metabolic flux, by variation in the amount of active enzyme, is a common mechanism used by cells to control the biosynthesis of important cellular components. Thus, the level of acetyl-CoA carboxylase in liver is varied according to the requirements for fatty acid biosynthesis.[1] Similarly, the uptake of lipoproteins into cells and the rate of cholesterol biosynthesis are processes modulated at the level of gene expression.[1] Within this context, it was surprising to find that PC biosynthesis was rarely altered by changes in the expression of biosynthetic enzymes. Thus, we view PC biosynthesis largely as a constitutive process, and flux through the biosynthetic pathways appears to be altered by substrate supply and activation/ inactivation of preexisting enzymes. There are, however, some well-documented instances in which the amount of enzymes in PC biosynthesis has been altered. However, it should be remembered that a change in the amount of an enzyme could be due to other factors than increased synthesis of mRNA. Thus, there could be a change in the amount of an enzyme due to an effect on mRNA translation or rate of degradation of the enzyme.

1. Development

Changes in the specific activity and total activity of PC biosynthetic enzymes have been detected in liver and lung during the perinatal period of development.[2-8] These changes have been monitored by enzyme assay at optimal conditions, but in no instance has the amount of enzyme mass been estimated. With the recent purification of choline kinase (CK), cholinephosphate cytidylyltransferase (CT), and phosphatidylethanolamine-*N*-methyltransferase (PEMT) and preparation of antibodies to these proteins, immunotitration studies are now feasible. Nevertheless, it is clear that there are differences in the amounts of these enzymes during the course of development. In rat liver there was a 2- to 4.8-fold increase in CK during the last 5 d of gestation, followed by a return to adult levels 5 d after parturition.[2,4] It was speculated[2] that this rapid increase in CK was to ensure adequate amounts of cholinephosphate for the rapid increase in PC biosynthesis that occurs around birth.[5] The activities of cholinephosphotransferase (CPT) and PEMT gradually increased from 5 d before birth to maximal levels 10 d after birth.[2,4] This result is consistent with these enzymes being expressed constitutively in surplus so that the amounts of these two enzymes never limit the rate of PC biosynthesis. The activity of CT in liver cytosol 5 d before birth was several times higher than observed in livers from adult animals whereas microsomal CT activity was at the same level as found in adults.[2] During the parturition period, there was a dramatic increase of CT in microsomes which appeared to be derived from CT in the cytosol. By 3 d after birth, the microsomal CT activity had changed to adult levels and there was a rise in cytosolic CT activity over 10 d to adult levels or higher.[2] The translocation of CT was thought[2] to be responsible for the increase in PC biosynthesis in liver at birth.[4] The coincidence of the increase in CT activity on microsomes and an increased rate of PC synthesis in liver provides another *in vivo* model system which suggests CT translocation is a physiologically important mechanism for control of PC biosynthesis. As discussed in Chapter 3, there was a translocation from cytosol to microsomes of CT at birth in the lungs of newly born rat pups that was coincident with an increase in the fatty acid content of microsomes.[6] It would be interesting to know if a similar mechanism operates in the new born ratliver.

2. Hormones and Growth Factors

The effects of hormones on PC biosynthesis have been investigated in several systems. Diethylstilbestrol, a synthetic estrogen, when injected into roosters stimulated the synthesis of PC.[9-11] This increase appeared to be due to a two- to threefold induction of CK which correlated with a similar rise in the levels of cholinephosphate.[9,10] Antibody to the partially purified enzyme showed that there was a similar increase in the titratable amount of CK.[11] Microsomal CPT and cytosolic CT activities were not altered as a result of the treatment. The activity of microsomal CT was not measured,[10] but PEMT activity was increased twofold in the treated roosters after 3 d.[10] However, the rate of the methylation reaction in the roosters was not assessed. Nor was it possible at that time to measure the amount of PEMT protein. These studies provide a clear example of regulation of PC biosynthesis via the CDP-choline pathway at the level of gene expression. The best correlation with PC biosynthesis was with the induction of CK. Since CK in the liver is composed of several isoenzymes (Chapter 2), it would be interesting to know how diethylstilbestrol affected the isoenzyme profile. The effect of the hormone on CT activity on the microsomes still needs to be evaluated.

Glucocorticoids stimulate pulmonary PC synthesis,[12-16] apparently by increasing the amount of CT on the microsomes as demonstrated in fetal lung, fetal lung explants, and fetal type II cells. The effect does not appear to be directly on the type II cells, but rather on fetal lung fibroblasts that are stimulated to produce fibroblast-pneumonocyte factor which acts on the alveolar type II cells.[13-15] Antibody to this factor blocked its effect on type II cells and delayed lung maturation in intact animals.[13] Evidence suggests that the factor stimulates PC biosynthesis by increasing CT activity in type II cells.[14] There was no effect on CK or CPT activity.[14] Subsequent studies suggested the glucocorticoid treatment increased both the microsomal and cytosolic CT activities,[15,16] but the mechanism still needs to be investigated. These results support the idea that glucocorticoids induce the synthesis of fibroblast-pneumonocyte factor which stimulates the activation of CT in type II cells by an unknown mechanism. Estrogens have also been shown to stimulate PC synthesis in fetal rabbit lung, and this has been attributed to an increased amount of cytosolic phospholipid which activated cytosolic CT.[17]

Other hormonal effects on PC biosynthesis have been investigated. Vasopressin inhibited PC synthesis in freshly isolated hepatocytes by an effect on the CT reaction.[18] Similarly, inhibition by norepinephrine appeared to be via an effect on CT activity.[19] Thyroid hormone stimulated PC synthesis in cultured fetal rabbit lung.[20] PC synthesis in neonatal mouse bone was unaffected by several hormones known to stimulate other processes in bone.[21] Prolactin induced a threefold elevation of CK in a lymphoma cell line.[22] However, there was no effect on the rate of PC synthesis nor any change in CT or CPT activities.[22] Finally, as discussed in Chapter 3, activation of the cAMP system appears to inhibit PC synthesis in liver via phosphorylation of CT, causing its release from cellular membranes.

The effect of mitogenic growth factors on PC biosynthesis in 3T3 fibroblasts was investigated.[23,24] As much as an eightfold increase in PC biosynthesis was reported which appeared to be due to an induction of choline kinase with a concomitant increase in the cholinephosphate pool, which was probably limiting the rate of PC synthesis in these cells.[23,24] This study represents another example in which an effect at the level of gene expression on PC biosynthesis appears to be mediated by the induction of CK.

3. Induction of Choline Kinase by Polycyclic Aromatic Hydrocarbons and Carbon Tetrachloride

As discussed in Chapter 2, these agents cause a striking induction of CK in rat liver. The induced forms of CK were not immunoprecipitable with antibody to the pure kidney enzyme, yet this antibody cross-reacted with the normal isoenzyme recovered from rat liver.[25] Since a direct relationship between CK induction and PC synthesis has not been established,[25] the function, if any, of the CK isoenzyme induction remains an enigma.

4. Expression of PE Methyltransferase in Liver

As discussed in Chapter 7 the only known tissue in which PEMT occurs at high levels is the liver. Clearly, there is a mechanism for the expression of this gene in the hepatocyte which is greatly reduced in other cell types. How this control is mediated at the level of gene expression is a topic of current interest.

5. Conclusion

The question asked at the beginning of this section has now been answered. PC biosynthesis is regulated at the level of gene expression in selected cases. The regulation in the CDP-choline pathway appears to be largely with the CK reaction. Science has now progressed to the point that elucidation of the mechanism of the induction of this enzyme is a realistic goal. The effect of the fibroblast-pneumonocyte factor on the CT reaction in type II cells needs to be confirmed and whether or not this is an effect at the level of gene expression clearly established. Finally, we are now in a position to study the mechanism for the tissue specific expression of PEMT.

B. CONTROL BY COVALENT MODIFICATION OF ENZYMES

1. cAMP and Protein Phosphorylation

There is now considerable evidence to support the hypothesis that PC synthesis is reduced by the induction of cAMP in hepatocytes (Chapter 3). In summary, cAMP analogues cause a reduction in PC synthesis in cultured hepatocytes which correlates with a loss of CT from the cellular membranes. *In vitro* experiments with cytosol, postmitochondrial supernatant, and pure enzyme show that CT is a substrate for the cAMP-dependent protein kinase. Experiments show that alkaline phosphatase can reverse the phosphorylation effects. Critical experiments to show that the cAMP regulatory system is physiologically important for PC biosynthesis are still to be done. It is necessary to link increases in cAMP levels in hepatocytes or liver to phosphorylation of CT and to show reactivation of the enzyme after dephosphorylation of CT in a physiologically relevant system. We need to identify the site on CT that is phosphorylated and the nature of the physiologically relevant phosphatase. Once CT is cloned, it would be interesting to mutate specifically the phosphorylated sequence from the enzyme and demonstrate the loss of control of CT by the cAMP system.

There is a report that CK can be phosphorylated by cAMP-dependent kinase (Chapter 2), and we know that PEMT is an *in vitro* substrate for the cAMP kinase (Chapter 7). Presently, there is no clear indication that these covalent modifications are of regulatory significance.

2. Are There Other Covalent Modifications with Regulatory Significance?

The simple answer to this question is we don't know. Possibilities to be considered would be other protein kinases (protein kinase C, calmodulin-dependent kinase, etc.), fatty acylation, glycosylation, or even modification by a mevalonic acid derivative. An important point is that with the availability of pure CT, CK, and PEMT, these questions can be addressed in a rigorous manner.

C. HOW IMPORTANT IS THE SUPPLY OF SUBSTRATES?

There is no question that substrate supply can affect the rate of PC biosynthesis. This was demonstrated in 1932 when Best and Huntsman discovered and described the effects of choline deficiency.[26] The early studies on the requirement of choline for PC biosynthesis have been reviewed.[27] Interestingly, under conditions of choline depletion in which PC synthesis is reduced, CT is translocated from cytosol to cellular membranes.[28,29] Apparently, CT has received a signal that PC synthesis is required and the enzyme is activated. However, because of choline depletion and, as a result, lowered levels of cholinephosphate, PC synthesis is not increased. This is one example in which CT translocation to membranes is inversely correlated to PC biosynthesis. When choline is resupplied, the rate of PC synthesis and CT distribution in the cell returns to normal.[28,29] At the present time, the mechanism causing CT translocation is unknown. The activities of CK and CPT are unchanged in the choline deficient cells,[28,30] whereas PEMT activity was increased almost twofold in the livers from 2-d choline-deficient rats.[30] This apparent activation of PEMT observed *in vitro* appears to be due to an increased concentration of PE in the microsomal membranes and is not due to an increased amount of enzyme as determined by immunoblot experiments.[31] It should also be noted that the conversion of PE to PC in these hepatocytes is actually reduced by methionine deficiency, even though PEMT activity appears to be elevated *in vitro*.

Other substrates that may be of importance in controlling the rate of PC biosynthesis via the CDP-choline pathway are CTP and diacylglycerol (DG). The evidence that CTP may be important is limited to studies in HeLa cells[32,33] and the myopathic hamster heart.[34] Poliovirus infection of HeLa cells caused a stimulation of PC biosynthesis by a specific acceleration of the CT reaction, despite a nearly 50% reduction of microsomal and cytosolic CT activities.[32] There was a close correlation between increased levels of CTP and the increased activity of CT in the postmitochondrial fraction isolated from the HeLa cells.[33] It was, thus, postulated that the concentration of CTP was regulating the rate of PC synthesis in the poliovirus-infected HeLa cells.[33] Although there is no evidence to disprove this hypothesis, subsequent studies in HeLa cells showed a dramatic stimulation of PC synthesis and the CT reaction by supply of fatty acids in the medium.[35] If the supply or concentration CTP were rate-limiting for PC synthesis in HeLa cells and the supply of this nucleotide was not affected by the fatty acid supplementation, how did fatty acid activation of CT result in a stimulation of PC biosynthesis? Unfortunately, the supply or concentration of CTP was not measured in this study.[35]

Myopathic hamsters have a reduced level of CTP compared to normal animals (10.0 vs. 15.4 nmol CTP per gram heart).[34] Despite this, the rate of PC biosynthesis in myopathic hearts was normal, which was ascribed to a doubling of CT activity on the microsomes. If CTP were truly limiting the rate of PC synthesis in the heart, an increase in microsomal CT should not affect the rate of PC synthesis. Perhaps this is a case in which both CT activity and CTP concentration can be partially limiting. A doubling of CT activity on microsomes may have been required as compensation for the 30% reduction of the levels of CTP. It is also noteworthy that the total activity of CT was doubled in these hearts as determined by assay of the enzyme.[34] If immunotitration experiments showed a doubling in enzyme mass in the myopathic hearts, this would be a rare example of control of the amount of CT at the level of gene expression.

The supply of DG also seems to be a significant factor in regulation of PC biosynthesis. This conclusion is complicated by several pleiotropic effects. Addition of fatty acids to cells in culture causes an increase in DG formation, resulting in an increased rate of PC synthesis (Chapter 3) and triacylglycerol synthesis.[36] However, as discussed in Chapter 3, fatty acids also cause activation of CT by translocation to the cellular membranes. It is also quite clear that DG in cells exists in distinct metabolic compartments.[37-39] Evidence supports the idea that newly synthesized, rather than preexisting, DG is preferentially used for the synthesis of PC, and this biosynthetic precursor pool is separate from the bulk pool of microsomal DG. This was fully discussed in Chapter 4. Studies with permeabilized HeLa cells, and with microsomes isolated from these cells clearly showed that the rate of PC biosynthesis could be affected by the supply of DG.[40] In circumstances where the level of CDP-choline was below saturation (30 μM), increased synthesis of DG resulted in an increase in the rate of PC synthesis. Thus, the rate of PC biosynthesis can be regulated by both the rate of CDP-choline synthesis via CT and the rate of DG synthesis.[40]

Another mechanism by which substrate might limit PC synthesis that has been largely ignored is the supply of lyso-PC. CHO cells deprived of choline can make sufficient PC from lyso-PC available in the serum in the medium.[41] Similarly, choline- and methionine-deficient hepatocytes defective in PC synthesis and secretion of very low density lipoproteins[42] can be returned to normal by the supply of lyso-PC in the medium. The lyso-PC is transported into the cells and reacylated to PC.[43] Two major sources of lyso-PC in the serum are secretion by hepatocytes[44,45] and formation of lyso-PC as a result of the synthesis of cholesterol esters from PC by lecithin:cholesterol acyltransferase in the plasma.[1]

As thoroughly discussed in Chapter 7, the supply of methionine and the ratio of *S*-adenosylmethionine to *S*-adenosylhomocysteine appear to be of major importance in the regulation of PE methylation in liver.

In conclusion, the above synopsis shows that substrate supply is a major and dominating factor in the regulation of PC biosynthesis in animals.

D. ARE THERE NONCOVALENT ACTIVATORS OR INHIBITORS OF PHYSIOLOGICAL SIGNIFICANCE?

There is surprisingly little information available to implicate regulation of PC biosynthesis by activators or inhibitors. Even though there is an enormous interest in calcium as a second messenger, there is no solid evidence that this cation is or is not involved in the regulation of PC biosynthesis. As discussed in Chapter 5 the base exchange enzyme requires calcium for activity, but this reaction is thought to be of minor quantitative importance in the biosynthesis of PC. Another second messenger of current interest, DG, can act as an activator of CT and, therefore, PC biosynthesis, as summarized in Chapter 3. However, more evidence is required to show that DG actually functions as an activator of CT in a physiologically relevant setting. Also discussed in Chapter 3, fatty acids clearly are able to activate CT and could be physiologically important activators of PC biosynthesis.

There is evidence for coordination of cholesterol and PC biosynthesis. Inhibition of cholesterol synthesis in a myoblast cell line was coordinated with an inhibition of CT and reduced PC biosynthesis.[46] Similarly, feeding young rats a diet enriched in cholesterol and a bile acid resulted in increased CT activity on the microsomes and an increased rate of PC biosynthesis in the liver.[47] The mechanisms for the cholesterol effects on PC biosynthesis are unclear.

E. HOW EXTENSIVE IS THE RECIPROCAL REGULATION OF THE CDP-CHOLINE AND METHYLATION PATHWAYS?

There are two studies, summarized in Chapter 7, that show reciprocal effects on the two PC biosynthetic pathways in liver. Fatty acids stimulate the CDP-choline pathway and cause an inhibition of PE methylation, and 3-deazaadenosine causes an inhibition of the methylation pathway with a stimulation of the CDP-choline pathway. On the other hand, there is an example in which both pathways were inhibited by treatment of hepatocytes with cAMP analogues.[48] We have, therefore, hypothesized[48] that when energy is not limiting, inhibition of one pathway may be compensated by an increase in the other. In instances where energy supply is limited, PC biosynthesis by both pathways may be reduced. Further studies are required to test this hypothesis and elucidate the mechanisms involved in communication between the two pathways.

F. IS THE SUBCELLULAR LOCATION OF PC BIOSYNTHESIS IMPORTANT?

PC biosynthesis can occur in several organelles in animal cells, but the major quantitative site is the endoplasmic reticulum (ER).[49] This might be partially explained by the extent of the ER in the cell. For example, the ER contains approximately 20% of the protein in a liver cell.[50] However, other membrane-rich organelles contain a large percentage of cellular proteins (mitochondria 25%, nuclei 15%)[50] yet are much less enriched in PC biosynthetic enzymes.[49] We can only speculate on why nature has selected the ER as the major site for PC synthesis. One hypothesis relates the need for PC synthesis to expand the bilayer to accommodate integral membrane proteins made by ribosomes on the ER. These proteins require a phospholipid environment. Consistent with this idea, the phospholipid would need to be constantly supplied for the formation of vesicles which bud from the ER and transport integral membrane proteins to other organelles in the cell. Similarly, the expansion of the bilayer would be required for the budding of vesicles from the ER for transport of luminal proteins to other organelles or for secretion. Yet, the secretion of all secretory proteins from hepatocytes, except very low density lipoproteins, does not appear to be impaired in hepatocytes with reduced PC biosynthesis.[42] Perhaps there are also advantages to the cell to have phospholipid, cholesterol, and triacylglycerol synthesis occur on the same membrane. Moreover, with an effective (but poorly described mechanism) and rapid transport of PC to other membranes in the cell, there is no obvious reason why PC synthesis on the ER should not be largely sufficient for the cell's requirements.

The question could be turned around. Why is there synthesis of PC on other membranes in the cells when the rate of synthesis on the ER could very well be sufficient? It appears that *cis*, but not *trans*, Golgi has the ability to make PC.[49] Yet, both organelles are apparently involved in the formation of vesicles for transport of luminal proteins. Thus, the guess is that the site of synthesis of PC in the cell is probably important, but at the present time we simply do not understand why.

G. COORDINATION WITH THE BIOSYNTHESIS OF OTHER PHOSPHOLIPIDS

The literature on this topic is sparse. One of the reasons is that very little is known about the regulation of the biosynthesis of phospholipids other than PC. In one study it was shown that choline regulated the biosynthesis of PE in the hamster heart by competitive inhibition of ethanolamine kinase.[51] The physiological importance of this observation would depend on how much fluctuation is seen in the serum levels of choline. On the other hand, although ethanolamine inhibited choline uptake into the perfused hamster heart, there was no immediate effect on the rate of PC biosynthesis, probably because there was no effect on the pool sizes of the choline-containing precursors.[52] Choline was also shown to have no effect on incorporation of [^{3}H]glycerol into PE of isolated hepatocytes.[53] Similarly, the ethanolamine concentration in the hepatocyte medium did not seem to affect the rate of PC biosynthesis.[53] Thus, at the present time the supply of ethanolamine does not seem to have a major effect on PC biosynthesis via the CDP-choline pathway, nor does choline seem to alter PE synthesis.

There are several examples in which the biosynthesis of PE and PC respond in parallel to the same treatment. Fatty acids stimulate the biosynthesis of both phospholipids.[53] Fasting produced a pronounced decrease in the concentrations of the water-soluble precursors to PC with a more modest decrease in the water-soluble precursors to PE.[54] Phorbol esters have been known for some time to stimulate PC biosynthesis[55] and have recently been shown to stimulate the biosynthesis of PE in rat hepatocytes.[56] Finally, glucagon caused the inhibition of PC[57] and PE[58] biosynthesis in hepatocytes. Thus, from the limited information available at this time, PE and PC synthesis seem to respond in a similar manner to energy supply perturbations and phorbol esters. How these factors affect the biosynthesis of other phospholipids such as phosphatidylserine remains to be investigated. It is not known if the cell requires a fixed ratio of phospholipids for optimal growth and maintenance.

III. REGULATION OF PHOSPHATIDYLCHOLINE CATABOLISM

A. INTRODUCTION

There are enormous complexities in trying to understand PC catabolism. Rather than dealing with a couple of biosynthetic sequences, there are several routes for catabolism of PC as discussed in Chapter 8. The catabolic enzymes are found in the cytosol, as well as on the major membranes of the liver[59] and most other cells. Thus, the relationship among the various phospholipases A_1 in the same cell needs to be defined. The enzymes are often not specific for a class of phospholipids such as PC. In addition, the enzymes are present at low concentrations and attempts at purification of the cellular phospholipases have met with limited success.[59,60] Within the context of these difficulties it is not surprising that understanding the regulation of PC catabolism is still very much in its infancy.

Considerable progress has been made in elucidating the reasons for turnover of PC and other phospholipids in cells since Dawson in 1966 raised the issue of the physiological meaning of "phospholipid turnover".[61] This was discussed in Chapter 8 and will only be summarized briefly. A major function of PC degradation is to remodel the fatty acid composition of PC in the membranes generally by substituting arachidonic acid for a more saturated fatty acid in the SN-2 position as discussed in Chapter 6. Generation of the second messenger DG by phospholipase C and release of arachidonic acid from PC by phospholipase A_2 are clearly of functional importance as discussed more fully in Section IV of this chapter. Degradation of PC that contains peroxidized fatty acids will be of protective importance to the cell structure and function. Lysophospholipases are important for catabolism of lyso-PC to protect the cell against the lytic properties of this lipid. Thus, these functions of phospholipases are clearly delineated. A question that remains is: other than the above functions how important is the general catabolism of PC? Is such general turnover of PC of quantitative and physiological significance in the cell?

B. DO THE ANABOLIC AND CATABOLIC PATHWAYS COMMUNICATE?

In thinking about PC homeostasis in the cell we might postulate that it would be important to have signals that link catabolism to anabolism. If PC were being degraded for some reason, would there be a signal to activate PC synthesis? There is evidence for a PC cycle in cells[62] and that products of PC catabolism activate PC synthesis. Thus, as discussed in Section II of this chapter, two products of PC catabolism, fatty acids and DG, are both activators of PC biosynthesis via the CDP-choline pathway. Arachidonic acid also seems to be a feedback inhibitor of phospholipase A_2 from the P388D macrophage cell line.[63] Whether or not DG has an effect on phospholipase activity in cells is not known, but it shows both activation and inhibition in model systems (Chapter 8). Thus, at the present time it appears that both DG and fatty acids are likely modulators of PC synthesis and catabolism.

Also discussed in Chapter 8, lyso-PC is a possible inhibitor of phospholipase A_2 and choline is an inhibitor of glycerolphosphocholine phosphodiesterase. Whether or not these compounds have a regulatory function in cells or tissues is presently unknown. Certainly the supply of choline is fundamental for PC biosynthesis, and an abundance of choline inhibiting glycerolphosphocholine catabolism makes physiological sense.

It is obvious that progress in understanding the communication between PC synthesis and catabolism is limited. There should be considerable progress in this area in the next decade.

C. OTHER MECHANISMS FOR CONTROL OF PC CATABOLISM

There are two recently described systems for control of PC catabolism. Vitamin E has been shown to inhibit phospholipase A_2 in platelets[64] and deficiency of this vitamin caused an accumulation of lyso-PC in rat heart.[65] Vitamin E was a noncompetitive inhibitor of both phospholipases A_1 and A_2 from heart.[65] Prolactin increased the PC level in a lymphoma cell line by an inhibition of PC turnover, but neither the phospholipase involved nor the mechanism were described.[22]

Thus, as described in the introduction to this section, there are many problems and complexities associated with studies on the control of PC catabolism. However, recent developments bode well for considerable progress in the next decade.

IV. FUNCTIONAL ASPECTS OF PHOSPHATIDYLCHOLINE METABOLISM

The major emphasis of this book has been on the enzymes involved in the metabolism of PC and regulation of PC biosynthesis. Discussion on the functional aspects of PC has been mostly limited to the section on platelet activating factor in Chapter 9. Several years ago this would have probably been an appropriate end to the book. However, recent studies have shown that PC is required for cell survival and for lipoprotein secretion from liver. In addition, PC is a source of DG as a second messenger in cells and is involved in tumor promoter effects on cellular metabolism. This chapter will elaborate on recent progress in these areas.

A. PC IS REQUIRED FOR EUKARYOTIC CELL SURVIVAL

Scientists have suspected for many years that PC is essential for cell viability. One indication of the importance of PC is the lack of inherited metabolic defects of PC synthesis in humans and other animals. Putative genetic defects in PC biosynthesis would probably be lethal at an early stage of development. However, the isolation of an animal cell mutant defective in PC biosynthesis in 1981 provided the first unambiguous proof that cells could not survive without PC biosynthesis.[66] Raetz and co-workers isolated a mutant of Chinese hamster ovary (CHO) cells which was temperature sensitive for growth and defective in the biosynthesis of PC.[66] The mutant would grow at 33°, but not at 40°C. After 36 h at the higher temperature, cell lysis began, and continued incubation at that temperature eventually led to cell death. The defect was shown to be in the CT reaction.[66] The activities of CK and cholinephosphotransferase were unaffected. (This same mutant was also instrumental in showing that the CDP-choline pathway was probably of no significance in the biosynthesis of sphingomyelin[66] as discussed in Chapter 11.) Subsequent studies showed that this mutant could be rescued at the restrictive temperature by supply of exogenous PC or lyso-PC in the medium.[67] Phospholipids with different polar head groups, lipoprotein-bound PC, sphingomyelin, and glycerolphosphocholine would not support prolonged growth at the restrictive temperature. However, the dipalmitoyl derivatives of phosphatidyldimethylethanolamine, phosphatidylmonomethylethanolamine, *sn*-glycerol-2-phosphocholine and *sn*-glycerol-1-phosphocholine would substitute for PC. Thus, it was not the biosynthetic process that was required, but rather the PC itself that ensured cell survival at the restricted temperature. Interestingly, supplementation with monomethyl- but not dimethylethanolamine rescued the cells as well as phosphatidylmonomethylethanolamine. A likely explanation is that phosphomonomethylethanolamine is converted to the CDP-derivative by the phosphoethanolamine cytidylyltransferase whereas phosphodimethylethanolamine would be converted to the CDP-derivative by CT.

The isolation of another CHO mutant with several genetic lesions has been reported.[68] In this case PC biosynthesis was reduced because of a defect in choline kinase. However, the levels of PC were not altered since the rate of PC turnover was also reduced by twofold.[68]

B. PC BIOSYNTHESIS AND LIPOPROTEIN SECRETION

Although it has long been suspected that choline deficiency causes lowered levels of serum

TG in rats due to a defect in secretion of very low density lipoprotein (VLDL), direct evidence for this hypothesis has only recently been published.[42] Hepatocytes isolated from rats fed a choline- and methionine-deficient diet for 3 d secreted reduced amounts of VLDL, but not high density lipoproteins (HDL).[42] In addition, the rate of PC synthesis in the deficient cells was also reduced. New synthesis of PC appeared to be required for VLDL secretion; preexisting cellular PC would not suffice. This defect could be corrected within hours by supplementation of the cultured hepatocytes with choline, methionine, or lyso-PC.[42,43] The secretion of other proteins from the liver cells did not seem to be affected by reduced PC biosynthesis. Supplementation of the cells with ethanolamine, monomethyl-, or dimethylethanolamine did not correct the defect in VLDL secretion.[69] This requirement for PC biosynthesis is highly specific and is one of the few examples where biosynthesis of another phospholipid will not replace PC. The mechanism for the strict requirement of PC biosynthesis for VLDL secretion is presently unknown.

Another problem of current interest is where the PC is added to the lipoproteins during the assembly process. The general view is that phospholipids are added to the nascent lipoproteins during assembly in the ER and that some exchange could occur with the PC of the Golgi membranes.[70] Several recent reports suggest that the situation might be more complicated. First, an inhibitor of PE and other methylation reactions in liver, 3-deazaadenosine, potently blocked the incorporation of labeled ethanolamine into hepatocytes and secreted PC.[71] Deazaadenosine also greatly reduced the labeling of PC in the cells from [^{3}H]serine via the methylation reaction, but had no effect on the labeling of secreted PC.[72] This result suggested that there might be a deazaadenosine-insensitive methylation of PE involved in lipoprotein secretion. Second, it was subsequently shown that PC labeled from serine (via the methylation of PE) had a higher specific radioactivity in the secreted lipoproteins than in the cells.[73] On the contrary, the specific radioactivity of PC labeled from ethanolamine (via the CDP-ethanolamine pathway) was lower in the secreted lipoproteins than in the hepatocytes. [3-^{3}H]Serine will label the choline moiety of PC both in the ethanolamine moiety and in the methyl groups. Apparently, there is a compartmentalization of PC and PE synthesis for lipoprotein secretion that was unexpected. Several hypotheses to explain this localized PC biosynthesis and assembly into lipoproteins have been tested. It was postulated that the Golgi might be a specific site for synthesis and methylation of PE that would explain the above compartmentation.[49] The results showed that there were enzymes for the synthesis of PC via the CDP-choline and PC methylation pathway present in *cis* Golgi.[49] However, phosphatidylserine decarboxylation was not detectable in this organelle although ethanolaminephosphotransferase was very active. Thus, these results could not explain the specific labeling of PE and PC from serine in the secreted lipoproteins. Selection of specific molecular species was also tested and found not to explain the preferred secretion of serine-derived PC in lipoproteins.[74] Finally, synthesis and specific delivery of serine-derived PC to the Golgi for assembly into nascent lipoproteins was also not observed.[75] Hence, at the present time the preferred incorporation of serine-derived PC into secreted lipoproteins remains an enigma.

C. PC AS A SOURCE FOR SECOND MESSENGER DIACYLGLYCEROL

Over the past 5 years there has been considerable interest in the phosphatidylinositol cycle and how the catabolism of phosphatidylinositol-4,5-bisphosphate leads to the production of the second messengers inositol triphosphate and DG.[1,76] Recent studies have implicated the catabolism of PC as an alternative and, possibly more important, source of DG for activation of protein kinase C in liver.[76] The evidence is consistent with an agonist such as vasopressin binding to a receptor on the hepatocyte cell surface which activates, via a G-protein, a phospholipase C which degrades PC to DG and phosphocholine.[76] There is also evidence that PC in hepatocytes can be degraded by a phospholipase D which initially gives rise to phosphatidic acid and choline.[76] The phosphatidic acid might subsequently be degraded to DG by a specific phosphatase such as phosphatidic acid phosphohydrolase. There is involvement of a G protein in this process as well.[76] Further research should establish the importance of PC as a source of DG for activation of protein kinase C in hepatocytes.

D. PHORBOL ESTERS AND PC METABOLISM

There are many studies on the effects of phorbol esters on PC biosynthesis and catabolism. It has been known for over a decade that phorbol esters stimulate PC biosynthesis in a variety of different cells in culture.[77] The stimulation occurs within minutes, is not affected by actinomycin D or cycloheximide, and is readily reversible. The increased rate of PC biosynthesis appears to be due to a translocation of CT from the cytosol to cellular membranes, probably ER.[78] The mechanism by which the phorbol ester causes the translocation of CT is presently unknown. An increase in the fatty acid pool of phorbol-treated HeLa cells was ruled unlikely.[79] A potential role for phorbol ester-activated protein kinase C acting directly on CT was also investigated. A 6-fold purified preparation of the kinase showed some inhibition of a 15-fold purified CT from HeLa cells, but the effect was independent of calcium and DG, suggesting that the inhibition was not due to protein kinase C.[79] In another approach, both phorbol esters and thyrotropin-releasing hormone caused translocation of protein kinase C in GH_3 pituitary cells.[80] When protein kinase C activity was down-regulated by incubation of the GH_3 cells for 16 h with phorbol ester, the effect of the phorbol ester and releasing hormone on PC synthesis was abolished. These apparently divergent results with protein kinase C *in vitro* and in intact cells could be explained if CT activation were an indirect effect of protein kinase C. One interesting possibility is that the phorbol esters could activate protein kinase C, which in turn activates a PC specific phospholipase C. The generated DG on the plasma membrane could then activate CT on the ER, as was shown to be possible when Krebs II cells were treated with phospholipase C in the medium.[81] However, it has been shown that phorbol esters and two cell-permeant DGs seem to activate PC biosynthesis by different mechanisms.[82,83] The safe conclusion at this time is that we do not know the mechanism by which phorbol esters activate CT and PC biosynthesis.

Phorbol esters also stimulate the catabolism of PC in a variety of cells within minutes after addition to the culture medium.[84-90] Some laboratories report increased levels of DG[86-88] and/or phosphocholine[84] consistent with an activation of phospholipase C. Alternatively, the PC might be degraded to phosphatidic acid by a phospholipase D with the subsequent formation of DG by the action of a phosphatase. Such a mechanism appears to occur in REF52 cells, a rat embryo cell line, treated with phorbol esters or vasopressin.[90] Other laboratories report that phorbol esters cause a release of choline into the medium[85,89] which might be caused either by a phospholipase D directly or via phospholipase A with subsequent metabolism of the lyso-PC to choline which is released into the medium. Further work will hopefully differentiate among and explain the above possible mechanisms by which phorbol esters stimulate PC catabolism.

V. FUTURE DIRECTIONS

Where will the field of PC metabolism go before another edition of this book is contemplated? Many frontiers have been clearly identified in the earlier chapters. With the availability of three purified enzymes in PC biosynthesis, we should see major developments in understanding the structure of the enzymes, the structure of the genes, and the regulation of gene expression. Such progress will also be possible for the other enzymes of PC biosynthesis and catabolism when they are purified. We should have proof of an unequivocal role for protein phosphorylation of CT as an important physiological control mechanism. The possible regulation of other PC metabolic enzymes by protein phosphorylation and other covalent modifications should have been identified. The role of noncovalent modifiers such as calcium, DG, and fatty acids should be more clearly understood. We should also have an improved understanding of the role of substrate supply, particularly CTP, for regulation of PC biosynthesis.

A major frontier is PC catabolism. The enzymes involved need to be purified, the structures of the enzymes determined, and information about the relevant genes obtained. What are the important regulatory mechanisms for PC catabolism? Do the anabolic and catabolic pathways communicate? Finally, the function of PC in cells and tissues should be more clearly understood in the next decade. Even now it is quite clear that PC functions in many ways other than just the major component of membrane bilayers which act as a barrier for the cell and a solvent for membrane proteins.

Perhaps the most exciting prospect for the future is that we will undoubtedly uncover new and unexpected aspects of PC metabolism and function.

ACKNOWLEDGMENT

I am most grateful to the Medical Research Council of Canada and the Canadian Heart Foundation for the support of my research on regulation of PC metabolism and the role of PC biosynthesis in lipoprotein secretion. I wish to acknowledge and thank my many colleagues who have worked on these projects in my laboratory over the past 15 years. It is clear that your work has made a difference. I thank Jean Vance, Zemin Yao, and Neale Ridgway for their critical evaluation of the manuscript.

REFERENCES

1. **Vance, D. E. and Vance J. E.**, *Biochemistry of Lipids and Membranes*, Benjamin/Cummings, Menlo Park, CA, 1985.
2. **Pelech, S. L., Power, E., and Vance, D. E.**, Activities of the phosphatidylcholine biosynthetic enzymes in rat liver during development, *Can. J. Biochem. Cell Biol.*, 61, 1147, 1983.
3. **Coleman, R. A. and Haynes, E. B.**, Selective changes in microsomal enzymes of triacylglycerol and phosphatidylcholine synthesis in fetal and postnatal rat liver; induction of microsomal *sn*-glycerol 3-phosphate and dihydroxyacetonephosphate acyltransferase activities *J. Biol. Chem.*, 258, 450, 1983.
4. **Weinhold, P. A., Skinner, R. S., and Sanders, R. D.**, Activity and some properties of choline kinase, cholinephosphate cytidylyltransferase and cholinephosphotransferase during liver development in the rat, *Biochim. Biophys. Acta*, 326, 43, 1973.
5. **Stern, W., Kovac, C., and Weinhold, P. A.**, Activity and properties of CTP: cholinephosphate cytidylyltransferase in adult and fetal rat lung, *Biochim. Biophys. Acta*, 441, 280, 1976.
6. **Weinhold, P. A., Rounsifer, M. E., Williams, S. E., Brubaker, P. G., and Feldman, D. A.**, CTP: phosphorylcholine cytidylyltransferase in rat lung; the effect of free fatty acids on the translocation of activity between microsomes and cytosol, *J. Biol. Chem.*, 259, 10315, 1984.
7. **Weinhold, P. A., Quade, M. M., Brozowski, T. B., and Feldman, D. A.**, Increased synthesis of phosphatidylcholine by rat lung following premature birth, *Biochim. Biophys. Acta*, 617, 76, 1980.
8. **Weinhold, P. A., Feldman, D. A., Quade, M. M., Miller, J. C., and Brooks, R. L.**, Evidence for a regulatory role of CTP: cholinephosphate cytidylyltransferase in the synthesis of phosphatidylcholine in fetal lung following premature birth, *Biochim. Biophys. Acta*, 665, 134, 1981.
9. **Vigo, C. and Vance D. E.**, Effect of diethylstilboestrol on phosphatidylcholine biosynthesis and choline metabolism in the liver of roosters, *Biochem. J.*, 200, 321, 1981.
10. **Vigo, C., Paddon, H. B., Millard, F. C., Pritchard, P. H., and Vance, D. E.**, Diethylstilbestrol treatment modulates the enzymatic activities of phosphatidylcholine biosynthesis in rooster liver, *Biochim. Biophys. Acta*, 665, 546, 1981.
11. **Paddon, H. B., Vigo, C., and Vance, D. E.**, Diethylstilbestrol treatment increases the amount of choline kinase in rooster liver, *Biochim. Biophys. Acta*, 710, 112, 1982.
12. **Farrell, P. M. and Zachman, R. D.**, Induction of choline phosphotransferase and lecithin synthesis in the fetal lung by corticosteroids, *Science*, 179, 297, 1973.
13. **Post, M., Floros, J., and Smith, B. T.**, Inhibition of lung maturation by monoclonal antibodies against fibroblast-pneumonocyte factor, *Nature (London)*, 308, 284, 1984.

14. **Post, M., Barsoumian, A., and Smith, B. T.**, The cellular mechanism of glucocorticoid acceleration of fetal lung maturation; fibroblast pneumonocyte factor stimulates choline-phosphate cytidylyltransferase activity, *J. Biol. Chem.*, 261, 2179, 1986.
15. **Post, M.**, Maternal administration of dexamethasone stimulates choline-phosphate cytidylyltransferase in fetal type II cells, *Biochem. J.*, 241, 291, 1987.
16. **Rooney, S. A., Dynia, D. W., Smart D. A., Chu, A. J., Ingleson, L. D., Wilson, C. M., and Gross, I.**, Glucocorticoid stimulation of choline-phosphate cytidylyltransferase activity in fetal rat lung: receptor-response relationships, *Biochim. Biophys. Acta*, 888, 208, 1986.
17. **Chu, A. J. and Rooney, S. A.**, Stimulation of cholinephosphate cytidylyltransferase activity by estrogen in fetal rabbit lung is mediated by phospholipids, *Biochim. Biophys. Acta*, 834, 346, 1985.
18. **Tijburg, L. B. M., Schuurmans, E. A. J. M., Geelen, M. J. H., and van Golde, L. M. G.**, Effects of vasopressin on the synthesis of phosphatidylethanolamines and phosphatidylcholines by isolated rat hepatocytes, *Biochim. Biophys. Acta*, 919, 49, 1987.
19. **Haagsman, H. P., van den Heuvel, J. M., van Golde, L. M. G., and Geelen, M. J. H.**, Synthesis of phosphatidylcholines in rat hepatocytes. Possible regulation by norepinephrine via an α-adrenergic mechanism, *J. Biol. Chem.*, 259, 11273, 1984.
20. **Ballard, P. L., Hovey, M. L., and Gonzale, L. K.**, Thyroid hormone stimulation of phosphatidylcholine synthesis in cultured fetal rabbit lung, *J. Clin. Invest.*, 74, 898, 1984.
21. **Stern, P. H. and Vance, D. E.**, Phosphatidylcholine metabolism in neonatal mouse calvaria, *Biochem. J.*, 244, 409, 1987.
22. **Ko, K. W. S., Cook, H. W., and Vance, D. E.**, Reduction of phosphatidylcholine turnover in a Nb 2 lymphoma cell line after prolactin treatment; a novel mechanism for control of phosphatidylcholine levels in cells, *J. Biol. Chem.*, 261, 7846, 1986.
23. **Warden, C. H. and Friedkin, M.**, Regulation of phosphatidylcholine biosynthesis by mitogenic growth factors, *Biochim. Biophys. Acta*, 792, 270, 1984.
24. **Warden, C. H. and Friedkin, M.**, Regulation of choline kinase activity and phosphatidylcholine biosynthesis by mitogenic growth factors in 3T3 fibroblasts, *J. Biol. Chem.*, 260, 6006, 1985.
25. **Tadokoro, K., Ishidate, K., and Nakazawa, Y.**, Evidence for the existence of isozymes of choline kinase and their selective induction in 3-methylcholanthrene- or carbon tetrachloride-treated rat liver, *Biochim. Biophys. Acta*, 835, 501, 1985.
26. **Best, C. H. and Huntsman, M. E.**, The effects of the components of lecithin upon deposition of fat in the liver, *J. Physiol. (London)*, 75, 405, 1932.
27. **Kuksis, A. and Mookerjea, S.**, Choline, *Nutr. Rev.*, 36, 201, 1978.
28. **Tesan, M., Ancheschi, M. M., and Bleasdale, J. E.**, Regulation of CTP:phosphocholine cytidylyltransferase activity in type II pneumonocytes, *Biochem. J.*, 232, 705, 1985.
29. **Yao, Z., Jamil, H., and Vance, D. E.**, unpublished results.
30. **Schneider, W. J. and Vance, D. E.**, Effect of choline deficiency on the enzymes that synthesize phosphatidylcholine and phosphatidylethanolamine in rat liver, *Eur. J. Biochem.*, 85, 181, 1978.
31. **Ridgway, N. D., Yao, Z., and Vance, D. E.**, Phosphatidylethanolamine levels and regulation of phosphatidylethanolamine-*N*-methyltransferase, *J. Biol. Chem.*, 264, 1203, 1989.
32. **Vance, D. E., Trip, E. M., and Paddon, H. B.**, Poliovirus increases phosphatidylcholine biosynthesis in HeLa cells by stimulation of the rate-limiting reaction catalyzed by CTP:phosphocholine cytidylyltransferase, *J. Biol. Chem.*, 255, 1064, 1980.
33. **Choy, P. C., Paddon, H. B., and Vance, D. E.**, An increase in cytoplasmic CTP accelerates the reaction catalyzed by CTP:phosphocholine cytidylyltransferase in poliovirus-infected HeLa cells, *J. Biol. Chem.*, 255, 1070, 1980.
34. **Choy, P. C.**, Control of phosphatidylcholine biosynthesis in myopathic hamster hearts, *J. Biol. Chem.*, 257, 10928, 1982.
35. **Pelech, S. L., Cook, H. W., Paddon, H. B., and Vance, D. E.**, Membrane-bound CTP:phosphocholine cytidylyltransferase regulates the rate of phosphatidylcholine synthesis in HeLa cells treated with unsaturated fatty acids, *Biochim. Biophys. Acta*, 795, 433, 1984.
36. **Sundler, R., Åkesson, B., and Nilsson, Å.**, Effect of different fatty acids on glycerolipid synthesis in isolated rat hepatocytes, *J. Biol. Chem.*, 249, 5102, 1974.
37. **Binaglia, L., Roberti, R., Vecchini, A., and Porcellati, G.**, Evidence for a compartmentation of brain microsomal diacylglycerol, *J. Lipid Res.*, 23, 955, 1982.
38. **Rüstow, B. and Kunze, D.**, Diacylglycerol synthesized *in vitro* from *sn*-glycerol 3-phosphate and the endogenous diacylglycerol are different substrate pools for the biosynthesis of phosphatidylcholine in rat lung microsomes, *Biochim. Biophys. Acta*, 835, 273, 1985.
39. **Rüstow, B. and Kunze, D.**, Further evidence for the existence of different diacylglycerol pools of the phosphatidylcholine synthesis in microsomes, *Biochim. Biophys. Acta*, 921, 552, 1987.

40. **Lim, P., Cornell, R., and Vance, D. E.**, The supply of both CDP-choline and diacylglycerol can regulate the rate of phosphatidylcholine synthesis in HeLa cells, *Biochem. Cell Biol.*, 64, 692, 1986.
41. **Esko, J. D. and Matsuoka, K. Y.**, Biosynthesis of phosphatidylcholine from serum phospholipids in Chinese hamster ovary cells deprived of choline, *J. Biol. Chem.*, 258, 3051, 1983.
42. **Yao, Z. and Vance, D. E.**, The active synthesis of phosphatidylcholine is required for very low density lipoprotein secretion from rat hepatocytes, *J. Biol. Chem.*, 263, 2998, 1988.
43. **Robinson, B. S., Yao, Z., Baisted, D. J., and Vance, D. E.**, Lysophosphatidylcholine metabolism and lipoprotein secretion by cultured rat hepatocytes deficient in choline, *Biochem. J.*, in press.
44. **Graham, A., Zammit, V. A., and Brindley, D. N.**, Fatty acid specificity for the synthesis of triacylglycerol and phosphatidylcholine and for the secretion of very-low-density lipoproteins and lysophosphatidylcholine by cultures of rat hepatocytes, *Biochem. J.*, 249, 727, 1988.
45. **Baisted, D. J., Robinson, B. S., and Vance, D. E.**, Albumin stimulates the release of lysophosphatidylcholine from cultured rat hepatocytes, *Biochem. J.*, 253, 693, 1988.
46. **Cornell, R. B. and Goldfine, H.**, The coordination of sterol and phospholipid synthesis in cultured myogenic cells. Effect of cholesterol synthesis inhibition on the synthesis of phosphatidylcholine, *Biochim. Biophys. Acta*, 750, 504, 1983.
47. **Lim, P. H., Pritchard, P. H., Paddon, H. B., and Vance, D. E.**, Stimulation of hepatic phosphatidylcholine biosynthesis in rats fed a high cholesterol and cholate diet correlates with translocation of CTP:phosphocholine cytidylyltransferase from cytosol to microsomes, *Biochim. Biophys. Acta*, 753, 74, 1983.
48. **Vance, D. E. and Ridgway, N. D.**, The methylation of phosphatidylethanolamine, *Prog. Lipid Res.*, 27, 61, 1988.
49. **Vance, J. E. and Vance, D. E.**, Does rat liver Golgi have the capacity to synthesize phospholipids for lipoprotein secretion?, *J. Biol. Chem.*, 263, 5898, 1988.
50. **Jacobson, G. R. and Saier, M. H.**, Structure and assembly of biological membranes, in *Biochemistry*, 2nd ed., Zubay, G., Ed., Macmillan, New York, 1988, chap. 6.
51. **Zelinski, T. A. and Choy, P. C.**, Choline regulates phosphatidylethanolamine biosynthesis in isolated hamster heart, *J. Biol. Chem.*, 257, 13201, 1982.
52. **Zelinski, T. A. and Choy, P. C.**, Ethanolamine inhibits choline uptake in the isolated hamster heart, *Biochim. Biophys. Acta*, 794, 326, 1984.
53. **Sundler, R. and Åkesson, B.**, Regulation of phospholipid biosynthesis in isolated rat hepatocytes. Effect of different substrates, *J. Biol. Chem.*, 250, 3359, 1975.
54. **Tijburg, L. B. M., Houweling, M., Geelen, M. J. H., and van Golde, L. M. G.**, Effects of dietary conditions on the pool sizes of precursors of phosphatidylcholine and phosphatidylethanolamine synthesis in rat liver, *Biochim. Biophys. Acta*, 959, 1, 1988.
55. **Paddon, H. B. and Vance, D. E.**, Tetradecanoyl-phorbol acetate stimulates phosphatidylcholine biosynthesis in HeLa cells by an increase in the rate of the reaction catalyzed by CTP:phosphocholine cytidylyltransferase, *Biochim. Biophys. Acta*, 620, 636, 1980.
56. **Tijburg, L. B. M., Houweling, M., Geelen, M. J. H., and van Golde, L. M. B.**, Stimulation of phosphatidylethanolamine synthesis in isolated rat hepatocytes by phorbol 12-myristate 13-acetate, *Biochim. Biophys. Acta*, 922, 184, 1987.
57. **Pelech, S. L., Pritchard, P. H., Sommerman, E. F., Percival-Smith, A., and Vance, D. E.**, Glucagon inhibits phosphatidylcholine biosynthesis via the CDP-choline and transmethylation pathways in cultured rat hepatocytes, *Can. J. Biochem. Cell Biol.*, 62, 196, 1984.
58. **Tijburg, L. B. M., Houweling, M., Geelen, M. J. H., and van Golde, L. M. G.**, Inhibition of phosphatidylethanolamine synthesis by glucagon in isolated rat hepatocytes, *Biochem. J.*, in press.
59. **Waite, M.**, *The Phospholipases*, Plenum Press, New York, 1987.
60. **Ulevitch, R. J., Watanabe, Y., Sano, M., Lister, M. D., Deems, R. A., and Dennis, E. A.**, Solubilization, purification and characterization of a membrane-bound phospholipase A_2 from the P388D_1 macrophage-like cell line, *J. Biol. Chem.*, 263, 3079, 1988.
61. **Dawson, R. M. C.**, The metabolism of animal phospholipids and their turnover in cell membranes, *Essays Biochem.*, 2, 69, 1966.
62. **Pelech, S. L. and Vance, D. E.**, Signal transduction via phosphatidylcholine cycles, *Trends Biochem. Sci.*, 14, 28, 1989.
63. **Lister, M. D., Deems, R. A., Watanabe, Y., Ulevitch, R. J., and Dennis, E. A.**, Kinetic analysis of the Ca^{2+}-dependent, membrane-bound, macrophage phospholipase A_2 and the effects of arachidonic acid, *J. Biol. Chem.*, 263, 7506, 1988.
64. **Douglas, C. E., Chan, A. C., and Choy, P. C.**, Vitamin E inhibits platelet phospholipase A_2, *Biochim. Biophys. Acta*, 876, 639, 1986.
65. **Cao, Y. Z., O, K., Choy, P. C., and Chan, A. C.**, Regulation by vitamin E of phosphatidylcholine metabolism in rat heart, *Biochem. J.*, 247, 135, 1987.
66. **Esko, J. D., Wermuth, M. M., and Raetz, C. R. H.**, Thermolabile CDP-choline synthetase in an animal cell mutant defective in lecithin formation, *J. Biol. Chem.*, 256, 7388, 1981.

67. **Esko, J. D., Nishijima, M., and Raetz, C. R. H.**, Animal cells dependent on exogenous phosphatidylcholine for membrane biogenesis, *Proc. Natl. Acad. Sci.*, 79, 1698, 1982.
68. **Nishijima, M., Kuge, O., Maeda, M., Nakano, A., and Akamatsu, Y.**, Regulation of phosphatidylcholine metabolism in mammalian cells. Isolation and characterization of a Chinese hamster ovary cell pleiotropic mutant defective in both choline kinase and choline-exchange reaction activities, *J. Biol. Chem.*, 259, 7101, 1984.
69. **Yao, Z. and Vance, D. E.**, Head group specificity in the requirement of phosphatidylcholine biosynthesis for VLDL secretion from cultured hepatocytes, *J. Biol. Chem.*, in press.
70. **Vance, J. E. and Vance, D. E.**, The role of phosphatidylcholine biosynthesis in the secretion of lipoproteins from hepatocytes, *Can. J. Biochem. Cell Biol.*, 63, 870, 1985.
71. **Vance, J. E., Nguyen, T. M., and Vance, D. E.**, The biosynthesis of phosphatidylcholine by methylation of phosphatidylethanolamine derived from ethanolamine is not required for lipoprotein secretion by cultured rat hepatocytes, *Biochim. Biophys. Acta*, 875, 501, 1986.
72. **Vance, J. E. and Vance, D. E.**, A deazaadenosine-insensitive methylation of phosphatidylethanolamine is involved in lipoprotein secretion, *FEBS Lett.*, 204, 243, 1986.
73. **Vance, J. E. and Vance, D. E.**, Specific pools of phospholipids are used for lipoprotein secretion by cultured rat hepatocytes, *J. Biol. Chem.*, 261, 4486, 1986.
74. **Vance, J.**, Compartmentalization of phospholipids for lipoprotein assembly on the basis of molecular species and biosynthetic origin, *Biochim. Biophys. Acta*, 963, 70, 1988.
75. **Vance, J.**, Compartmentalization and differential labeling of phospholipids of rat subcellular membranes, *Biochim. Biophys. Acta*, 963, 10, 1988.
76. **Exton, J. H.**, Mechanism of action of calcium mobilizing agonists — some variations on a young theme, *FASEB J.*, 2, 2670, 1988.
77. **Pelech, S. L. and Vance, D. E.**, Regulation of phosphatidylcholine biosynthesis, *Biochim. Biophys. Acta*, 779, 217, 1984.
78. **Pelech, S. L., Paddon, H. B., and Vance, D. E.**, Phorbol esters stimulate phosphatidylcholine biosynthesis by translocation of CTP:phosphocholine cytidylyltransferase from cytosol to microsomes, *Biochim. Biophys. Acta*, 795, 447, 1984.
79. **Cook, H. W. and Vance, D. E.**, Evaluation of possible mechanisms of phorbol ester stimulation of phosphatidylcholine synthesis in HeLa cells, *Can. J. Biochem. Cell Biol.*, 63, 145, 1984.
80. **Kolesnick, R. N.**, Thyrotropin-releasing hormones and phorbol esters induce phosphatidylcholine synthesis in GH_3 pituitary cells. Evidence for stimulation via protein kinase C, *J. Biol. Chem.*, 262, 14525, 1987.
81. **Tercé, F., Record, M., Ribbes, G., Chap, H., and Douste-Blazy, L.**, Intracellular processing of cytidylyltransferase in Krebs II cells during stimulation of phosphatidylcholine synthesis. Evidence that a plasma membrane modification promotes enzyme translocation specifically to the endoplasmic reticulum, *J. Biol. Chem.*, 263, 3142, 1988.
82. **Kolesnick, R. N. and Paley, A. E.**, 1,2-Diacylglycerols and phorbol esters stimulate phosphatidylcholine metabolism in GH_3 pituitary cells. Evidence for separate mechanisms of action, *J. Biol. Chem.*, 262, 9204, 1987.
83. **Liscovitch, M., Slack, B., Blusztajn, J. K., and Wurtman, R. J.**, Differential regulation of phosphatidylcholine biosynthesis by 12-*O*-tetradecanoylphorbol-13-acetate and diacylglycerol in NG108-15 neuroblastoma x glioma hybrid cells, *J. Biol. Chem.*, 262, 17487, 1987.
84. **Guy, G. R. and Murray, A. W.**, Tumor promoter stimulation of phosphatidylcholine turnover in HeLa cells, *Cancer Res.*, 42, 1980, 1982.
85. **Liscovitch, M., Blusztajn, J. K., Freese, A., and Wurtman, R. J.**, Stimulation of choline release from NG108-15 cells by 12-*O*-tetradecanoylphorbol 13-acetate, *Biochem. J.*, 241, 81, 1987.
86. **Besterman, J. M., Duronio, V., and Cuatrecasas, P.**, Rapid formation of diacylglycerol from phosphatidylcholine: a pathway for generation of a second messenger, *Proc. Natl. Acad. Sci.*, 83, 6785, 1986.
87. **Daniel, L. W., Waite, M., and Wykle, R. L.**, A novel mechanism of diglyceride formation. 12-*O*-tetradecanoylphorbol-13-acetate stimulates the cyclic breakdown and resynthesis of phosphatidylcholine, *J. Biol. Chem.*, 261, 9128, 1986.
88. **Cabot, M. C., Welsh, C. J., Zhang, Z. C., Cao, H. T., Chabbott, H., and Lebowitz, M.**, Vasopressin, phorbol diesters and serum elicit choline glycerophospholipid hydrolysis and diacylglycerol formation in nontransformed cells: transformed derivatives do not respond, *Biochim. Biophys. Acta*, 959, 46, 1988.
89. **Mueller, H. W., Pritchard, P. H., and Vance, D. E.**, unpublished results.
90. **Cabot, M. C., Welsh, C. J., Cao, H. T., and Chabbott, H. M.**, The phosphatidylcholine pathway of diacylglycerol formation stimulated by phorbol diesters occurs via phospholipase D activation, *FEBS Lett.*, 233, 153, 1988.

INDEX

D

E

F

G

M

N

O

P

R

S

T

U

V

W

X

Y

Z